*László Poppe and*
*Mihály Nógrádi*

*Editors*

*László Poppe, József Nagy, Gábor*
*Hornyánszky and Zoltán Boros*

*Contributing Authors*

**Stereochemistry and Stereoselective Synthesis**

*Edited by*
*László Poppe and Mihály Nógrádi*

*Contributing Authors*
*László Poppe, József Nagy, Gábor Hornyánszky and Zoltán Boros*

# Stereochemistry and Stereoselective Synthesis

## An Introduction

Verlag GmbH & Co. KGaA

**Editors**

*Dr. László Poppe*
Budapest Univ. of Technology & Economics
Dept. of Organic Chemistry & Technology
Szt. Gellért tér 4
1111 Budapest
Hungary

*Dr. Mihály Nógrádi*
Budapest Univ. of Technology & Economics
Dept. of Organic Chemistry & Technology
Szt. Gellért tér 4
1111 Budapest
Hungary

**Contributing Authors**

*Dr. László Poppe*
Budapest Univ. of Technology & Economics
Dept. of Organic Chemistry & Technology
Szt. Gellért tér 4
1111 Budapest
Hungary

*Dr. József Nagy*
Budapest Univ. of Technology & Economics
Dept. of Organic Chemistry & Technology
Szt. Gellért tér 4
1111 Budapest
Hungary

*Dr. Gábor Hornyánszky*
Budapest Univ. of Technology & Economics
Dept. of Organic Chemistry & Technology
Szt. Gellért tér 4
1111 Budapest
Hungary

*Zoltán Boros*
H-ION Research, Development & Innovation Ltd.
Konkoly-Thege Miklós út 29-33
1121 Budapest
Hungary



**Library of Congress Card No.:** applied for

**British Library Cataloguing-in-Publication Data**
A catalogue record for this book is available from the British Library.

**Bibliographic information published by the Deutsche Nationalbibliothek**
The Deutsche Nationalbibliothek lists this publication in the Deutsche Nationalbibliografie; detailed bibliographic data are available on the Internet at <http://dnb.d-nb.de>.



**Print ISBN:** 978-3-527-33901-3
**ePDF ISBN:** 978-3-527-69630-7
**ePub ISBN:** 978-3-527-69631-4
**Mobi ISBN:** 978-3-527-69632-1
**oBook ISBN:** 978-3-527-34117-7

**Cover Design** Adam-Design, Weinheim, Germany
**Typesetting** SPi Global, Chennai, India
**Printing and Binding** Markono Print Media Pte Ltd, Singapore

Printed on acid-free paper

# Contents

## About the Companion Website

This book is accompanied by a companion website:

**www.wiley.com/go/poppe/stereochemistry**

The website includes animations about the software developer.

# Introduction

Nowadays, the development of molecular sciences brings about a revolutionary change of our world, life, and culture as once did the industrial revolution laying the foundations of our modern world. This molecular revolution, one of the milestones of which was the elucidation of the structure of the hitherto largest known natural compound, the human genome, extends in a dimension hitherto unheard of our knowledge of both ourselves and the universe. A key element of this molecular revolution is chemistry, and within it organic chemistry, contributing a lion's share in the twentieth and twenty-first centuries to the significant achievements of biology, medical, material, and environment sciences.

Similar to other sciences, organic chemistry plays a key role in our knowledge of the universe, and within chemistry a special place is allotted to the study of organic molecules. Apart from the potential of synthetic organic chemistry to construct molecules to be found in nature, it is capable to construct molecules not produced in nature.

A key problem of organic synthesis is selectivity, and within this domain stereoselectivity, a capacity to prepare selectively just one of the possible stereoisomeric structures. The importance of stereochemistry has been recognized in the very early period of organic chemistry: J. B. Biot observed in 1815 that certain organic compounds and their solutions rotate the plane of planar polarized light. L. Pasteur (1948) separated (resolved) the optically inactive tartaric acid to two optically active forms and made one of the most important hypotheses in stereochemistry, namely that the two forms are related as mirror images. J. A. LeBel and J. J. van't Hoff (1874) recognized the tetrahedral bond structure of carbon and that this structure enables, in the case of four different ligands, the existence of two nonidentical mirror image structures (enantiomers). H. E. Fischer after having identified and synthesized most of the 16 possible stereoisomeric forms of aldohexoses (1891) suggested a representation of three-dimensional structures in two dimensions, while M. A. Rosanoff (1905) proposed the conventional absolute configuration of D-(+)-glyceraldehyde.

Stereochemistry and stereoselective synthesis received a significant impetus in the middle of the past century when J. M. Bijvoet (1951) determined the actual absolute configuration of (+)-tartaric sodium rubidium salt with the aid of anomalous scattering in X-ray diffraction.

The relevance of stereochemical studies was recognized by awarding a series of Nobel Prizes. The foundation of modern stereochemistry was laid down in the monograph of M. S. Newman (1956). D. H. R. Barton and O. Hassel were awarded the Nobel Prize (1969) for conformational studies, while V. Prelog and J. W. Cornforth (1975) for analyzing the stereochemistry of enzyme-catalyzed reactions. Nobel-Prize-winning studies were carried out by D. J. Cram, J. M. Lehn, and C. J. Pedersen (1987) of selective interactions in supramolecular systems; W. S. Knowles, R. Noyori, and B. Sharpless (2001) for elaborating stereoselective synthetic methods.

The practical importance of stereochemistry is accentuated by the fact that nowadays almost exclusively enantiopure drugs can be registered and the inactive enantiomers are regarded as "contaminants." It is therefore not surprising that manufacturing enantiopure compounds is a multibillion dollar business increasing about 10% per year. Accordingly, development of stereoselective methodology of manufacturing and analyzing pure enantiomers is becoming a central issue for the pharmaceutical, pesticide, cosmetic, and even of household chemical industry.

To write a textbook on any field of science is always challenging, especially about stereochemistry and stereoselective methodology, which is now in extremely fast development. The present work is intended to serve not only students of chemistry but also a wider circle of readers, namely to those whose main interest is outside stereochemistry or even organic chemistry, but who wish to have an overview about the problems, the scope, and potentials of this highly interesting field of chemistry. We hope to find among our potential readers biochemists, polymer chemists, pharmacologists, pharmacists, biologists, and workers in other branches of biosciences.

# Part I
# Basic Concepts at the Molecular Level

The inherent difficulty of correlating structure with properties is that structure is a concept at the molecular (*microscopic*) level, while properties are in general "*macroscopic*" manifestations. Difficulties of comparing the two levels can be attributed to two factors: the quantity of material and the time required for the determination of properties.

This part deals with the basic concepts of stereochemistry focusing at the molecular (*microscopic*) level.

*Stereochemistry and Stereoselective Synthesis: An Introduction*, First Edition.
László Poppe and Mihály Nógrádi.

Companion Website: www.wiley.com/go/poppe/stereochemistry

# 1 Structure and Properties

A general opinion of chemists adopted by textbooks is that in chemistry, structure[1] is a central concept: the key to everything. Properties[2] of a given substance depend on the number and nature of its constituting atoms and their mode of connection (*connectedness*). It is less obvious but quite important that looking at structural formula, a well-versed chemist should be able to deduce many properties of a substance. In order to exploit this possibility, in this chapter, we are going to define some basic principles associated with structure and properties. After having defined the concept of structure, we will proceed to discuss correlations of structure and properties, with special emphasis on the spatial features of structure and properties derived from changes thereof. First of all, we will discuss properties derived from the spatial structure of typical organic compounds; but for the sake of completeness, the stereochemical features of inorganic compounds will be dealt briefly as well.

## 1.1 The Covalent Bond and the Octet Rule

The covalent nature of the chemical bond, assuming a shared pair of electrons, was first proposed by G. N. Lewis in 1916 [1]. According to this concept, by sharing two electrons, two hydrogen atoms can establish a stable bond by forming a closed shell of electrons similar to that of the noble gas helium (Figure 1.1).

In the ***Lewis* structures,** electrons are symbolized by dots. The amount of energy required to dissociate a hydrogen molecule to two hydrogen atoms is called the bond dissociation (or bond) energy. In the case of $H_2$, this is quite high: 435 kJ mol$^{-1}$.

1) Under *structure*, chemists often understand a single state of a single molecule; therefore, at this level, structure is a *microscopic* concept (to be discussed in more detail later).
2) *Properties* of a material are generally defined not on the basis of a given state of a single molecular structure but on an assembly of a large number of molecules, on a timescale relatively long compared to the timescale of molecular events, and therefore these properties are generally *macroscopic* concepts.

*Stereochemistry and Stereoselective Synthesis: An Introduction*, First Edition.
László Poppe and Mihály Nógrádi.

Companion Website: www.wiley.com/go/poppe/stereochemistry

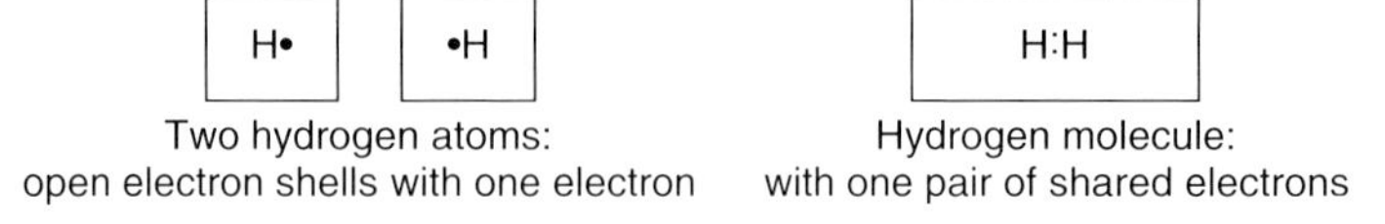


**Figure 1.1** Formation of the covalent bond of the hydrogen molecule.

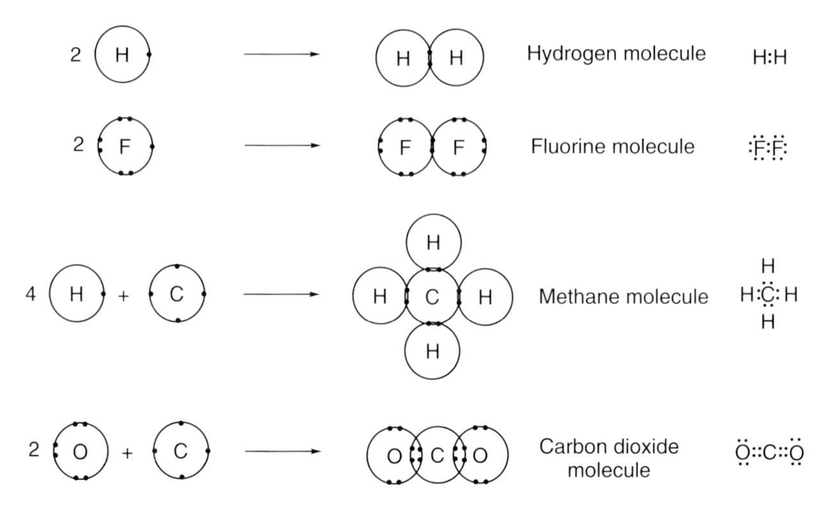


**Figure 1.2** Covalent bonds in some simple molecules (*Lewis* representation).

While with hydrogen molecule the number of electrons present in the valence forming shell is limited to two, in the *Lewis* model, molecules composed from elements of the second row (Li, Be, B, C, N. O, F, and Ne) in the valence shells contain eight (shared and unshared) electrons. Most organic compounds follow the octet rule: on formation of their compounds, electrons are taken up, and elements are shared or removed in a way that they should assume a stable structure involving eight valence electrons. When in compounds of carbon, nitrogen, oxygen, and fluorine, the octet rule is valid, their electron configuration is analogous to that of the noble gas neon. *Lewis* representation of some simple molecules is shown in Figure 1.2.

Such structures showing the distribution of electrons (*Lewis* structures) are useful aids for understanding covalent bond formation, but it is simpler to use the s.c. ***Kekulé* formulas**.[3] The latter are derived from *Lewis* formulas by replacing a shared pair of electrons with a line connecting the corresponding atomic symbols. Nonbonding electrons are shown by dots. Examples for **structural formulas** drawn according to this principle are shown in Table 1.1. As a further

3) Based on the work of A. M. Butlerov, A. Couper, and F. A. Kekulé but at variance with their original formulas, bonds between elements are shown by lines (Kekulé formulas).

**Table 1.1** Lewis and Kekulé formulas of some simple molecules.

| Name | Molecular formula | *Lewis* formula | *Kekulé* formula |
|---|---|---|---|
| Water | $H_2O$ | H:Ö: / H | H–Ö: / H |
| Ammonia | $H_3N$ | H:N̈:H / H | H–N̈–H / H |
| Methane | $CH_4$ | H / H:C:H / H | H / H–C–H / H |
| Methanol | $CH_4O$ | H / H:C:Ö:H / H | H / H–C–Ö–H / H |
| Methylamine | $CH_5N$ | HH / H:C:N: / H | H H / H–C–N–H / H |
| Ethene | $C_2H_4$ | H. .H / C::C / H˙ ˙H | H H / C=C / H H |
| Formaldehyde | $CH_2O$ | H. / C::Ö: / H | H / C=Ö: / H |
| Acetylene | $C_2H_2$ | H:C⋮⋮C H | H–C≡C–H |
| Hydrogen cyanide | CHN | H:N⋮⋮C | H–C≡N: |

simplification of *Kekulé* formulas, nonbonding electrons are not shown since these can be readily calculated following the octet rule.

As illustrated in Table 1.1 by ethene, formaldehyde, acetylene, and hydrogen cyanide, atoms may share more than one pair of electrons forming in this way **multiple bonds**. Compounds of boron, such as $BH_3$ or $BF_3$, are exceptional in a way that the valence shell of boron is not filled up with electrons as would be required by the octet rule. Accordingly, these compounds have a high affinity to electrons and are very reactive.

**Valence** is the number of those valence electrons, which must be taken up or shed that the valence shell should attain the octet state (Table 1.2). In their covalent compounds, the number of bonds adjoined to an atom is equal to the valence of the given atom. Valences listed in Table 1.2 are typical of atoms common in organic compounds.

**Table 1.2** Valence states of selected elements common in organic compounds.

| Atom | H | C | N | O | F | Cl | Br | I |
|---|---|---|---|---|---|---|---|---|
| Valence | 1 | 4 | 3 | 2 | 1 | 1 | 1 | 1 |

## 1.2
## Representation of Chemical Structures

For the representation of simpler compounds, condensed *Kekulé* formulas are suitable; but this is a cumbersome way to represent more complex compounds. Thus, cyclic compounds are best represented by further simplified, s.c. **linear formulas**. Application of linear formulas is exemplified by formulas for cycloalkanes (Figure 1.3).

Linear formulas are also convenient to depict open chain compounds (Figure 1.4).

Cyclopropane    Cyclobutane    Cyclopentane    Cylohexane    Cycloheptane

**Figure 1.3** Representation of cycloalkanes by linear formulas.

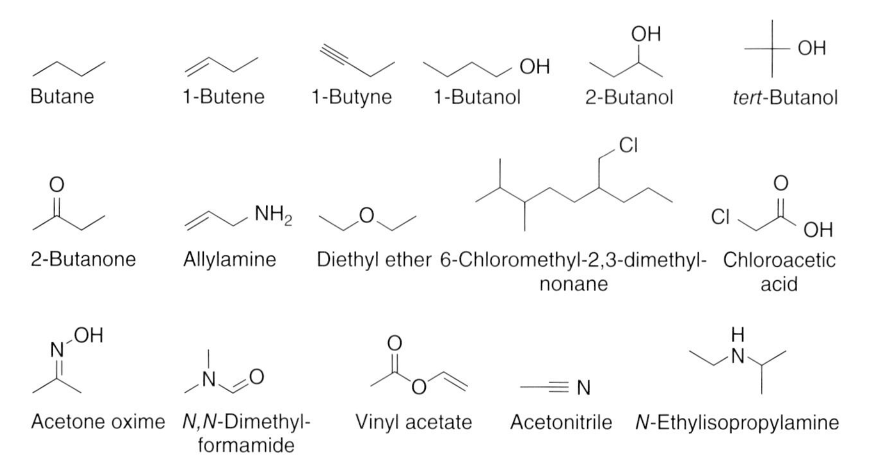


**Figure 1.4** Representation of open-chain compounds with linear formulas.

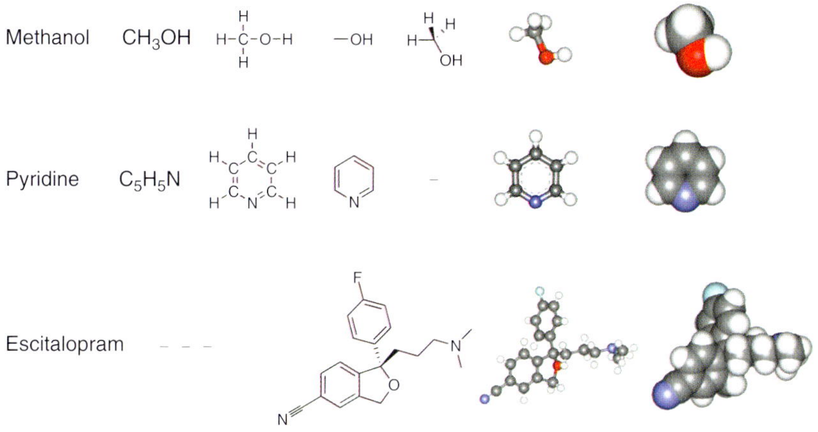

**Figure 1.5** Various modes of representation for methanol, pyridine, and the antidepressant escitalopram.

When drawing linear formulas, for simplicity, symbols of carbon atoms, the pairs of electrons, and hydrogen atoms attached to carbons are omitted. Accordingly, all end points, breaking points, and points of branching represent a carbon atom. Multiple bonds are shown by an appropriate number of parallel lines. Heavy atoms other than carbon are shown by their atomic symbols together with the hydrogen atoms attached.

Additional ways of representation are shown in Figure 1.5 by the example of methanol, pyridine, and the antidepressant escitalopram. Representations include the following:

- **A**: Name (trivial or systematic)
- **B**: Condensed formulas (generally suitable for printing in a single line)
- **C**: *Kekulé* formulas (showing all the atoms and bonds)
- **D**: Linear formulas (hydrogens omitted, breaking, branching, and end points represent carbon atoms; heavy atoms are shown with the hydrogen attached)
- **E**: **Stereoformulas**[4] (depicting the spatial orientation of bonds)
- **F**: Ball and stick model (often used to depict X-ray structures or computed models)
- **G**: Space-filling model (approaches the electron distribution).

Figure 1.5 well demonstrates that for smaller open-chain organic compounds (e.g., methanol), it is the condensed or *Kekulé* formula that is the most appropriate for the demonstration of the two-dimensional features of the structure. In case of smaller cyclic compounds (e.g., pyridine), the linear formulas are most often used. To depict more complex structures (e.g., escitalopram), condensed

4) The details of the use of stereo formulas are exactly defined by IUPAC recommendations [2]. Easily perceptible stereo formulas have only become generally used in the second half of the twentieth century. Earlier, various other representations were used.

or *Kekulé* representations are practically useless, and therefore a combination of linear and stereo representation is recommended. The complete stereostructure of molecules can be best depicted by a simplified representation of 3D structures. In Figure 1.5, two possible modes of representation of molecular models that is, ball and stick and space-filling models are shown.[5] As can be seen with more complex molecules, space-filling models refer to their overall shape, while the details of the structure are less apparent.

## 1.3 Description of Chemical Structure

Chapters of the rules of International Union of Pure and Applied Chemistry (IUPAC) for naming compounds [3] do not give a precise definition of what should be understood under "chemical structure." Therefore, we define in a general way as **chemical structure** as it is understood by crystallographers: chemical structure is an accurate description of the spatial arrangement of the constituting atoms (atomic nuclei) in space.

The exact spatial arrangement of atoms can be described, for instance, by their **Cartesian coordinates**.[6] From a structural point of view, however, it is not the absolute position and orientation of the molecule that is decisive; therefore, it is often more useful to describe the relative position of atoms by s.c. **internal coordinates**[7] (Figure 1.6). Internal coordinates for the description of molecules are the bond lengths ($r$), the bond angles ($\alpha$), and the torsion angles ($\omega$).[8]

For the description of molecules containing two atoms (**I**), besides defining the type of atoms (A, B), it is sufficient to give the bond length as a single internal coordinate. In organic molecules, the value of **bond lengths** varies in a relatively narrow range (Table 1.3).

The structure of a triatomic molecule (**II**) is characterized by the type of atoms (A, B, and C), two bond lengths ($r_1$ and $r_2$), and one **bond angle** ($\alpha$).[9] In real triatomic molecules, bond lengths and bond angles depend on the type of constituting atoms, their hybridization state, and the order of the bonds (Figure 1.7).

It is apparent from Figure 1.7 that in triatomic molecules, bond angles vary in a wide range. The molecules of hydrogen sulfide and water form an angular

5) Computer programs suitable to visualize chemical structures offer further possibilities (e.g., wire models, ball and stick models, covering surfaces according to various properties and combinations thereof).
6) A molecule containing $n$ atoms can have a maximum of $3n - 6$ internal degrees of freedom and can be described by $3n - 6$ Cartesian coordinates. Use of Cartesian coordinates also defines its absolute position, and therefore their use is uncomfortable.
7) A molecule containing $n$ atoms can have a maximum of $3n - 6$ internal degrees of freedom and can be described by $3n - 6$ internal coordinates. Internal coordinates reflect to local bonds and symmetries; they can be composed by linear combinations of Cartesian coordinates.
8) Bond angles are often denoted by the symbol $\Theta$. In polymer chemistry, it is usual to denote bond angles as $\tau$, while the symbol $\Theta$ is for torsion angles.
9) Bond angles are sometimes called valence angles.

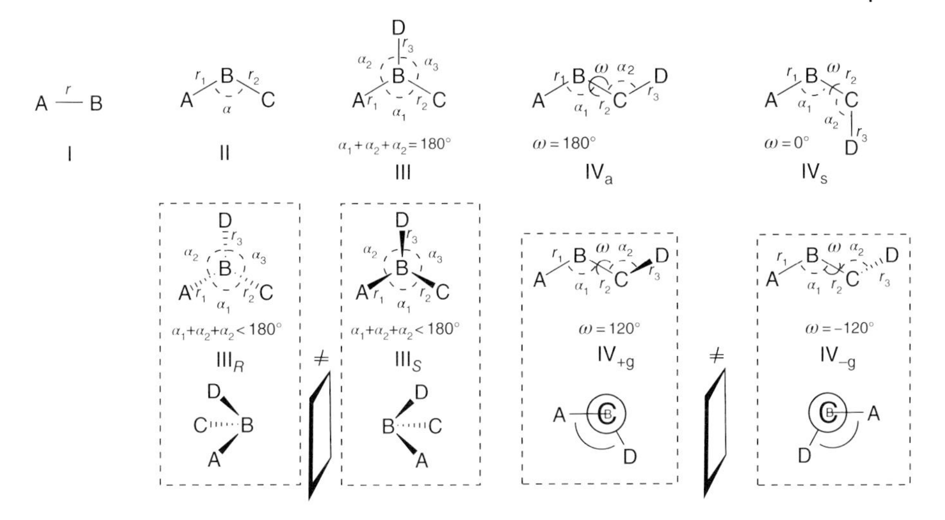


**Figure 1.6** Characterization of molecules consisting of two, three, and four atoms with internal coordinates.

**Table 1.3** Characteristic values of bond lengths commonly occurring in organic compounds.

| | C–C | C=C | C≡C | C–H | C–O | C=O | C–N | C–S | C–F | C–Cl | C–Br | C–I |
|---|---|---|---|---|---|---|---|---|---|---|---|---|
| pm | 154 | 133 | 120 | 109 | 143 | 122 | 147 | 182 | 135 | 176 | 192 | 212 |
| Å | 1.54 | 1.33 | 1.20 | 1.09 | 1.43 | 1.22 | 1.47 | 1.82 | 1.35 | 1.76 | 1.92 | 2.12 |

$\alpha = 92.1°$ Hydrogen sulfide

$\alpha = 103.9°$ Water

$\alpha = 180°$ Carbon dioxide

$\alpha = 180°$ Hydrogen cyanide

**Figure 1.7** Bond angles in selected triatomic molecules.

structure owing to the presence of two nonbonding pairs of electrons in an approximately tetrahedral arrangement. (The difference in the bond angles is caused by the different strengths of repulsion between the pairs of electrons.) On the other hand, hydrogen cyanide and carbon dioxide are linear due to the *sp* hybridization state of the central atom. In bond angles between atoms of given types, the value of bond angles varies in a narrow range and therefore when discussing structures, organic chemists take them as constant.

Molecules consisting of four atoms (A, B, C, and D) present a new situation (Figure 1.6). If the fourth atom (D) is attached to the central atom (B) of the chain formed by the other three, we are dealing with a branched structure (Figure 1.6, **III**, **III**$_R$, and **III**$_S$). When the four atoms (A, B, C, and D) are coplanar (the sum of the three bond angles $\alpha_1$, $\alpha_2$, and $\alpha_3$ is 180°), the structure is planar (**III**). When, however, the sum of the three bond angles $\alpha_1$, $\alpha_2$, and $\alpha_3$ is less than 180°, pyramidal structures (**III**$_R$ and **III**$_S$) are formed. It can be seen that while the internal coordinates of the two structures ($\alpha_1$, $\alpha_2$, and $\alpha_3$) are identical, the structures themselves are nonidentical.

If the atoms in a molecule consisting of four atoms form a continuous chain, in addition to bond lengths ($r_1$, $r_2$, and $r_3$) and bond angles ($\alpha_1$ and $\alpha_2$), an additional type of internal coordinate, the **torsion angle** ($\omega$),[10] has to be taken into consideration (Figure 1.6). With the variation of torsion angles, a further multiplication of structural alternatives arises (Figure 1.6, **IV**$_S$, **IV**$_R$, **IV**$_{+g}$, and **IV**$_{-g}$). While in organic molecules, as discussed before, the values of bond lengths ($r_1 \cdots r_n$) and bond angles ($\alpha_1 \cdots \alpha_n$) are relatively constant, torsion angles ($\omega$) for single bonds can take up any value between 0° and 180°. Coplanarity of the four atoms is possible in two different ways: $\omega = 0°$ (**IV**$_s$) or $\omega = 180°$ (**IV**$_a$) (Figure 1.6). While in structures built up from the same kind of atoms, bond lengths and bond angles may be identical, it is apparent that owing to the difference of torsion angles, the two structures are different. Besides the two coplanar arrangements (**IV**$_s$ and **IV**$_a$), an infinite number of noncoplanar structures is possible with the alteration of torsion angles in the range of 0–180°. It can be seen that the arrangement of a chain of atoms A–B–C–D assuming a given torsion angle (e.g., $\omega = 120°$: **IV**$_{+g}$) and the corresponding arrangement characterized by an identical but negative torsion angle ($\omega = -120°$: **IV**$_{-g}$) are nonidentical mirror images.

## 1.4
## Problems of Correlating Chemical Structure with Properties

For a representation of chemical structures **bonding matrices**, describing the order of attachment of atoms and bonds (connectedness) may be useful (Figure 1.8).

Figure 1.8 clearly demonstrates that while the chemical formula of ethanol and dimethyl ether ($C_2H_5O$) is identical, their bonding matrices are different. We can therefore reasonably assume that these molecules are in fact different, in other words **isomers**.[11]

While acetaldehyde and “vinyl alcohol” have the same chemical formula ($C_2H_4O$, Figure 1.9), both their structural formulas and their bonding matrices are apparently different.

10) A torsion or dihedral angle ($\omega$) is defined in case of a consecutive nonlinear chain of atoms A, B, C, and D as the angle formed by the planes of atoms ABC and BCD. If the plane is seen along the line BC, the torsion angle is taken as positive if the bond AB ought to be moved clockwise (in an angle <180°) to superpose bond CD. If AB ought to be rotated counterclockwise, the torsion angle is taken as negative.

11) According to their most common definition, **isomers** have the same molecular formula but are nevertheless different. This definition is inexact, since it lacks the definition of the circumstances (thermal energy level, environment, state, and time interval) under which the compounds are viable as stable entities (cf. Section 2.2.2).

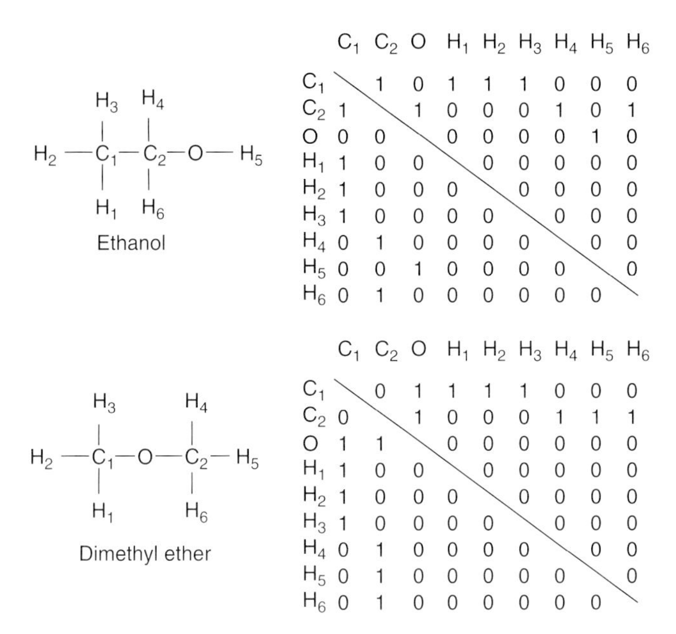


| | $C_1$ | $C_2$ | O | $H_1$ | $H_2$ | $H_3$ | $H_4$ | $H_5$ | $H_6$ |
|---|---|---|---|---|---|---|---|---|---|
| $C_1$ | | 1 | 0 | 1 | 1 | 1 | 0 | 0 | 0 |
| $C_2$ | 1 | | 1 | 0 | 0 | 0 | 1 | 0 | 1 |
| O | 0 | 0 | | 0 | 0 | 0 | 0 | 1 | 0 |
| $H_1$ | 1 | 0 | 0 | | 0 | 0 | 0 | 0 | 0 |
| $H_2$ | 1 | 0 | 0 | 0 | | 0 | 0 | 0 | 0 |
| $H_3$ | 1 | 0 | 0 | 0 | 0 | | 0 | 0 | 0 |
| $H_4$ | 0 | 1 | 0 | 0 | 0 | 0 | | 0 | 0 |
| $H_5$ | 0 | 0 | 1 | 0 | 0 | 0 | 0 | | 0 |
| $H_6$ | 0 | 1 | 0 | 0 | 0 | 0 | 0 | 0 | |

| | $C_1$ | $C_2$ | O | $H_1$ | $H_2$ | $H_3$ | $H_4$ | $H_5$ | $H_6$ |
|---|---|---|---|---|---|---|---|---|---|
| $C_1$ | | 0 | 1 | 1 | 1 | 1 | 0 | 0 | 0 |
| $C_2$ | 0 | | 1 | 0 | 0 | 0 | 1 | 1 | 1 |
| O | 1 | 1 | | 0 | 0 | 0 | 0 | 0 | 0 |
| $H_1$ | 1 | 0 | 0 | | 0 | 0 | 0 | 0 | 0 |
| $H_2$ | 1 | 0 | 0 | 0 | | 0 | 0 | 0 | 0 |
| $H_3$ | 1 | 0 | 0 | 0 | 0 | | 0 | 0 | 0 |
| $H_4$ | 0 | 1 | 0 | 0 | 0 | 0 | | 0 | 0 |
| $H_5$ | 0 | 1 | 0 | 0 | 0 | 0 | 0 | | 0 |
| $H_6$ | 0 | 1 | 0 | 0 | 0 | 0 | 0 | 0 | |

**Figure 1.8** Kekulé formula and the bonding matrix of ethanol and dimethyl ether.

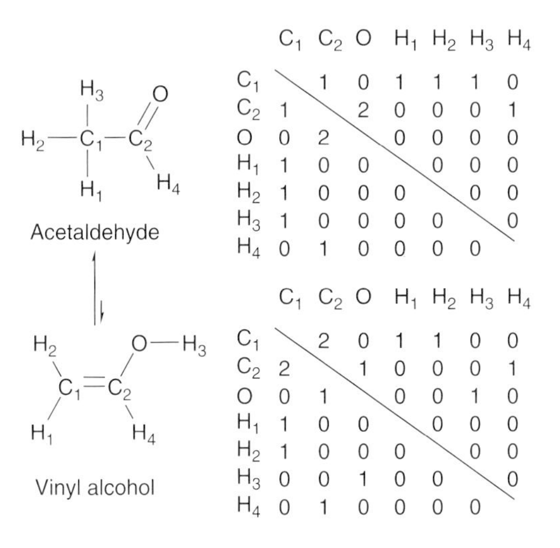


| | $C_1$ | $C_2$ | O | $H_1$ | $H_2$ | $H_3$ | $H_4$ |
|---|---|---|---|---|---|---|---|
| $C_1$ | | 1 | 0 | 1 | 1 | 1 | 0 |
| $C_2$ | 1 | | 2 | 0 | 0 | 0 | 1 |
| O | 0 | 2 | | 0 | 0 | 0 | 0 |
| $H_1$ | 1 | 0 | 0 | | 0 | 0 | 0 |
| $H_2$ | 1 | 0 | 0 | 0 | | 0 | 0 |
| $H_3$ | 1 | 0 | 0 | 0 | 0 | | 0 |
| $H_4$ | 0 | 1 | 0 | 0 | 0 | 0 | |

| | $C_1$ | $C_2$ | O | $H_1$ | $H_2$ | $H_3$ | $H_4$ |
|---|---|---|---|---|---|---|---|
| $C_1$ | | 2 | 0 | 1 | 1 | 0 | 0 |
| $C_2$ | 2 | | 1 | 0 | 0 | 0 | 1 |
| O | 0 | 1 | | 0 | 0 | 1 | 0 |
| $H_1$ | 1 | 0 | 0 | | 0 | 0 | 0 |
| $H_2$ | 1 | 0 | 0 | 0 | | 0 | 0 |
| $H_3$ | 0 | 0 | 1 | 0 | 0 | | 0 |
| $H_4$ | 0 | 1 | 0 | 0 | 0 | 0 | |

**Figure 1.9** Kekulé formulas and the bond matrices of acetaldehyde and "vinyl alcohol."

Looking at the structures, we might assume that also in this case the compounds are isomers, but in reality, the two structures cannot be separated because they form a dynamic equilibrium mixture of **tautomers** (cf. Chapter 6).[12]

12) A relatively exact definition of **tautomers** may be that these are inseparable isomers in a dynamic equilibrium.

Examples in Figures 1.7–1.9 already demonstrate that without the examination of additional features, structures in themselves do not give information about the nature of differences of the entities in question. The inherent difficulty of correlating structure with properties is that structure is a concept at the molecular (*microscopic)* level, while properties are in general "*macroscopic*" manifestations. Difficulties of comparing the two levels can be attributed to two factors: the quantity of material and the time required for the determination of properties.

For the determination of the **amount of substance** ($n$, mol), it is necessary to know the number ($N$) of the units (generally atoms or molecules) constituting the substance (Figure 1.10). This can be determined based on the ***Avogadro* constant** ($N_A$) using the equation $N_A = N/n$. The best value of the experimentally determined value of the *Avogadro* constant is $6.022140857(74) \times 10^{23}$ units/mol [4]. This number is the link between microscopic (at the atomic or molecular level) and macroscopic level of observing natural phenomena.

In order to demonstrate the significance of **time**, it is useful to compare the timescale of molecular vibrations and rotations with the actual time needed to determine the properties of a given substance. On the molecular level, it is the former that characterizes and determines the dynamics of chemical bonds and reactions. The timescale of these motions is in the range of $10^{-10}$–$10^{-13}$ s. Femtosecond laser techniques are probably the fastest method of measurement.

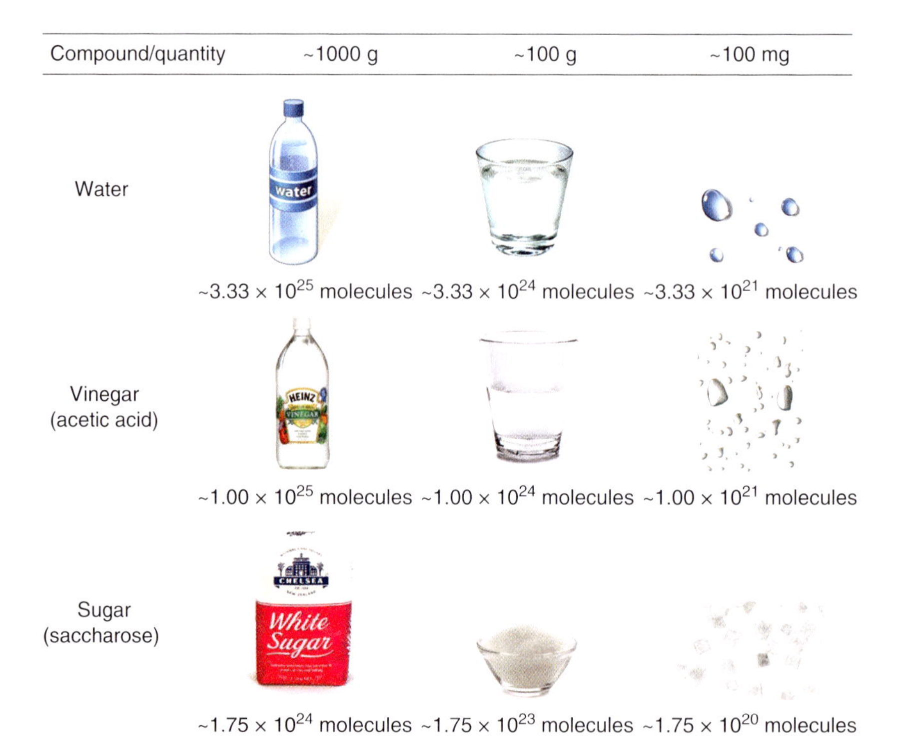


**Figure 1.10** Number of molecules in a customary amount of various substances.

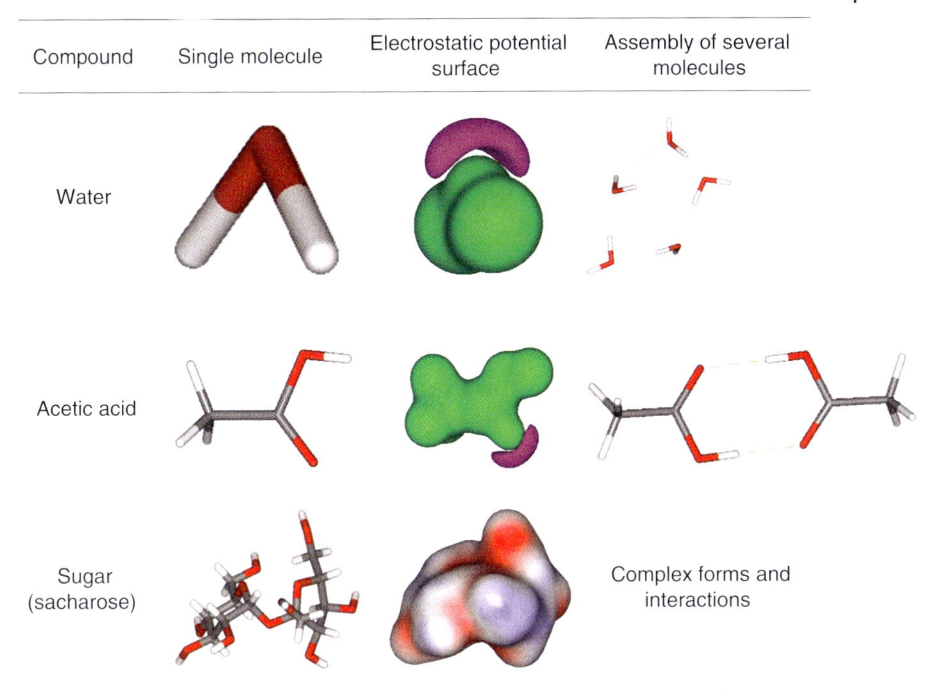


**Figure 1.11** Properties of certain well-known compounds at the molecular level.

For elaborating a method capable to study the response of molecules brought into a coherent state of laser irradiation for a few femtoseconds,[13] A. H. Zewail was awarded the Nobel Prize (1999).

For the determination of a property on the macroscopic scale (e.g., boiling point, melting point, dielectric constant, heat of combustion, optical rotation, infrared spectrum, and $pK_a$), usually at least 10 mg of material and 100 s of time are required. It is very enlightening to consider that these quantities and time mean dealing with an assembly of at least $10^{19}$ molecules and each of this vast number of individual entities undergoes, during the time of measurements, at least $10^{12}$–$10^{15}$ changes of rotational and vibrational states.

At the same time, when defining our concepts on the microscopic (molecular) level, most often we investigate the properties of a single isolated molecule (Figure 1.11, second column) in vacuum using the methods of quantum chemistry providing structural (energetic) properties, for example, electrostatic potential surfaces (Figure 1.11, third column).

When interpreting structure–property correlations, what sort of molecular assemblies should be considered as uniform is important from the point of view of properties. In general, molecules are taken as the smallest unit of a chemical entity. This interpretation assumes that molecules included in the assembly of

13) $1\ \text{fs} = 10^{-15}\ \text{s}$.

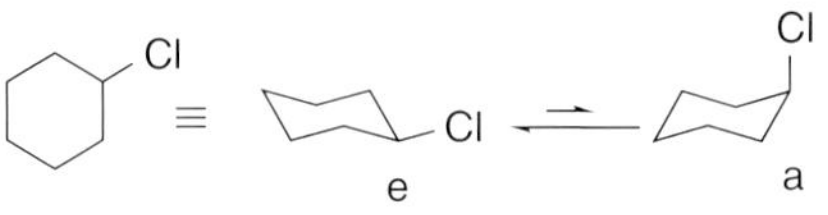

Figure 1.12 Structure of chlorocyclohexane.

a given material are identical/indistinguishable. This assumption is not fulfilled under all circumstances, as illustrated by the example of chlorocyclohexane [5] (Figure 1.12).

Under **normal conditions,**[14] we consider chlorocyclohexane as a single molecule. It is well known, however, that owing to conformational motions[15] chlorocyclohexane is a mixture of conformational states containing the chlorine atom either in an axial or in an equatorial disposition, respectively, as can be demonstrated by observing two different C–Cl bond vibration bands in the infrared spectrum.[16] Nevertheless, from most practical points of views – such as distillation, density, chromatographic behavior, and chemical reactions – chlorocyclohexane behaves as a homogeneous material. Some features of chlorocyclohexane are, however, temperature dependent. For example, at room temperature its $^{13}$C NMR spectrum contains, as expected for a single molecule, four different signals, while at −100 °C, eight signals, indicating the coexistence of the two conformations. At this temperature, chlorocyclohexane exists in two clearly discernable molecular states. Whether these molecular states can be distinguished is thus dependent on the temperature and the technique used for their differentiation (isolation, IR, or NMR spectroscopy). At low temperature, for example, −150 °C, the two forms of chlorocyclohexane behave as isomers[17] and can be isolated [6, 7]. It was proposed [8, 9] that substances with interconvertible molecular states should be considered as uniform when the energy barrier separating them is less than $kT$,[18] while different when it exceeds $kT$.

An additional difficulty of interpreting structure–property correlations is that looking at individual molecular structures is often insufficient for the assessment of even fundamental macroscopic properties. Characteristics of materials may be significantly influenced by interactions of entities of the same structure and produce assemblies (Figure 1.11, fourth column) significantly differing from the properties of the isolated molecules (Figure 1.11, second column).

14) Under *normal conditions*, generally, room temperature and atmospheric pressure are understood. Conditions of observation may be, however, much influenced by latitude and height above sea level.

15) For a more precise definition of conformation, see Section 2.2.1.

16) The energy barrier separating the axial and equatorial conformations of chlorocyclohexane is relatively high (~45 kJ mol$^{-1}$).

17) For a more precise definition of isomerism, see Section 2.2.2.

18) In macroscopic systems, with large numbers of molecules, $RT$ value is commonly used. Its SI units are joules per mole (J mol$^{-1}$): ($RT = kT \cdot N_A$). At room temperature 25 °C (298 K), $kT$ is equivalent to $4.11 \times 10^{-21}$ J; thus, $kT \cdot N_A = RT = 2.479$ kJ mol$^{-1}$.

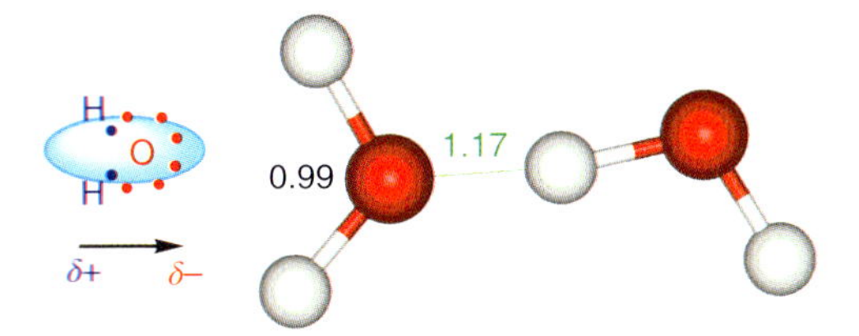


**Figure 1.13** Dipolar character of water and the hydrogen bond.

The molecular mass of water ($M = 18$) is, for instance, much lower than that of molecular nitrogen ($M = 28$) or oxygen ($M = 32$), and on the basis of molecular mass, we might infer that at room temperature water is, like air, a gas. This assumption might be confirmed by the fact that hydrogen sulfide having a similar structure but a much higher molecular mass ($M = 34$) than water is under normal conditions a gas. A similar anomaly is encountered when regarding a single isolated molecule of acetic acid. Comparing its molecular mass ($M = 60$) with that of diethyl ether ($M = 74$), we may infer that acetic acid should be more volatile than diethyl ether (bp = 34.6 °C). Similarly, methoxyethane of the same molecular mass ($M = 60$) or $n$-butane with only a slightly lower molecular mass ($M = 58$) are both gases. Acetic acid ($M = 60$) is, however, a liquid with a rather high boiling point (118 °C) and a freezing point of 16.7 °C.

When instead of individual molecular structures (Figure 1.11, second column) we regard their assembly (Figure 1.11, fourth column), the importance of interactions between individual molecules becomes apparent. Water is namely a neutral molecule but with unevenly distributed positive and negative charges (Figure 1.13). As a consequence, a hydrogen bond[19] is established between the oxygen atom with an excess of negative charge and the hydrogen atom of a second water molecule (Figure 1.13).

Formation of hydrogen bonds is possible between molecules in an intra- or an intermolecular way[20] (Figure 1.11, fourth column). Formation of intermolecular hydrogen bonds explains the surprisingly high melting and boiling points relative to their low molecular mass of water and acetic acid. Intramolecular hydrogen bonds contribute significantly to the steric structure of peptides and proteins. For instance, a hydrogen atom attached to a nitrogen atom in a peptide bond often forms a hydrogen bond with the carbonyl oxygen of another peptide bond.

19) Among secondary chemical bonds, *hydrogen bond* (*hydrogen bridge*) is the strongest. Its energy is about a tenth of that of a primary chemical bond. For its formation, an element of small dimensions and displaying high electronegativity (first of all O and N) with a nonbonding pair of electrons and a directly attached hydrogen atom is necessary. The hydrogen bond is established between this hydrogen atom and the nonbonding electron pair on the aforementioned other atom.

20) A pragmatic definition of a chemical bond by Pauling [10] is as follows: "… there is a chemical bond between two or more atoms when forces acting between them are such that they result in the formation of aggregates which are sufficiently stable that chemists should deal with them as independent molecular entities."

## References


1. Lewis, G.N. (1916) The Atom and the Molecule. *J. Am. Chem. Soc.*, **38** (4), 762–785.
2. Brecher, J. (2006) Graphical representation of stereochemical configuration [IUPAC Recommendations 2006]. *Pure. Appl. Chem.*, **78** (10), 1897–1970.
3. Cross, L.C. and Klyne, W. (1976) Rules for the nomenclature of organic chemistry. Section E: stereochemistry (recommendations 1974). *Pure Appl Chem*, **45**, 11–13.
4. CODATA (2014): http://physics.nist.gov/cuu/Constants/ (link retrieved: 31.03.2016)
5. Eliel, E.L., Wilen, S.H., and Mander, L.N. (1994) *Stereochemistry of Organic Compounds*, John Wiley & Sons, New York.
6. Jensen, F.R. and Bushweller, C.H. (1966) Separation of Conformers. I. Axial and Equatorial Isomers of Monosubstituted Cyclohexanes. *J. Am. Chem. Soc.* **88** (18), 4279–4281.
7. Jensen, F.R. and Bushweller, C.H. (1969) Separation of conformers. II. Axial and equatorial isomers of chlorocyclohexane and trideuteriomethoxycyclohexane. *J. Am. Chem. Soc.* **91** (12), 3223–3225.
8. Eliel, E.L. (1976) Stereochemistry since LeBel and van't Hoff. Part I. *Chemistry* **49** (1), 6–10; Stereochemistry since LeBel and van't Hoff. Part II. *Chemistry* **49** (3), 8–13.
9. Eliel, E.L. (1977) On the concept of isomerism. *Israel J. Chem.* **15** (1–2), 7–11.
10. Pauling, L.C. (1960) *The Nature of the Chemical Bond and the Structure of Molecules and Crystals. An Introduction to Modern Structural Chemistry.* 3rd ed, Cornell University Press, Ithaca, N.Y.

# 2
# Concepts of Stereochemistry

After clarification of the most important aspects of the relationship of macroscopic and microscopic levels, it is necessary to classify and exactly define the most important stereochemical concepts. In the present work, we mostly use the definitions proposed by Eliel and associates [1].

## 2.1
## Elements and Groups of Symmetry

Since the basic concepts of stereochemistry are much dependent on the symmetry relationships of the object (molecule) studied, before discussing them in detail it is appropriate to discuss briefly the symmetry elements and symmetry groups.

The concept of symmetry is used in more than one meaning. According to one school of thought, a symmetrical form is proportional and in equilibrium and symmetry is the harmony unifying the individual parts to a whole. A second meaning is derived from the Greek words *sum* (συμ) and *metros* (μετροσ) and is strictly speaking the common measure of things. Symmetry is thus at the same time an esthetic category in the everyday sense and also a scientific concept in an exact mathematical sense, which is suitable to differentiate between a variety of regular patterns and enables to classify them. We encounter symmetry both in natural phenomena and human creations.

In chemistry, molecular symmetry is describing symmetry relationships of molecules and classifying them according to their symmetry. Molecular symmetry is of basic importance in chemistry because it helps to understand and predict several molecular properties, such as dipole moments and some direction-dependent interactions of molecules. It is therefore not surprising that almost all textbooks of chemistry at the university level, should they deal with quantum chemistry, spectroscopy, inorganic, organic, or physical chemistry, devote at least a short section to discussing symmetry.

*Stereochemistry and Stereoselective Synthesis: An Introduction,* First Edition.
László Poppe and Mihály Nógrádi.

Companion Website: www.wiley.com/go/poppe/stereochemistry

Three basic concepts are associated with symmetry:

**Element of symmetry**: An element of symmetry is composed of the totality of points of a symmetrical phenomenon, the topological parameters of which remain invariant during a transformation.

**Symmetry operation**: An operation that transforms the structural elements of a symmetrical phenomenon in itself (superposition operation).

**Member of symmetry**: A member of symmetry is a structural element of a symmetrical phenomenon. It behaves as a unit in a symmetry operation (i.e., it is the totality of those particular points that are displaced in a symmetry operation). The relationship of members of symmetry remains unchanged after the transformation.

The character of symmetry elements depends on whether they refer to an object of finite dimensions (stereochemistry deals with such cases) or also to objects repeated in infinite extension (this is the subject of crystallography[1)]). In the following, we are going to review only symmetry elements associated with objects of finite dimensions, those important from the aspect of stereochemistry. These are the axis of symmetry $C_n$, the plane of symmetry $\sigma$, the center of symmetry $i$, and the alternating axis of symmetry $S_n$.[2)]

The axes of symmetry $C_n$, which require only rotation for symmetry transformations, are the **symmetry elements of first order**, while the plane of symmetry $\sigma$, the center of symmetry $i$, and the alternating axis of symmetry $S_n$ are **symmetry elements of second order.** An exact discussion and classification of molecular symmetries is possible by an appropriate adaptation of mathematical group theory.

A $C_n$ axis of symmetry is, by definition, an axis around which by a rotation of $360/n$ degrees the original object (molecule) results in an object that is indistinguishable and can be brought to superposition with the original one. A complete rotation is equivalent with an identity transformation ($C_1 = E$). Figure 2.1 shows examples of various objects (molecules) possessing an axis of symmetry $C_n$.

Rotations according to axes of symmetry $C_n$ are "real" transformations in the sense that points of molecules brought into superposition are real material points. Inspecting each of the objects in Figure 2.1 possessing symmetry elements $C_2$ (a), $C_3$ (b), $C_4$ (c), $C_6$ (d), and $C_\infty$ (e), it can be observed that in the course of rotation around an axis of symmetry $C_n$ by an angle of $360°/n$, every mass point is taking up a real mass position indistinguishable from the original one. Until the discovery of fullerenes (f) it seemed that no object of a $C_5$ symmetry axis does exist among real molecules of finite dimensions. Conical (e) or cylindrical objects possess a rotational symmetry ($C_\infty$).

1) With objects repeated in infinite extension, a translational element of symmetry also contributes to symmetry conditions. Rotation and translation (twisting) result in *helicogyrs*. Reflection and translation produce *glide planes*.

2) For the theory of point groups, it is important to take into account the identity transformation $E$ too.

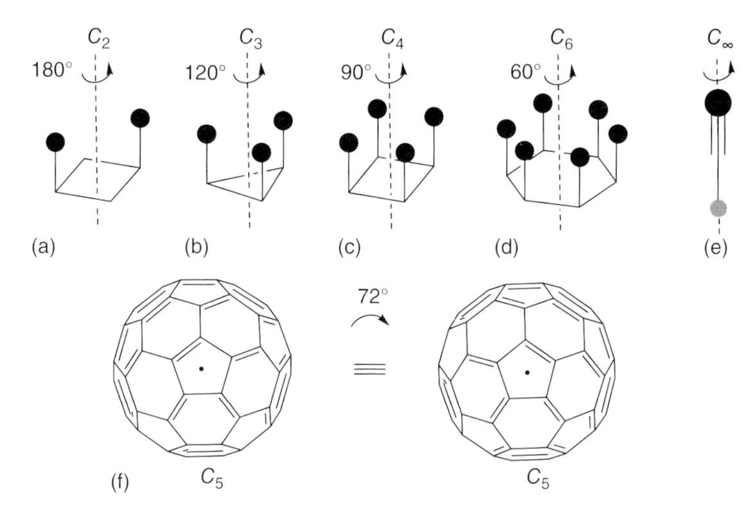


**Figure 2.1** Examples of objects having axes of symmetry ($C_n$).

Among real molecules, a $C_2$ axis of symmetry can be found in water, a $C_3$ axis in ammonia, and a $C_6$ axis in benzene. A molecule can have more than one axis of symmetry. Thus, in benzene, three $C_2$ axes can be found coplanar with the benzene ring, for example, the one transecting carbon atoms 1 and 4. The axis having the highest $n$ number is the **main axis**.

Reflecting an object through a plane of symmetry $\sigma$, an object indistinguishable from the original one is obtained. In other words, along a straight line perpendicular to the $\sigma$ plane from a given mass point on the other side of the mirror plane, a mass point indistinguishable from the original can be found at identical distance. Therefore, the plane of symmetry $\sigma$ is often called a **mirror plane**. In Figure 2.2, two objects possessing a plane of symmetry are shown. In object Figure 2.2a (a molecule), the plane of symmetry $\sigma$ touches centers of masses (atoms), while in objects (molecules) b and c, $\sigma$ dissects bonds.

Turning to real molecules, in water, two planes of symmetry can be found. One of them dissects all three atoms of the molecule, while the other one is

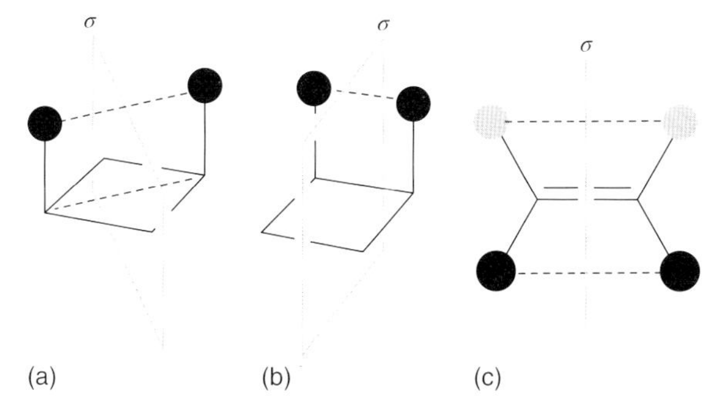


**Figure 2.2** Examples of objects with a plane of symmetry $\sigma$.

perpendicular to the first one and includes only the oxygen atom. The plane of symmetry coplanar with the main axis is the *vertical* plane of symmetry ($\sigma_v$), while the one perpendicular to the former is the *horizontal* plane of symmetry ($\sigma_h$).[3] Accordingly, in benzene ($C_6H_6$), the plane of symmetry incorporating the six carbon atoms is a horizontal one ($\sigma_a$), while the plane of symmetry perpendicular to the former and including atoms $C_1$–$C_4$ is a perpendicular one ($\sigma_v$).

The center of symmetry $i$ is a point inside an object, which satisfies the condition that along a straight line drawn from any point of the object passing the point of symmetry at the other end of the line at equal distance, another mass point indistinguishable from the former can be found (Figure 2.3). The center of symmetry is often also called a center of inversion. The center of symmetry $i$ can be situated either at an "*empty*" part of the space, where no real atom can be found (a and b in Figure 2.3) or coincide with a real atom (c in Figure 2.3). Among real molecules, xenon-tetrafluoride ($XeF_4$) is an example when the center of symmetry $i$ coincides with the center of the xenon atom, while in benzene ($C_6H_6$), the center of symmetry resides in the "empty" space, at the center of the aromatic ring.

The **alternating axis of symmetry** (or **rotation–reflection axis**) $S_n$ is such a complex element of symmetry according to which the object is first rotated around an axis by $360°/n$ followed by reflecting the original object through a mirror plane perpendicular to this axis. The resulting object (molecule) can be brought into superposition with the original object. It can be easily understood that a rotational-reflection operation involving an alternating axis of symmetry by complete rotation (360°) can only produce an object identical with the original when $n$ is an even number. Note that the alternating axis of symmetry $S_1$ is identical with the mirror plane ($S_1 = \sigma$), while $S_2$ is identical with the center of symmetry $i$ ($S_2 = i$) .

In Figure 2.4, some objects with an alternating axis of symmetry $S_n$ are shown. Object a exhibits an $S_2$ rotation–reflection axis (i.e., a center of symmetry $i$), objects b and c an $S_4$ type alternating axis of symmetry, and objects d and e an $S_6$ rotation–reflexion axis. Examples of real molecules with an alternating axis of symmetry $S_n$ are methane ($CH_4$) or tetrafluoromethane ($CF_4$) possessing three $S_4$ rotation–reflection axes (structure c). Structure e is realized in the open conformation[4] of ethane ($C_2H_6$) and hexafluoroethane ($C_2F_6$), which possesses an $S_6$ type alternating axis of symmetry.

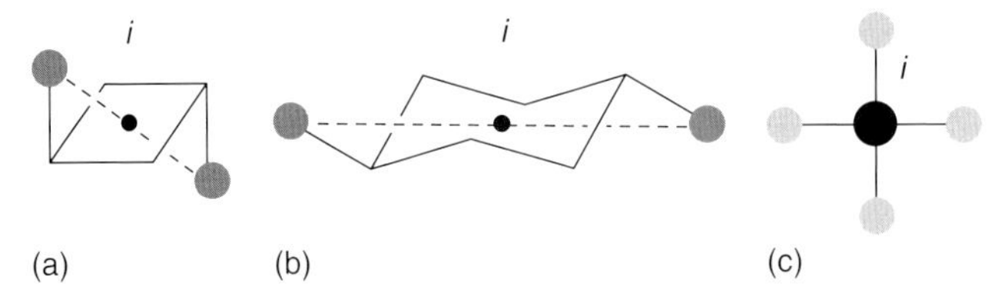


**Figure 2.3** Examples of objects having a center of symmetry *i*.

3) The plane of symmetry can also be characterized by its orientation in Cartesian coordinates, for example, ($xz$) or ($yz$).

4) For the definition of conformation, see Section 2.2.1.

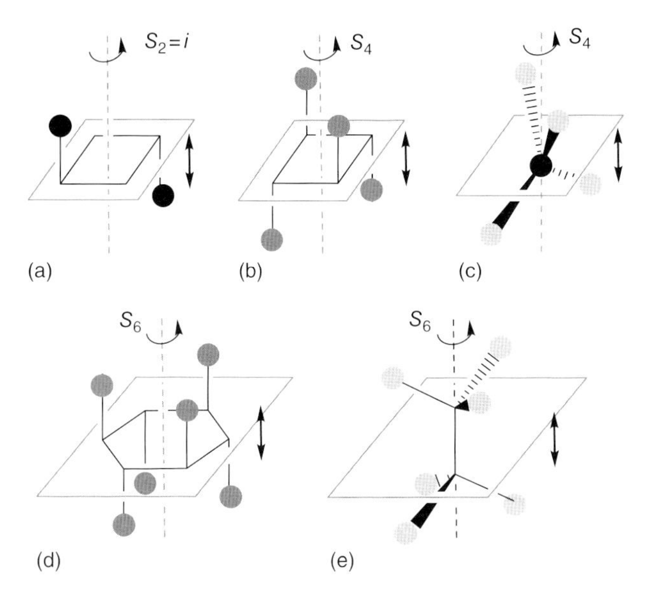


**Figure 2.4** Examples of objects with an alternating axis of symmetry $S_n$.

A **point group** is such a group of symmetry transformations in the mathematical sense that can be carried out with the aid of symmetry elements. In the course of symmetry transformations permitted within the given group, at least one point of the object remains fixed.[5] From the point of view of basic features connected with symmetry, objects belonging to the same point group behave similarly.

## 2.2 Classification of the Basic Concepts of Stereochemistry

After having clarified symmetry relationships, for interpretation and systematization of the basic concepts of stereochemistry, it is helpful if we are conscious of the level at which these concepts are valid: microscopic level → structure and macroscopic level → properties. One way of classification of concepts is whether they refer (i) to a single object (or part of it), (ii) to the relationship of two objects, or (iii) to the identical parts of the single object, and (iv) to the possibilities of approaching the object (cf. Table 2.1).

When studying structural features of objects with the purpose of classification and interpreting the meaning of concepts at the microscopic level, we regard them as practically homogeneous and constant in time. Note that as shown

5) In the case of molecules with finite dimensions, the number of point groups relevant for stereochemistry is smaller (30), since crystallographic point groups (32) permit also translational symmetries in 3D.

Table 2.1 Classification of some basic concepts of stereochemistry.

| Concepts referring to a single object (molecule) or part of it  | Concepts referring to two objects (molecules) (*isomeric relations*)  | Concepts referring to the relationship of parts[a] of or mode of approach[b] to a single object (molecule) (*topicity*) 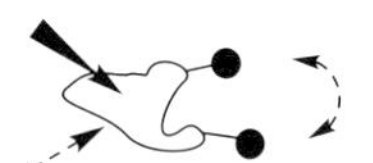 |
|---|---|---|
| **Constitution**[c)]: Order of connection of atoms and bonds building up the molecule (connectedness). It can be described by a 2D bonding matrix | **Constitutional isomerism**: Constitution of the two molecules compared is different (and they can be separated) | **Groups**[a)] **attached to sites of different constitution** |
| **Configuration**[d)]: Under well-defined conditions stable arrangement of parts of the molecule, which can only be described by steric characteristics (stereodescriptors) | **Stereoisomerism:** The two molecules compared are of the same constitution, they are nevertheless nonidentical and can be separated May belong to two groups: diastereomers or enantiomers | **Stereoheterotopic groups**[a)]**/faces**[b)]**:** (i) Distinguishable groups attached to sites of identical constitution (ii) Distinguishable modes of approach of a face Their relationship may be twofold: diastereotopic or enantiotopic |
| **Conformation**[c)]: Steric (3D) arrangement of the atoms of a molecule that is energetically permissible under the given conditions | **Diastereomers:** Nonenantiomeric stereoisomers | **Diastereotopic groups/faces:** Related neither by an axis ($C_n$) nor by symmetry elements of second kind ($S_n$) |
| **Chirality**[c)]: A chiral object (molecule) is nonidentical with its mirror image | **Enantiomers:** Stereoisomers related as nonidentical mirror images | **Enantiotopic groups/faces:** Not related by an axis of symmetry ($C_n$) but by a symmetry element of second kind ($S_n$) |
| **Achirality**[c)]: An achiral object (molecule) is identical with its mirror image | | **Homotopic groups/faces:** Related by an axis of symmetry ($C_n$) |

a) **Groups**: Parts of a molecule of identical (homomorphic) structure.
b) **Faces**: Two sides of a plane defined by a planar part of a molecule [a = X(b)c], where to a central atom X, three ligands (a, b, and c) are linked.
c) Refers strictly to a single molecule.
d) Refers only to a well-defined part of the molecule, with a stable steric arrangement under the given energy conditions.

before, properties linked to structural features can generally only be established at the microscopic level because in a multitude assembly, individual molecules can interact with each other even in case of "pure" substances. Therefore, when defining concepts related to a given structural state, we have to know when and how they contain elements that can be correlated to properties measurable on a macroscopic level.

Concepts important for the description of the state of a given molecule can be assigned at the structural level to a single object (molecule) or to a part of it. Accordingly, to the first group (cf. Table 2.1, column 1) concepts were listed for the interpretation of which it is sufficient to regard a single molecule.

It is often important how much individual molecular states differ from each other, since such differences determine the separability of an assembly of molecules, important both from a synthetic and an analytical point of view. This requires examining the degree of similarity of two molecules and interpretation of related concepts (Table 2.1, column 2).

It is also quite common that in the course of a chemical synthesis or applying a method based on physical, chemical, or biological interactions, a molecule can behave in more than one manner. This means that the molecule can participate in a given process showing different "facets." Examination of the relationship of different parts of a single molecular entity and definition of concepts describing these intramolecular relationships is also of prime importance (Table 2.1, column 3).

About the definition of concepts related to a molecular entity and parts thereof, there is no consensus in the literature, both older and more recent. Therefore, it is useful to become acquainted with the principles along which the individual concepts have been elaborated. Where definitions according to IUPAC rules [2–4] were unequivocal, we preferred them, where – in our opinion – more accurate ones should be introduced, we indicated and explained the differences suggested.[6] In the following, we list the concepts summarized in Table 2.1 inclusive of their definitions as interpreted both by us and others.

### 2.2.1 Concepts Related to a Single Object (Molecule) or Parts of Thereof

**Constitution:** Description of the quality and connectedness of atoms constituting a molecule (inclusive bond orders) but disregarding any differentiation concerning spatial arrangement.[7]

**Configuration:** Arrangement of a well-defined and, under the given conditions, stable part of the molecule that can only be characterized by spatial features (stereodescriptors).

[According to IUPAC recommendations [2–4], the concept of configuration refers to such an arrangement of the atoms of a molecule, which permits to

6) Interpretation of concepts was largely based on discussions in Ref. [1].

7) This definition is in accordance with the IUPAC recommendations [2–4]. *Constitution* practically corresponds to the description of molecules with bond matrices (cf. Figures 1.8 and 1.9).

distinguish stereoisomers, the isomerism between which is not due to conformation differences. Note that the IUPAC definition of isomerism (stereoisomerism) does not include the criterion of isolability of isomers, and therefore the concepts configuration and conformation are not clearly distinguished.]

**Conformation:** Energetically permissible spatial (3D) arrangements of the atoms of a molecule under the given conditions (generally taken as normal conditions[8])).

[According to IUPAC recommendations [2–4], conformation is such an arrangement of atoms, which produces stereoisomers differing from each other formally by rotation(s) around single bonds. This interpretation is restrictive since individual conformational states may differ not only by torsion angles but also by bond angles or even by bond lengths].

**Chirality:** An object (molecule) is chiral when it is not identical with its mirror image.

[According to IUPAC recommendations [2–4], chirality is that particular geometrical property of a rigid object (or spatial arrangement of points or atoms) that cannot be brought into superposition with its mirror image. Such objects lack symmetry elements of the second kind ($\sigma = S_1$ (plane of symmetry) $i = S_2$ (center of symmetry; $S_n$ (alternating axis of symmetry))].

**Achirality:** Any object (molecule) is achiral when it is identical with its mirror image. The simplest definition can be derived from the very name: an object is achiral when it is not chiral.

[According to IUPAC recommendations [2–4], achiral objects possess a symmetry element of second kind.]

Interpretation of the above concepts is problematic, first of all for *configuration* and *conformation*. If we interpret them rigidly and are not aware that the concept *isomerism*, as defined by the IUPAC rules, is inexact insofar, it lacks the criterion of separability based on energetics. Misinterpretations are possible because in certain molecules, rotation around single bonds is hindered and therefore stereoisomerism based on conformational differences, for example, *atropisomerism* is possible. On the other hand, there are cases when a molecule is chiral but cannot be separated into pure enantiomers. For example, amines substituted with three different groups are chiral, but the energy barrier between the enantiomers is so low that enantiomers cannot be separated (apart from some special cases such as Tröger's base that exhibits chirality due to the presence of two bridgehead stereogenic nitrogen atoms).

Although the criterion of isolability, similar to "normal conditions"[9] cannot be precisely defined, it is nevertheless, quite useful. According to one of the suggestions [5], two rapidly interconverting molecular species can only be regarded as distinct when the energy barrier separating them is higher than $kT$[9]) (Figure 2.5).

8) Under *normal conditions*, generally room temperature and atmospheric pressure are understood. Conditions of observation may be, however, much influenced by latitude and height above sea level.

9) In macroscopic systems, with large numbers of molecules, $RT$ value is commonly used. Its SI units are joules per mole (J mol$^{-1}$): ($RT = kTN_A$). At room temperature 25 °C (298 K), $kT$ is equivalent to $4.11 \times 10^{-21}$ J; thus, $kTN_A = RT = 2.479$ kJ mol$^{-1}$.

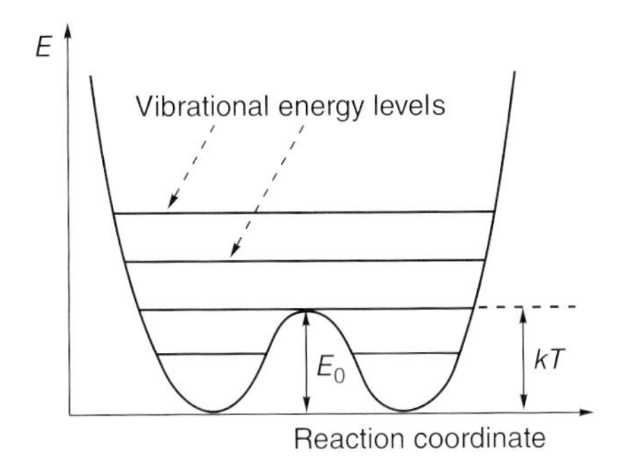


**Figure 2.5** Double potential energy minima. If $E_0 < T$, the first vibrational energy level is higher than the barrier separating the two energy minima [5].

Should the energy barrier be less than $kT$, then we regard the system as a single species.

Under normal conditions (25 °C), a transformation barrier of 84 kJ mol$^{-1}$ (20 kcal mol$^{-1}$) free enthalpy corresponds to a transformation rate of $1.3 \times 10^{-2}$ s$^{-1}$ (i.e., $t_{1/2} = 1$ min, which means that the compounds cannot be separated by conventional methods), while a value of 105 kJ mol$^{-1}$ (25 kcal mol$^{-1}$) corresponds to a transformation rate of $1.3 \times 10^{-2}$ s$^{-1}$ (i.e., $t_{1/2} = 66$ h) permitting the separation of the isomers by preparative methods.

Despite the problems associated with the criterion of separability, it has a sound thermodynamic foundation and it is of consequence whether a given molecular state can be regarded as an isolable entity. According to Gibb's rule, $P + DF = C + 2$ (where $P$ = the number of phases, $DF$ = the number of degrees of freedom, and $C$ = the number of components), since $C$ is equal to the number of *isolable* components.

### 2.2.2
### Concepts Referring to the Relationship of Two Objects (Molecules) (Isomerism, Isomeric Relationship)

The domain of the concept of **isomerism** refers to the relationship of two molecular objects, and its importance is associated with the separability of substances. As pointed out earlier, definition of isomerism is not straightforward[10] and for an exact definition of **isomerism**, recommendations of IUPAC rules [2–4] are of not much help either. Consequently, along with explanations, occasionally we recommend definitions at variance with IUPAC recommendations.

**Isomerism:** relationship between **isomers** [2–4].

**Isomer:** one among compounds (or molecular entities) that have identical chemical formulas but can be ***isolated*** and exhibit different physical and/or chemical properties.

10) According to their most common definition, **isomers** have the same molecular formula but are nevertheless different. This definition is inexact, since it lacks the definition of the circumstances (thermal energy level, environment, state, and time interval) under which the compounds are viable as stable entities (cf. Section 2.2.2).

[According to IUPAC recommendations [2–4], an isomer is one of the several species (or molecular entities) that have the same atomic composition (molecular formula) but different line formulae or different stereochemical formulae and hence different physical and/or chemical properties.]

We have to admit that even our proposed concept of isomerism including isolability is not completely unambiguous,[11] since the criterion for isolability cannot be defined precisely. Nevertheless, our proposition is nearer to the everyday praxis of organic chemist than the IUPAC recommendations, since when taking them strictly, then labile, but chiral forms of tertiary amines, the tautomeric pair of acetaldehyde and vinyl alcohol, and further even different conformations of *n*-butane (e.g., its eclipsed and open conformations) should be regarded as isomers.

Isomers can arise (taking into account the criterion of isolability) for multiple reasons (Figure 2.6). To distinguish the types of isomers different terms are given.

**Constitutional isomer:** one among isomers, that is, molecules that have different constitutions but the same chemical (molecular) formula. In constitutional isomers, the *connectedness* of the atoms is different.
[According to IUPAC [2–4] recommendations, constitutional isomerism is a phenomenon associated with structures differing in their constitution and represented by different linear formulas, for example, $CH_3OCH_3$ and $CH_3CH_2OH$.][12]

**Stereoisomer:** one among isomers that display the same constitution.

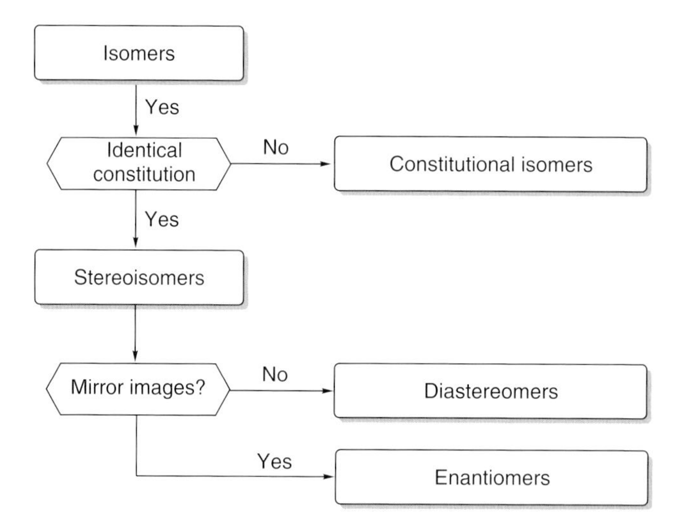


**Figure 2.6** Possible forms of isomerism.

11) A relatively exact definition of **tautomers** may be that these are inseparable isomers in a dynamic equilibrium.

12) See Figure 1.8. According to the definition of *constitutional isomerism* by IUPAC, which lacks the criterion of isolability, acetaldehyde and vinyl alcohol (see Figure 1.2) should be considered as constitutional isomers, while in fact these are tautomers.

[According to of IUPAC recommendations, stereoisomers [2–4] are isomers of the same constitution but having different spatial arrangement of their atoms. Accordingly, stereoisomerism [2–4] is a kind of isomerism that originates from the different spatial arrangement of atoms, without differing in the connectivity or bond order of isomers.][13)]

**Diastereomer:** one among stereoisomers not related as mirror images. *Diastereomers differ in all of their physical properties* as well as in their chemical behavior toward both chiral and achiral reagents. *Diastereomers can be either chiral or achiral.*

[This definition is in concordance with the recommendations of IUPAC [2–4] stating that diastereoisomerism is a kind of stereoisomerism different from enantiomerism. Diastereomers are stereoisomers that are not related as mirror images.][14)]

**Enantiomer:** one among those nonsuperposable stereoisomers that are related as mirror images (enantiomers thus form pairs of mirror images). *Enantiomers must be chiral. Scalar physical properties of enantiomers are identical; they only differ in vectorial physical properties*, such as optical rotation. Their chemical behavior toward achiral agents is identical, while it may differ toward chiral agents [2–4].

[This definition is in accord with IUPAC recommendation [2–4] on stereochemistry stating that enantiomers are pairs of molecular entities, which are not superposable and are related as mirror images.][15)]

Conformation/configuration and isomeric relationships are illustrated by the isomers of cyclohexane ($C_6H_{12}$) (Figure 2.7). It follows from the definitions and examples in Figure 2.7 that it depends on the type of isomerism under which conditions components of a mixture of isomers can be separated. Constitutional isomers and diastereomers are differing in most of their chemical and physical properties, while enantiomers can only be differentiated by direction-dependent (vectorial) properties and exhibit different chemical reactivities only toward chiral partners. Examples in Figure 2.7 well illustrate the necessity of giving exact definitions and justify taking into consideration the criterion of isolability.

By the example of cyclohexane shown in Figure 2.7, it can be seen that its different conformational states cannot be isolated and are not separable molecular entities. Here, the imprecision of the IUPAC definition of *conformation* becomes apparent since the three states differ not only by rotation around single bonds. In the chair and the half-chair conformations, C–C–C bond angles are also different.

13) According to the IUPAC definition of *stereoisomerism* that lacks the criterion of isolability, the structures of chlorocyclohexane shown in Figure 1.12 have also be considered as *stereoisomers*, while in fact they are only different conformational states containing the chlorine atom in equatorial and axial orientation, respectively. Under normal conditions, chlorocyclohexane behaves as a homogeneous material.

14) The IUPAC definition of *diastereoisomerism* seems to be correct, but since the basic definitions of isomer/stereoisomer fail to contain the criterion of isolability, this concept is also inexact.

15) The IUPAC definition of *enantiomeric* relationship seems to be correct, but since the basic definitions of isomer/stereoisomer fail to contain the criterion of isolability, this concept is also inexact.

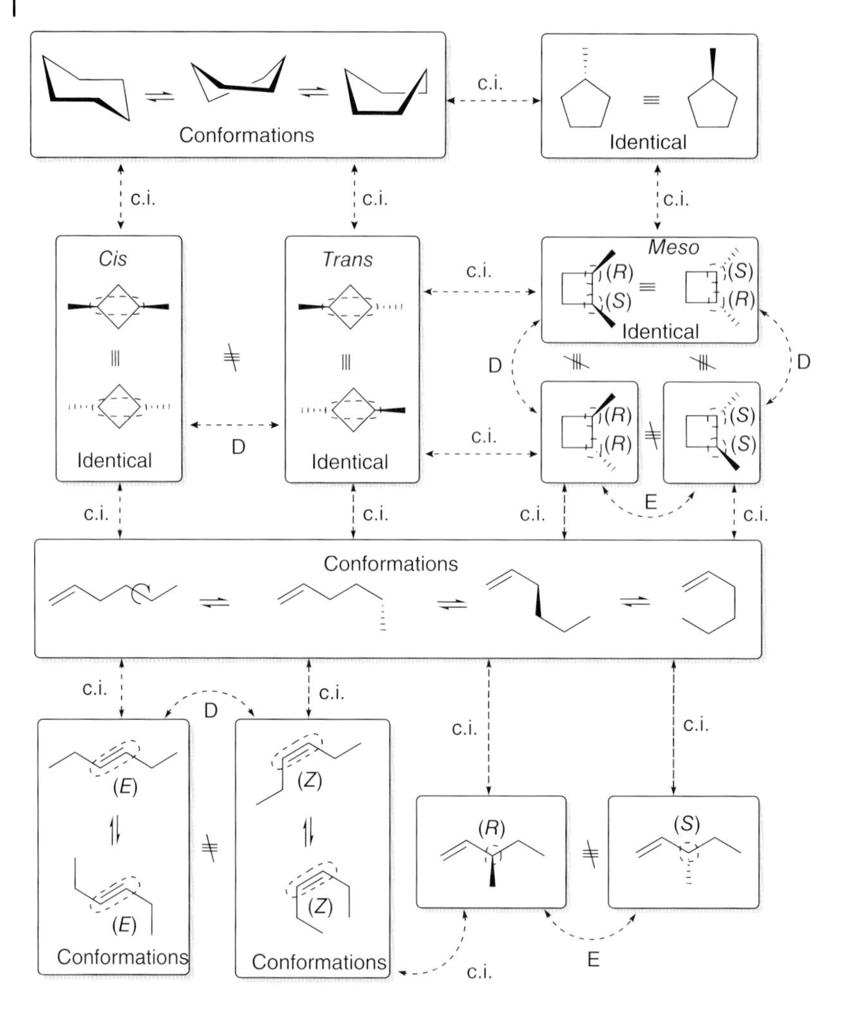


**Figure 2.7** Conformation, configuration, and various types of isomerism. The molecules shown all have the same molecular formula ($C_6H_{12}$). Molecules placed into a common box are either identical (≡) or exist as a set of conformations capable to interconversion (⇌). (c.i.: constitutional isomers; D: diastereomers; and E: enantiomers).

The two states of methylcyclopentane shown are related as mirror images, but these are superposable and therefore *achiral.*

Figure 2.7 also helps to understand the concept of *configuration.* In the figure, several parts of the molecule are circumscribed by dashed lines. The spatial arrangement of these is constant for some reason, that is, it can only be changed by transgressing a considerable energy barrier (e.g., by bond rupture). Configuration refers therefore not to the entire molecule, but characterizes only the arrangement of a part of it fixed in space. This arrangement can only be described by ***stereodescriptors*** referring to orientation in space (such as *cis/trans* in case of 1,3-dimethylcyclobutane isomers, (*R*)/(*S*) with 1,2-dimethylcyclobutane and

3-methylpent-1-ene isomers, and of (*E*)/(*Z*)-hex-3-ene isomers, respectively). It is apparent from Figure 2.7 that in case of (*E*)- and (*Z*)-3-hexene, the *configuration* of the parts marked is indifferent to the *conformational* state of the mobile parts of the molecule. We can notice that stereoisomers always contain some stereogenic element[16] that exhibits some feature fixed in space, that is, a *configuration.*

Figure 2.7 also clearly demonstrates that while every molecule has a *constitution* and resides in a certain *conformational* state, there exist wide ranges of molecules that lack a part to which a *configuration* can be assigned (e.g., cyclohexane, methylcyclopentane, and 1-hexene).

*Chirality* and *achirality* can be interpreted at the microscopic level as the property of a fixed molecule. The two properties are exclusive: any of the molecular objects shown in Figure 2.7 are either *chiral* or *achiral*. Extension of the concepts *chiral/achiral* to a material assembly is not trivial; IUPAC rules suggest avoiding it when discussing the concept "chiral" [2–4]. Problem of extending the concept "*chirality*" to material assemblies is apparent in the case of 1-hexene, shown in Figure 2.7, where both chiral and achiral conformational states at the microscopic level are shown; while at the macroscopic level, 1-hexene is treated as an *achiral* compound.

A similar problem arises when the concepts *chiral/achiral* are applied not to the complete assembly of molecules but to a part of it. This use is also discouraged by the IUPAC recommendations [2–4] discussing the concept "*chiral.*"

One should not be misled by the way the IUPAC rules [2–4] define the concept of a *center of chirality.*[17] These do not imply that the presence of a *center of chirality* does necessarily mean that the molecule itself is chiral. This is well exemplified by the *meso* stereoisomer of 1,2-dimethylcyclobutane, which is achiral, while its diastereomer (*R*,*R*)-1,2-dimethylcyclobutane is chiral. Diastereomers of a compound can even all be achiral, such as *cis*- and *trans*-1,3-dimethylcyclobutane *or* (*E*)- and (*Z*)-hex-3-ene.

Note that there exists no universal and completely exact definition of isomerism and consequently none exists for stereoisomerism either. Connectedness in molecules is generally clearly defined and describing constitution, and thus when describing a molecule, single and higher order bonds are taken into consideration. As a cautionary note, Figure 2.8 quotes situations when the presence of bonds with an order of less than one generates molecular associates related as stereoisomers.

Since definitions are never absolutely exact and universal, in cases where differences cannot be completely covered by definitions, it is advised to give precise descriptions. Accordingly, with the carboxylic acids shown in Figure 2.8 we can treat them as dimers in diastereomeric relationship, although regarding the criterion of separability they are not diastereomers. Complexes

16) *Stereogenic elements* will be discussed in detail in Section 2.3.

17) A *center of chirality* is an atom bearing ligands in such a spatial arrangement, which cannot be brought into superposition with its own mirror image. This means an extension of the concept of *asymmetric carbon atom* to atomic centers of any kind, as for example, $P^+$*abcd* or to $N^+$*abcd* [2–4].

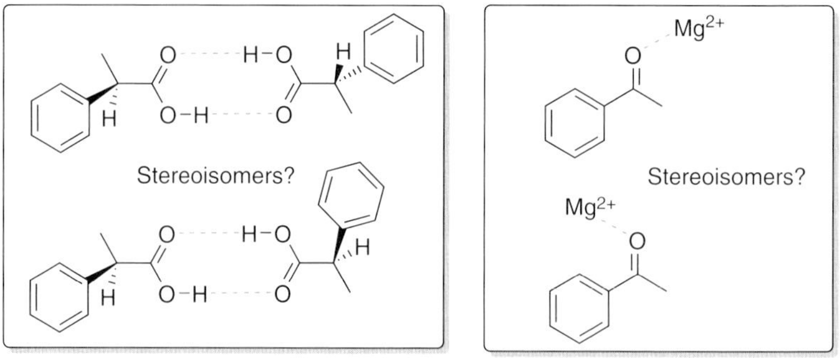

Figure 2.8 Molecules related as stereoisomers owing to bonds of an order of less than one.

of acetophenone with magnesium are not identical at the molecular level and although their relationship is diastereomeric, we do not treat them as diastereomers.

Misunderstandings associated with concepts of the relationship of two objects (molecules) are caused most often if level and timescale are disregarded. At the molecular level, for instance, we can examine the relationship of two molecules in a fixed conformation, but even then we should keep in mind that comparison of a chemical structure in a given state (a concept at the microscopic level) and its properties (a concept associated with macroscopic assemblies of material) may cause problems.[18)]

Therefore, supported by Figure 2.9, we suggest that

(i) if two molecules are compared at the *microscopic (molecular) level,* without taking into account properties not directly derived from their structure, then when making comparisons we should use terms referring to the *isomeric relationship of the given type.* In this sense, it is true for the forms shown on the left of Figure 2.9 that they are related as constitutional isomers (a and d), diastereomers (b), or enantiomers (c and e), respectively.

(ii) When substances consisting of molecules are related as isomers and we characterize them *at the macroscopic level* based (also) on properties derivable from those associated with an isolable substance, we should use the term for the *isomer* of the given type.
[In this sense, it is correct to state that ethanol and dimethyl ether (Figure 2.9) are constitutional isomers (a), (*Z*)- and (*E*)-hex-3-ene are diastereomers (b), and (*R*)- and (*S*)-3-methylpent-1-ene are enantiomers (c). However, in the sense of the criterion of isolability, we cannot state that acetaldehyde and vinyl alcohol are constitutional isomers (in fact they are a mixture of tautomers (d), further owing to the low-energy barrier of interconversion, the (*R*)- and (*S*)-forms of ethyl-phenyl-methylamine cannot be isolated as enantiomers (e)).]

18) See Section 1.4.

It can be seen therefore that it may lead to confusion if we apply the concept of isomerism according to IUPAC recommendations [2–4], without taking into account isolability at the macroscopic level. In this sense, we would regard the *R* and *S* forms of ethyl-phenyl-methylamine as enantiomers and not merely as structures in an enantiomeric relationship. A similar misunderstanding would be to treat the various conformations of 1-hexene, among which there can be found both chiral and achiral states as stereoisomers (since they do not fulfill the criterion of isolability). At the same time, we may state unambiguously that the first achiral conformational state shown in Figure 2.7 on the left and the chiral conformational state shown to the right are related as diastereomers (despite being nonisolable).

Unambiguous formulation is aided by the above considerations. Only when stating the relationship of two objects at the microscopic level, the term *isomeric relationship* should be used. While referring to a macroscopic assembly of molecules in an isomeric relationship, or to a property that can be determined only at the macroscopic level, then (and only then) the term *isomer* should be used.

### 2.2.3 Concepts Referring to the Relationship of Parts of an Object (Molecule) or Modes of Approach to an Object (Topicity)

From the point of view of the behavior of a molecule, the study of the relationship between parts of the same constitution (connectedness) may be of relevance. These can be arranged within the molecule in different ways (Figure 2.10). Such relationships are described by different terms and subject to different symmetry relationships.

***Homotopic groups*** [2–4]: Homotopic groups within a molecule are atoms or groups having the same constitution and related by an $n$-fold axis of symmetry ($C_n$, $n = 2, 3, \ldots$).

As examples, the chiral form of tartaric acid ($C_2$ axis, two homotopic carboxyl groups), chlorines in chloroform ($C_3$ axis, three homotopic chlorine atoms) or $\alpha$-cyclodextrin ($C_6$ axis, six homotopic glucose units) can be quoted (Figure 2.11). Homotopic groups can often be found in chiral molecules such as the two methyl groups in (4*R*,5*R*)-2,2-dimethyl-4,5-diphenyl-1,3-dioxolane (Figure 2.11). Homotopic groups cannot be distinguished by any physical or chemical methods. Replacing any member of a set of homotopic groups gives the same product (Figure 2.11, example of chloroform).

***Enantiotopic groups*** [2–4]: Enantiotopic groups are atoms or groups of identical constitution, which are exclusively related by some symmetry element of second kind, that is, a plane of symmetry ($\sigma$) or an alternating axis of symmetry ($S_n$).

An example may be the two groups "c" in a molecule of type Cabcc, the hydroxyl groups in glycerol ($\sigma$), or in *meso*-tartaric acid ($i$) marked gray in Figure 2.12. In an achiral environment, physical and chemical properties of enantiotopic groups are identical. They give identical signals in their nuclear magnetic resonance

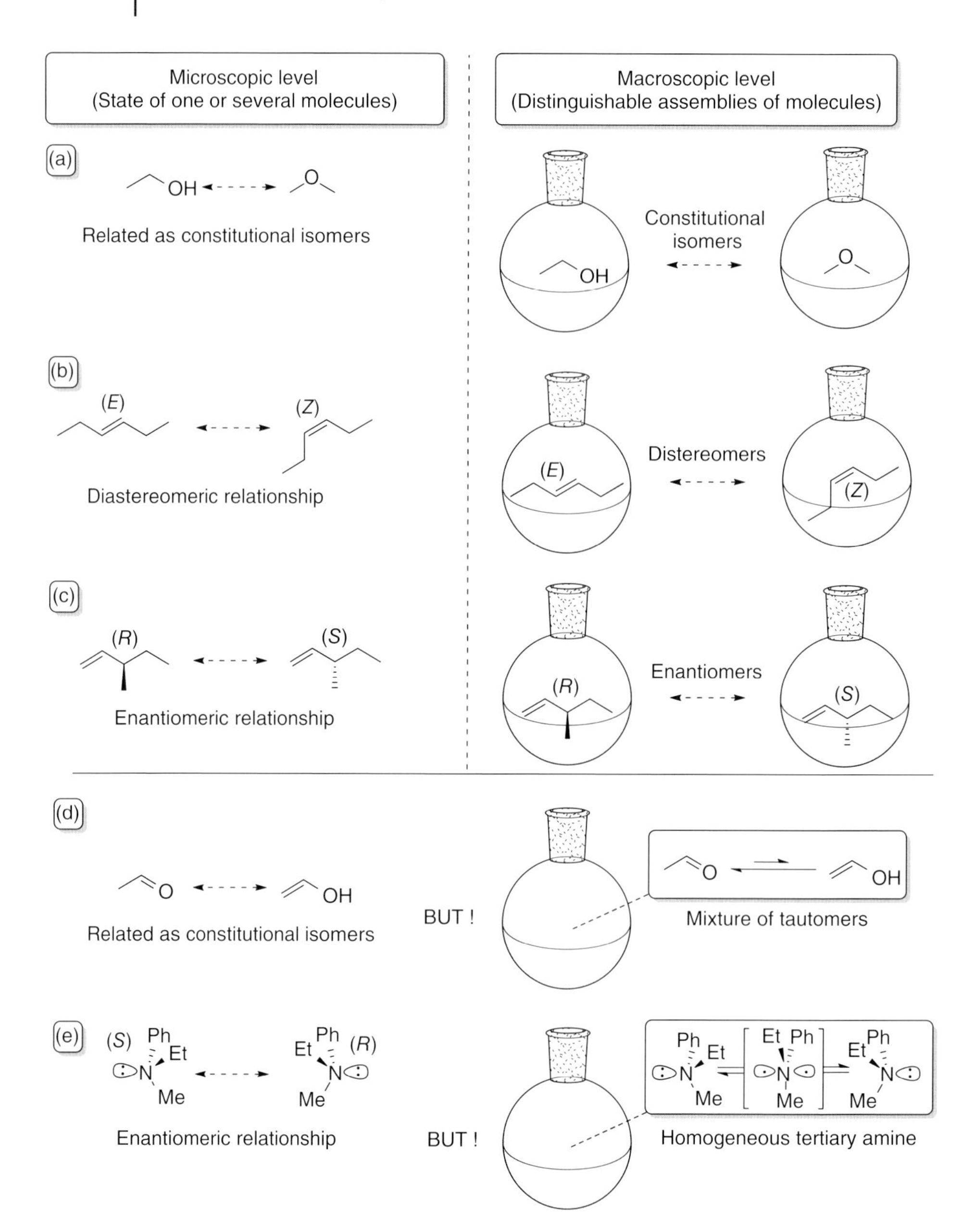


**Figure 2.9** Significance of differentiation between an isomeric relationship (microscopic level) and isomerism (macroscopic level).

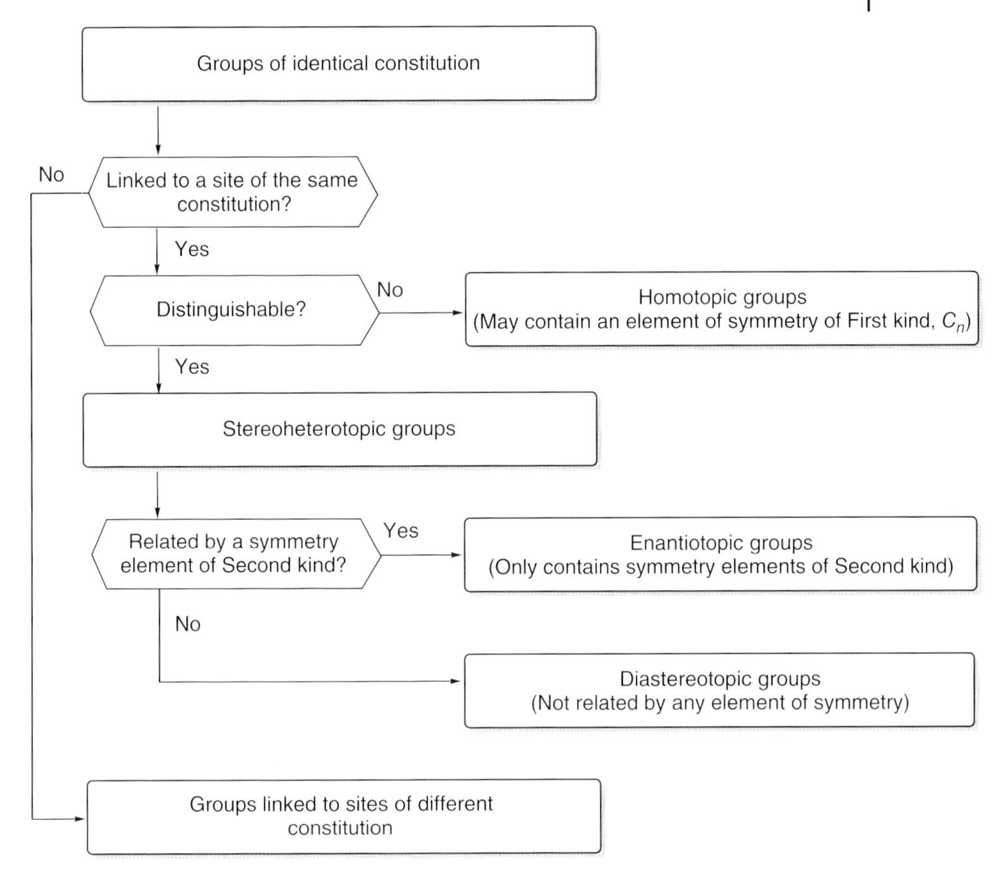


**Figure 2.10** Possible intramolecular relationships of arrangements of groups of identical constitution.

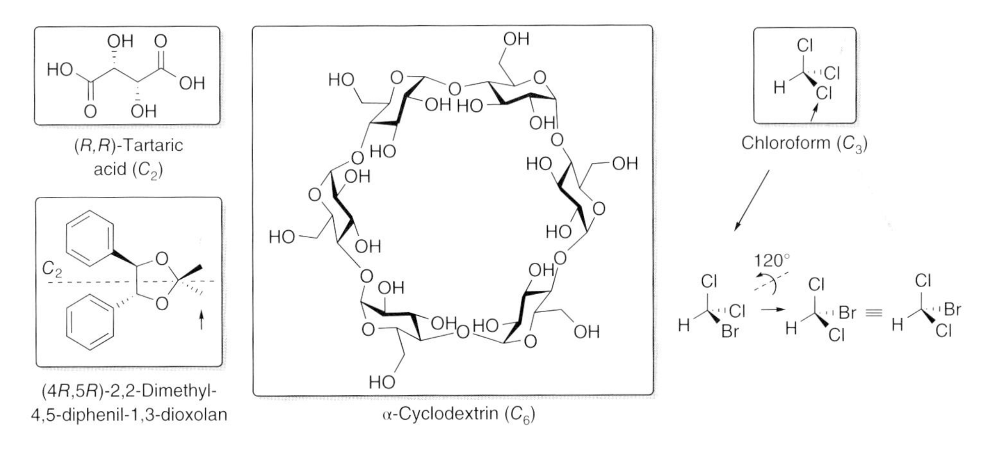


**Figure 2.11** Molecules containing homotopic atoms or groups.

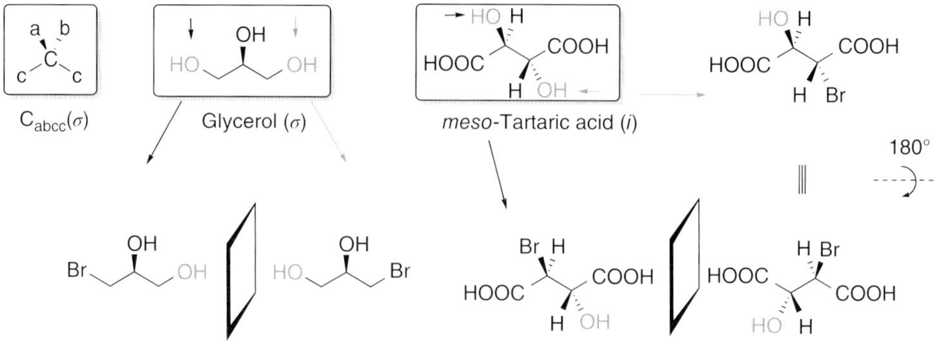

Figure 2.12 Molecules with enantiotopic atoms or groups.

(NMR) spectra and react with achiral reagents with the same ratio and same rate. In a chiral environment, however, they behave differently (e.g., may give different NMR signals when associated with a chiral shift reagent) and react with different rates. Replacement of one or the other of an enantiotopic pair of groups gives different enantiomers (Figure 2.12), the ratio of which is 1 : 1 in an achiral environment, while unequal in a chiral environment.

**Diastereotopic groups** [2–4]: Diastereotopic groups are atoms or groups within a molecule attached to sites of the same constitution but not related by any element of symmetry.

Examples are the hydrogen atoms of (*S*)-2-butanol or malic acid marked gray in Figure 2.13. Physical and chemical properties of diastereotopic groups are different in any environment: they may give different NMR signals and react differently. Replacement of one or the other member of a diastereotopic pair of atoms or groups gives different diastereomers. The ratio of the products is in principle, other than 1 : 1.

It is apparent that groups attached to sites of different constitution cannot be related by any symmetry element. They are distinguishable by any of their

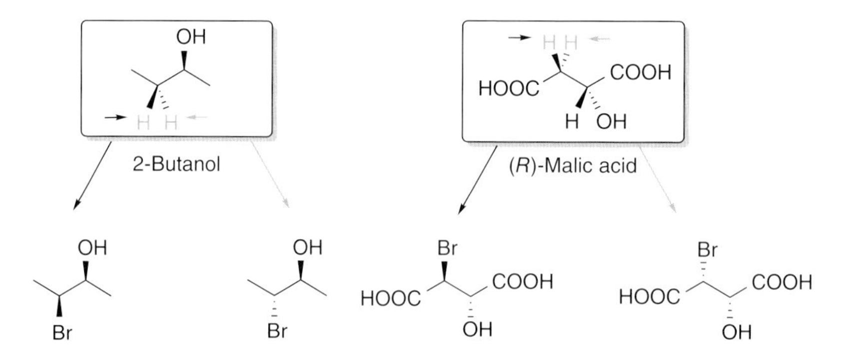


Figure 2.13 Molecules containing diastereotopic atoms or groups.

properties, and replacement of the individual groups provides constitutional isomers in a ratio other than 1 : 1.

Within a molecule independent of each other, more than one kind of topicity can be assigned to different parts, as, for instance, to the hydrogen atoms in glycerol (Figure 2.14).

It is important to note that when examining topicity we often take unwittingly into consideration free rotation (Figure 2.14). In the case of glycerol in its most stable conformation (framed in the figure), $H_a$, $H_d$ and $H_b$, $H_c$ are pairwise in an enantiotopic relationship, while $H_a$, $H_b$ and $H_c$, $H_d$ are pairwise diastereotopic. Glycerol can take up innumerable conformations in which $H_a$, $H_b$, $H_c$, and $H_d$ are all diastereotopic. Note that while the relationship of $H_a$, $H_d$ and $H_b$, $H_c$ is dependent pairwise on the conformational state of the molecule, the diastereotopic relationship of the pairs $H_a$, $H_b$ and $H_c$, $H_d$ is independent of the actual conformation.

Similarly, in the most stable conformation of ethanol (framed in figure), the relation of $H_a$ and $H_b$ is enantiotopic, while that of $H_b$ and $H_c$ is diastereotopic, nevertheless owing to free rotation around the C–C bond, the relationships of $H_a$, $H_b$, and $H_c$ are, regarded practically, homotopic.

Hydrogens of $CH_2$ groups in cyclobutanol in Figure 2.14 illustrate also diastereotopic and enantiotopic relationships. The $H_a$ hydrogens are enantiotopic, while the relation of hydrogens $H_a$ and $H_b$ is diastereotopic. Hydrogens $H_c$ and $H_d$ while also diastereotopic are in a constitutionally different position and have no symmetry relationship to hydrogens $H_a$ and $H_b$.

The relationship of given parts of a molecule depends on energy and the timescale. In the case of the 9-methyl-9,10-dihydro-9,10-[1,2]benzenoanthracene derivative shown in Figure 2.14, the hydrogens of the methyl group marked gray

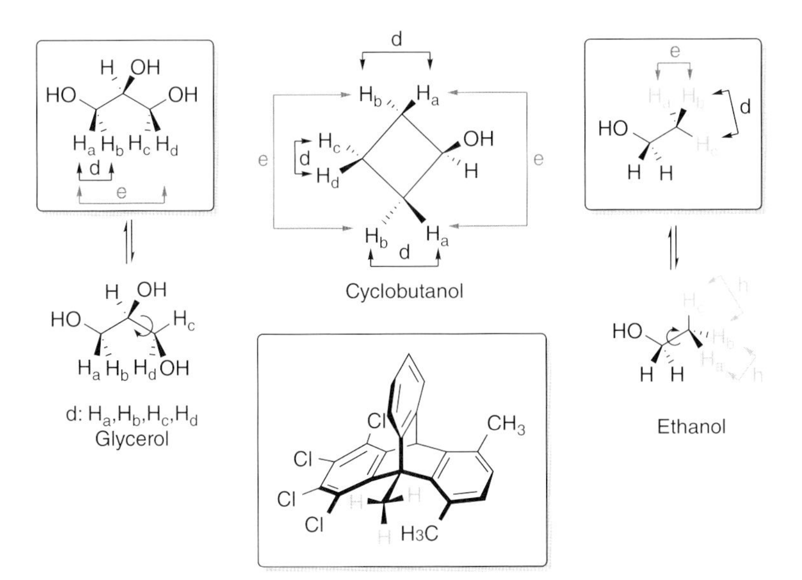


Figure 2.14 More than one type of topicity within the same molecule.

give at −90 °C, owing to hindered rotation three signals in the $^1$H spectrum indicating their diastereotopic relationship. Generally, however, in practice, we treat the hydrogens of methyl groups as homotopic.

From the point of view of molecular properties, not only the relationships of the individual parts of the molecule may be of relevance, but also the differences of how given parts of the molecule can be approached in addition reactions. These modes can be classified using symmetry considerations (Figure 2.15). Should a planar group contain more than one atomic center, topicity should be determined for each center separately.

**Homotopic faces:** Homotopic faces are modes of approach to parts of a molecule, which are related by an $n$-fold axis of symmetry ($C_n$). An example is the two faces of a center $C$ of a molecule of type X = $Caa$ ($C_2$ axis) or the two faces of the double bond in 2-methylpropene for substituted carbon atom of the double bond. Homotopic faces cannot be differentiated by any means. Addition of any of the two homotopic faces produces the same product (see the example of HCN addition to acetone in Figure 2.16).

**Enantiotopic faces:** Enantiotopic faces are modes of approach to parts of a molecule that are related exclusively by an element of symmetry of the second order (a plane of symmetry $\sigma$). An example is the two faces in a molecule of type X = $Cab$, such as vinyl bromide in respect of the substituted carbon atom or the two faces of the carbonyl group in acetophenone (Figure 2.17). Chemical properties of enantiotopic faces are identical in an achiral environment (they react in the same proportion), while different in a chiral environment (reacting in a different

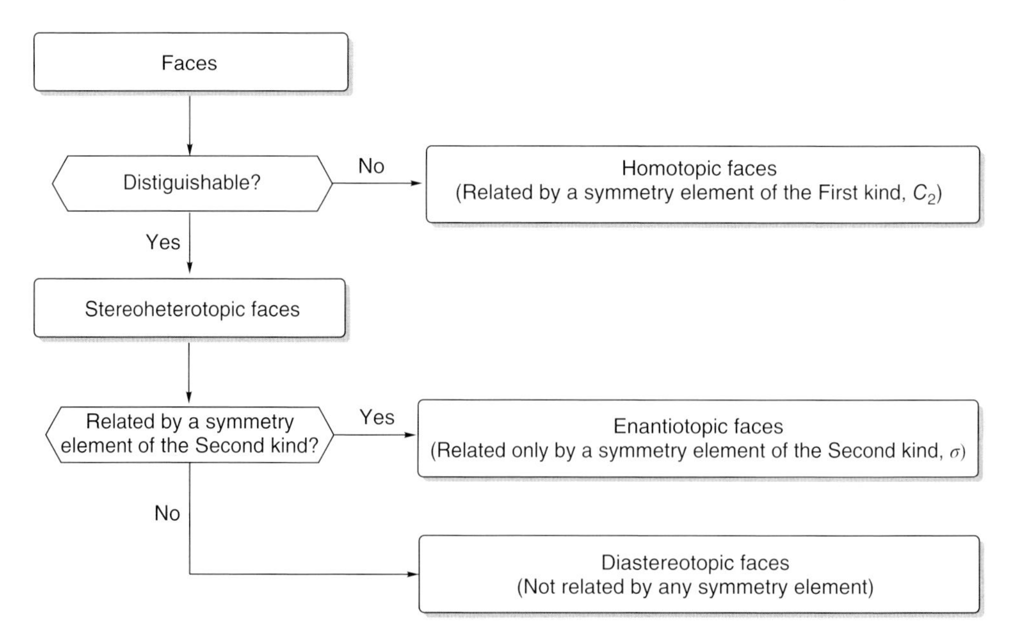


**Figure 2.15** Possible modes of approach to a planar part of a molecule.

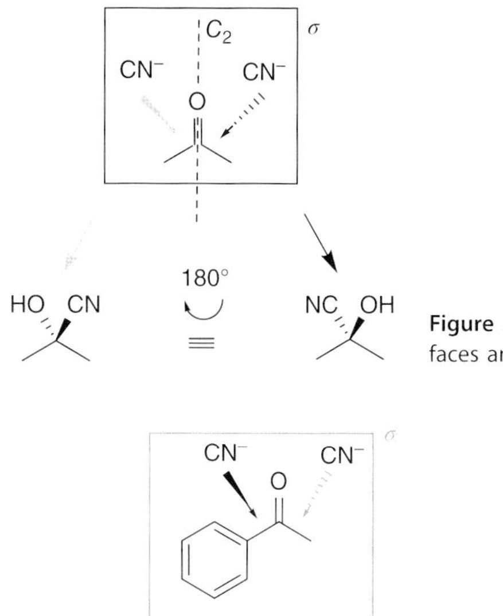


**Figure 2.16** A molecule containing homotopic faces and addition of HCN to it.

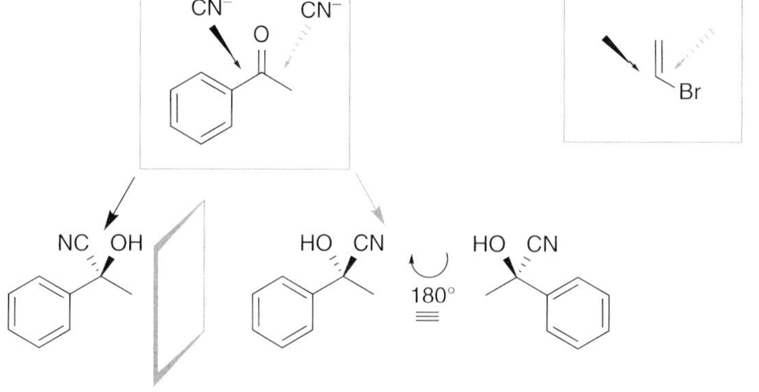


**Figure 2.17** Molecules containing enantiotopic faces.

ratio). Addition to one or the other enantiotopic face gives rise to different enantiomers (e.g., acetophenone in Figure 2.17), the ratio of which is 1 : 1 in an achiral environment but different from 1 : 1 in a chiral environment.

**Diastereotopic faces:** Diastereotopic faces are modes of approach to parts of a molecule that are not related by any element of symmetry. Diastereotopic faces are the two faces of a carbon center in a molecule of type X=C*aR** in which *R** contains a chiral element. Examples are the two faces of the carbonyl group in (*S*)-2-hydroxypropanal or of the methylene group in 5-methylene-tetrahydro-2*H*-pyran-2-one (Figure 2.18). Chemical properties of diastereotopic faces are different in any environment. Reaction from the direction of one or the other of diastereotopic faces produces different diastereomers and in principle in a ratio other than 1 : 1.

## 2.3
## Stereodescriptors

In connection with the concept of configuration, we stated that this could only be associated with a definite part/parts of the molecule having fixed steric properties

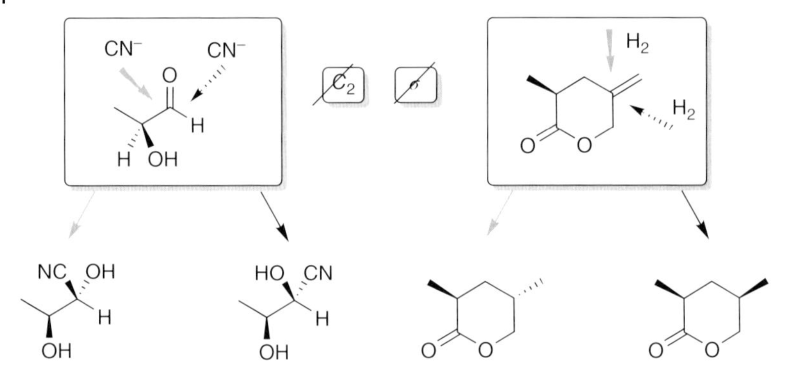


Figure 2.18 Molecules containing diastereotopic faces.

and can be characterized by *stereodescriptors* linked to the s.c. stereogenic center(s) (Figure 2.19).

**Stereogenic center:** A part of a molecule that can be the source of stereoisomerism. Interchange of a pair of atoms or groups (not necessarily any of them) produces a new stereoisomer. There must be present at least one stereogenic

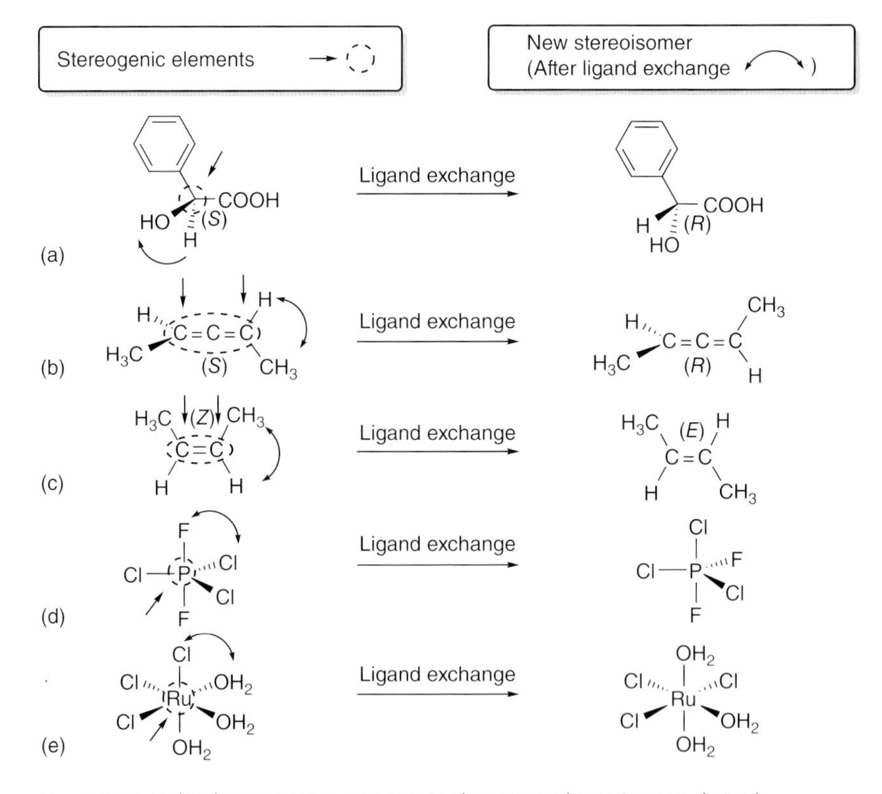


Figure 2.19 Molecules containing stereogenic elements and new isomers derived therefrom.

element in every enantiomer, while the presence of stereogenic elements does not mean that the given molecule, as a whole, is chiral. For systems comprising atoms attached to not more than four ligands, fundamental cases of stereogenic elements (Figure 2.19) are as follows:

a) Group of atoms, comprising a central atom and its ligand with the condition that interchange of any two of the ligands generates stereoisomers (e.g., Figure 2.19a). A typical example of such a stereogenic element is an *asymmetric atom* (*center of chirality*).
b) A stable conformation of four noncoplanar atoms (or rigid groups) in which (hindered) rotation around the central bond (resulting in a change of torsion angle) produces a stereoisomer (*axial chirality*) (e.g., Figure 2.19b).
c) A group of atoms containing a double bond that generates (*E*)–(*Z*) isomers (e.g., Figure 2.19c).

In systems containing atoms attached to more than four ligands, further types of stereogenic centers are possible (e.g., Figure 2.19d and e).

In the present section, stereodescriptors associated with stereogenic elements will be discussed disregarding those linked to more than four ligands.

### 2.3.1
### The D/L System, Fischer's Projection

Example (a) illustrates the case when four ligands are attached to a central atom in a way that interchanging any pair of ligands results in a new stereoisomer. A typical example of such a stereogenic element is the *asymmetric atom* (*center of asymmetry*). This kind of asymmetry has been recognized early,[19)] but it was not followed immediately by a simple system of graphic representation and the introduction of readily applicable generally valid stereodescriptors. Stereoformulas in Figure 2.20 are drawn with broken and solid lines for the sake of better understanding, noting that this usage started much later than the introduction of *Fischer* formulas.

The *Fischer–Rosanoff* convention, the first stereodescriptor system conceived, relates the description of the chirality to the structure of D- or L-glyceraldehyde. The projective representation of configuration was suggested by H. E. Fischer after elucidation of the stereostructure of the 16 stereoisomers of aldohexoses (1891). The *Fischer* projection enables to depict 3D structures in 2D (Figure 2.20). This was followed by the proposition of M. A. Rosanoff (1905) concerning the arbitrarily assigned absolute configuration of D-glyceraldehyde

19) In 1874, Jacobus Henricus van't Hoff and Joseph Achille Le Bel came, independently from each other, to the conclusion that optical activity (rotation of the plane of plan polarized light when passing a material) can be associated with that particular property of the carbon atom, that its ligands take up positions corresponding to the apexes of a tetrahedron. If the four ligands are different, they can be arranged in space in two ways. Thereby, the opposite optical rotations and the structure of certain pairs of isomers found in nature can be explained.

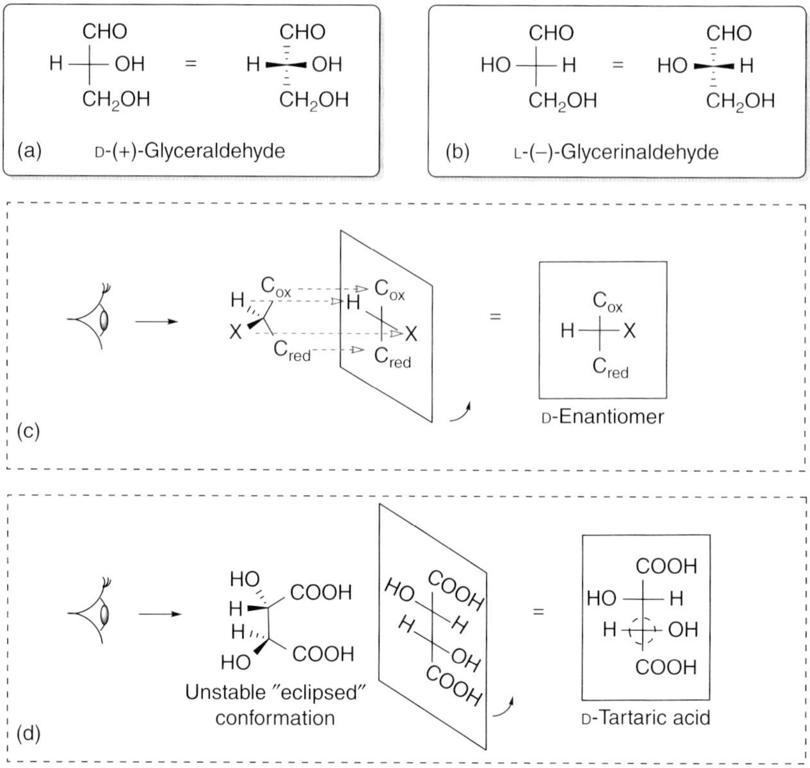

**Figure 2.20** Proposed absolute configuration of (+)-glyceraldehyde and the rules of *Fischer's* projection.

(Figure 2.20a and b)[20] and applied for the depiction of the stereostructure of molecules using the *Fischer* projection (Figure 2.20).

In a *Fischer* formula, horizontal lines show bonds emerging from the plane of paper, while vertical lines point away from the plane of the drawing (Figure 2.20a and b). When applying these projection rules, we select (i) the main carbon chain of the compound (Figure 2.20c), (ii) place the carbon chain end in a higher oxidation state to the top, the less oxidized one to the bottom end in a way that the chain ends should be closer to the plane of projection, and (iii) the ligands of the central carbon atom should draw coplanar with the projection plane. With molecules having more than one stereocenter (Figure 2.20d), we proceed in a way that the carbon chain is rotated to a conformation that all the substituents should be eclipsed. Thereafter, we perform arranging and projecting the molecule according to Figure 2.20c. In assigning a descriptor to the molecule, only the

20) According to this proposition, the graphical representation of D-glyceraldehyde should correspond to the configuration of D-glyceraldehyde, that is, to the form rotating the polarized light into (+)-direction. Later, this arbitrary assignment proved to describe, by pure chance, the actual situation.

configuration of the stereocenter that is most remote from the most oxidized end is considered, and if its ligand other the hydrogen projects to the right, the descriptor D is assigned both to this center and the whole of the molecule.

In this respect, the D/L system differs basically from the *R/S* and *E/Z* systems, since the descriptors D and L only describe the steric orientation of a single stereogenic center while referring to the whole molecule. The orientation of all the other stereogenic centers is defined by the trivial names of sugars since every sugar of a different name has a set of different steric arrangements along the carbon skeleton. In contrast, the *R/S* and *E/Z* systems define stereodescriptors for each and every stereogenic center separately. The D/L system originates from an early epoch of organic chemistry and is nowadays only used in carbohydrate chemistry and for natural amino acids. The latter are all of L configuration, while most natural sugars are of D configuration.[21)]

### 2.3.2
### The *R/S* System

A precondition for the unambiguous application of the *R/S* system was the development of an adequate representation of 3D structures in the plane of paper. Depicting bonds emerging or receding from the plane of paper by bold and broken lines, respectively, has become generally accepted only in the second half of the twentieth century. Earlier, other modes of representation (e.g., the *Fischer* projection) existed. Details of the use of stereoformulas were described in the 2006 recommendations of IUPAC.[22)]

The *R/S* system is a system of stereodescriptors typically serving to characterize *centers of chirality* (*centers of asymmetry*), that is, of stereogenic elements (as shown in Figure 2.19a). Interchange of ligands attached to such centers results in a stereoisomer different from the original. In contrast to the D/L-system, to be able to characterize the stereostructure of the complete molecule, in the *R/S* system, stereodescriptors must be assigned to each stereogenic element (Figure 2.21).

In order to describe a given stereogenic center, first the precedence of each ligand attached to the center must be established (Figure 2.21a).[23)] Main rules of precedence are as follows: (i) In the first domain of bonds, an atom of higher atomic number is given precedence, or in case of isotopes, the one of higher mass has precedence. (ii) If no decision is reached in the first domain of bonds, we go one bond farther. Here, besides precedence according to atomic number/mass number, the multiplicity of atoms of higher precedence also counts. Thus, geminal diol (two oxygens) > carbonyl group (one virtual oxygen and one real one linked by

21) Certain bacteria are producing D-amino acids having the function to disturb the building of the cell wall of other bacteria in their vicinity. It has also been found that there exist bacteria that are able to metabolize by alternative mechanisms certain L-sugars and grow in the presence of L-sugars as the sole carbon source.

22) The details of the use of stereo formulas are exactly defined by IUPAC recommendations [6]. Easily perceptible stereo formulas have only become generally used in the second half of the twentieth century. Earlier various other representations were used.

23) Sequence rules of the C. I. P. system are discussed in detail in Refs. [7] and [8].

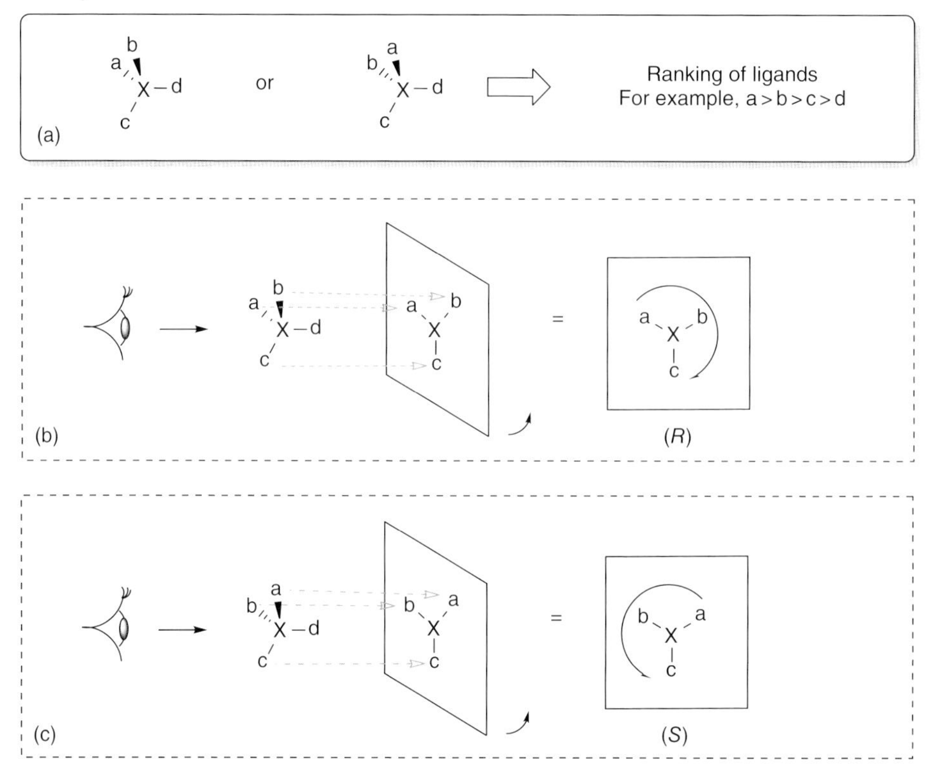


Figure 2.21 Rules to determine stereodescriptors according to the *R*/*S* system.

its second bond to carbon) > alcohol (one real oxygen). (iii) Other points of view, for example, stereochemistry [*R* ligand > *S* ligand].

After having completed the sequencing of ligands, the stereocenter is viewed along the bond connecting the central atom and the ligand of lowest precedence form the side of the central atom. When the sequence of ligands in the foreground is clockwise, the configuration of the center is *R* (Figure 2.21b). If the order of the same ligands is counterclockwise, configuration *S* is allotted to the center (Figure 2.21c).

The presence of centers of chirality in a molecule, which can be characterized by the stereodescriptors *R* or *S*, does not necessarily involve that the molecule as a whole is chiral. A classic example is the achiral (*R*,*S*)-tartaric acid (*meso*-tartaric acid) containing two *enantiomorphic* stereogenic elements.

Stereogenic elements resulting in chirality are manifested in manifold ways (Figure 2.22).

Among the stereogenic centers resulting in central chirality, the best known is the tetrahedral arrangement of four ligands around a central atom. The typical example is the *asymmetric carbon atom.* In Figure 2.22, it can be seen that the central atom can be other than carbon, ligands can be attached to the central atom

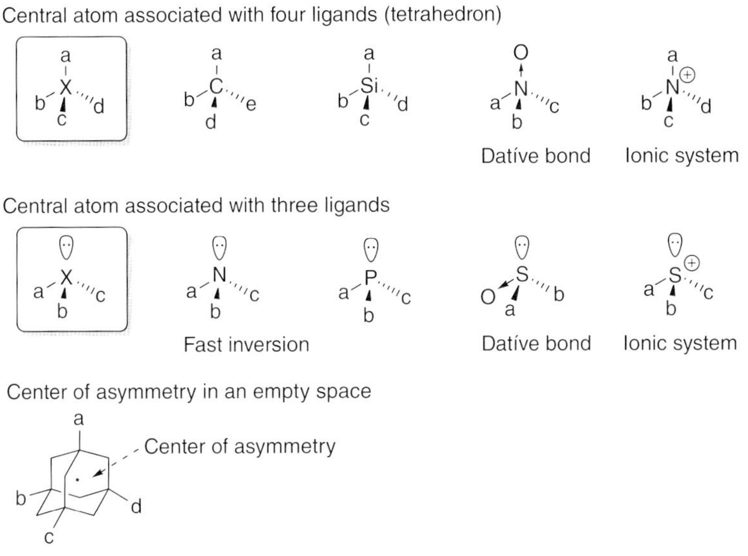

Central atom associated with four ligands (tetrahedron)

Datíve bond Ionic system

Central atom associated with three ligands

Fast inversion Datíve bond Ionic system

Center of asymmetry in an empty space

Center of asymmetry

**Figure 2.22** Stereogenic elements associated with central chirality.

by a dative bond, the central atom can bear a charge, and one of the ligands can even be a pair of nonbonding electrons.

The center of chirality is not necessarily a real atom, as exemplified by the chiral adamantane in Figure 2.22 in which the center of asymmetry is residing in the empty center of the molecule.

### 2.3.3
### Hindered Rotation around a Double Bond, *E/Z* Isomerism

The source of *E/Z* isomerism is the highly hindered rotation around a double bond. Apart from hindered rotation, it is also a condition that the substituents attached to the bridgehead atoms of the double bond should be different. For example, olefins of type $R^1R^2C{=}CR^3R^4$ show *E/Z* isomerism when $R^1 \neq R^2$ and $R^3 \neq R^4$. For the determination of *E/Z* stereodescriptors for olefins and similar systems containing double bonds, ligands attached to the bridgehead atoms of the double bond have to be sequenced separately following the rules of the *R/S* system (Figure 2.23a).

If the two higher ranking substituents at the terminals of the double bond are on the same side, then a *Z* configuration (zusammen in German) should be assigned to the double bond, while, if on opposite sides, the descriptor is *E* (entgegen). If a hydrogen atom is attached to each of the carbon atoms of the double bond, the use of the traditional *trans/cis* notation is still acceptable.

*E/Z* descriptors can also be used with structures when the bridgehead atoms are connected with a partial double bond owing to hindered rotation or when the bridgehead atom(s) are other than hydrogen, as for example, in oximes

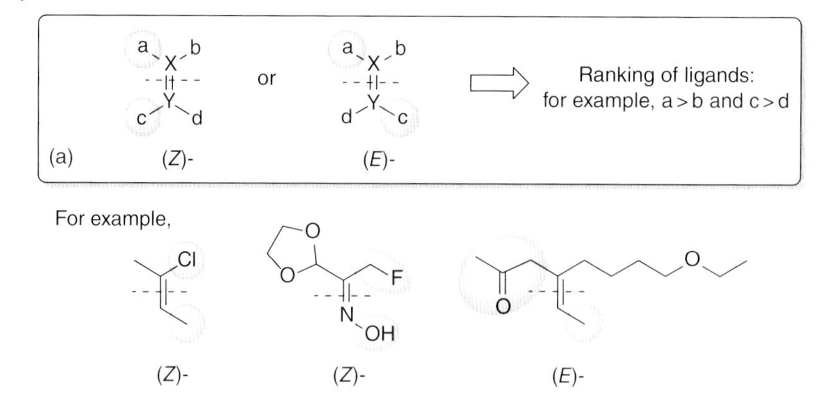


**Figure 2.23** Rules for the determination of the stereodescriptors for *E/Z* stereoisomers.

($R^1R^2C{=}NOH$). *E/Z* descriptors may serve to distinguish stereoisomeric cumulenes ($R^1R^2C[{=}C{=}C]_n{=}CR^3R^4$, if $R^1 \neq R^2$ and $R^3 \neq R^4$) and similar systems (e.g., $HON[{=}C{=}C]_n{=}NOH$).

## 2.3.4
## Axial Chirality, Helicity, *M/P* Descriptors

There are several chiral systems in which no center of chirality can be identified and which cannot therefore be characterized using the *R/S* or *E/Z* systems. These are typically helical or propeller-like structures.

**Axially chiral** molecules [2–4] are molecules in which along an axis four groups are situated in two noncoplanar planes, one pair in each. Rotation along the axis should be hindered and groups attached to the same plane should be different. Examples of axial chirality are shown in Figure 2.24. It is apparent that it is not

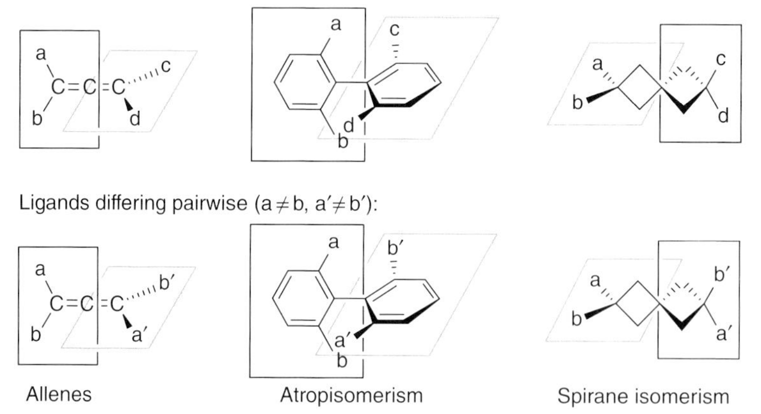


**Figure 2.24** Basic types of axial chirality.

required that all four ligands (a–d) should be different, but it suffices when ligands in a common plane (a, b and a′, b′, respectively) are different.

The reason for the chirality of the borane shown in Figure 2.26 is steric hindrance of rotation around the C–B bonds. The three aromatic rings avoid collision by taking up a propeller-like arrangement around the central boron atom. The sense of arrangement of the blades of the propeller can be right- or left-handed; consequently, two enantiomeric propellers, configured *P* and *M*, exist.

Stereogenic elements causing axial chirality can be characterized by the stereodescriptors *P*/*M* or $R_a/S_a$. A descriptor characterizing the sense of twisting is assigned by looking at the molecule along the axis of the helix and determining separately the precedence of substituents attached to the "near" and "remote" ends of the helix following the rules of the *R*/*S* system; according to a by-rule, the "near" end has precedence over the "remote" one (Figure 2.25). If viewing the molecule along its axis, the substituent of higher order on the "remote" end is situated clockwise relative to the substituent of higher precedence at the "near" end, then the helix has a right-hand twist and the descriptor *P* (plus) is assigned to the molecule. Otherwise, the descriptor *M* (minus) is assigned to the system. Note that it is indifferent from which end of the axis the molecule

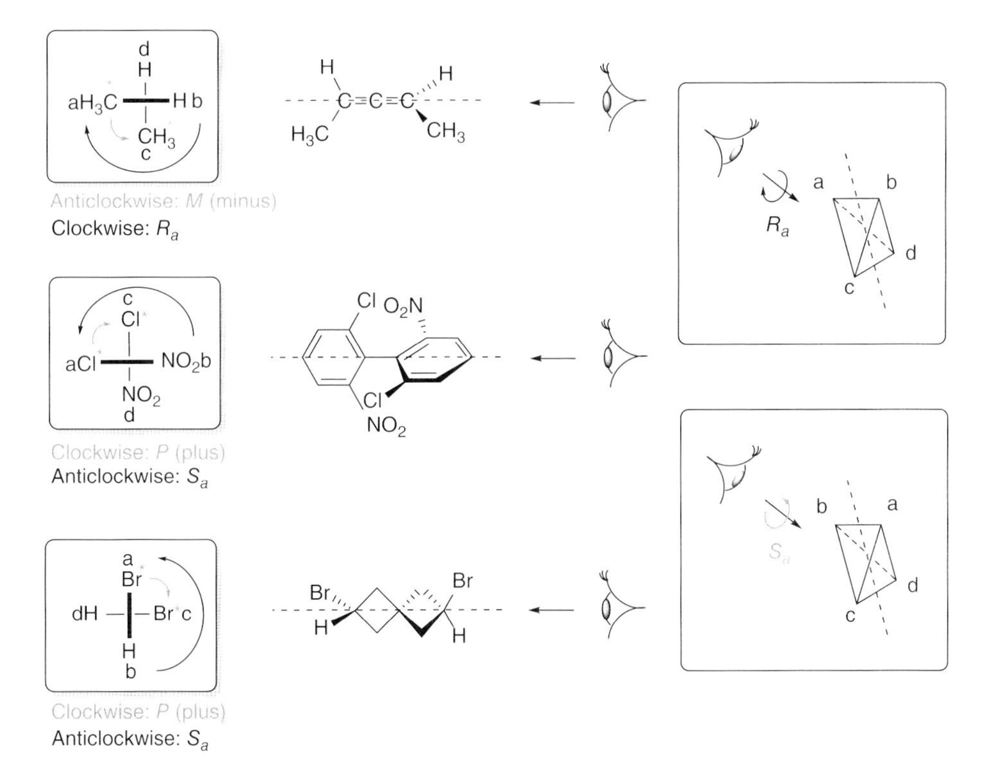


**Figure 2.25** Atropisomerism. Examples of the determination of the stereodescriptors *M*/*P* and $R_a/S_a$.

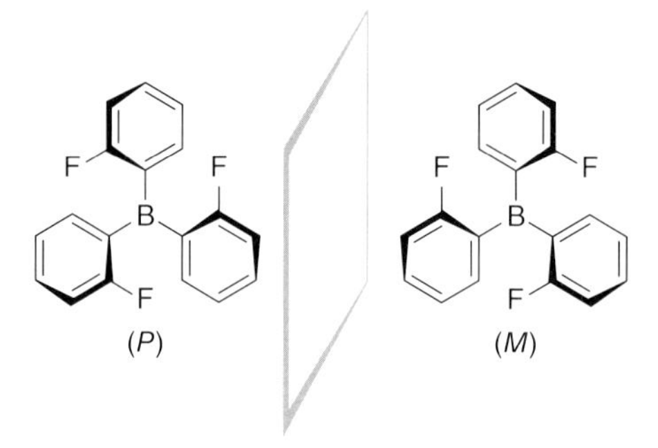


**Figure 2.26** Propeller-like enantiomers.

is viewed. Determination of helicity is illustrated by the examples in Figure 2.25 (substituents of higher precedence are marked with asterisks and the sense of twisting with arrows). In the case of axial chirality, generally, $P = S_a$ and $M = R_a$.

The source of the chirality of the hexahelicene in Figure 2.27 is also the tendency to avoid steric tensions. This is relaxed by taking up a right- or left-handed helical form of the polyaromatic ring system.

### 2.3.5 Planar Chirality, $R_p/S_p$-Descriptors

A stereogenic element can arise from the situation when a molecule contains a plane to which a direction of circumvention can be assigned based on the arrangement of groups of atoms, further there exists an element by which one face of the plane can be distinguished from the other (Figure 2.28).

Planar chirality can occur with cyclophanes (Figure 2.28 (a)). In *trans*-cyclooctene, the source of planar chirality is that the aliphatic chain is too short to permit to flip from one side of the double bond to the other (Figure 2.28). With metallocenes, complexation of an appropriately substituted aromatic ring by the metal atom at one face may also generate planar chirality (Figure 2.28, (d)).

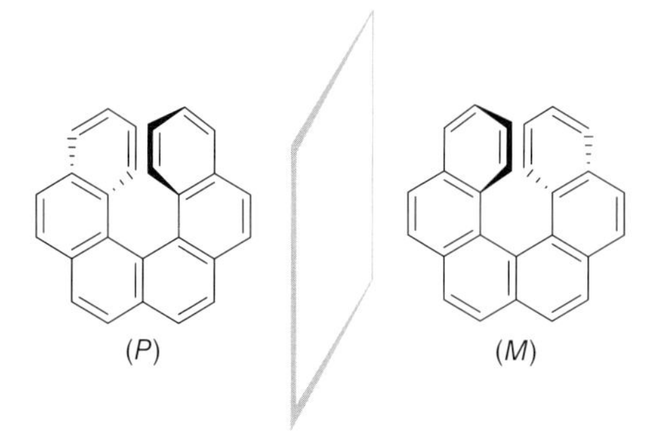


**Figure 2.27** Enantiomers of hexahelicenes.

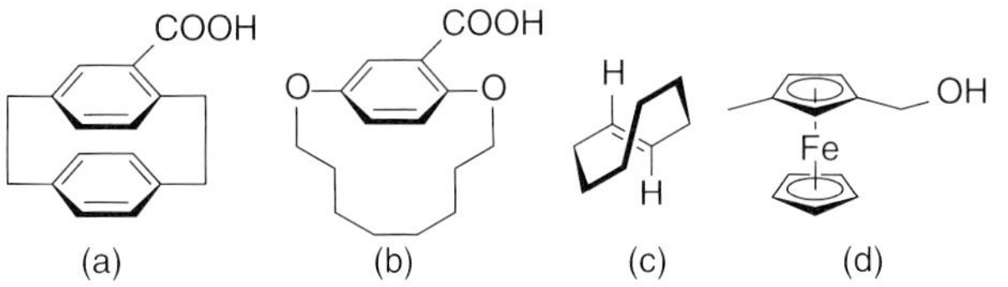

Figure 2.28 Molecules with a planar element of chirality.

Stereogenic elements resulting in planar chirality can be characterized by the stereodescriptors $R_p/S_p$ or $P/M$. In order to assign the proper descriptors, an s.c. *pilot atom* has to be selected being the out-of-plane atom or group of atoms closest to the plane (or in case of identity one of highest priority) (Figure 2.29).

For finding the stereodescriptors $R_p/S_p$, we look at the molecule from the direction of the pilot atom, where a, b, and c are the three groups following the pilot atom (Figure 2.29). After that, the stereodescriptor is determined by viewing the molecule from the direction of the pilot atom. If the route following the arrow in the direction $a \rightarrow b \rightarrow c$ is clockwise, configuration of the molecule is $R_p$ and $S_p$ when anticlockwise. Alternatively, when viewing the molecule from the direction of the pilot atom, the third group (c) counted from the pilot atom is situated clockwise, then the system is right-twisted and should be characterized by a $P$ (plus) descriptor. When the orientation is left-twisted, the descriptor $M$ (minus) should be allocated to the molecule. (The groups to be considered for the determination of $P/M$ descriptors were marked with asterisks, and the direction of twisting by an arrow.) In case of planar chirality, generally, $P = R_p$ and $M = S_p$.[24)]

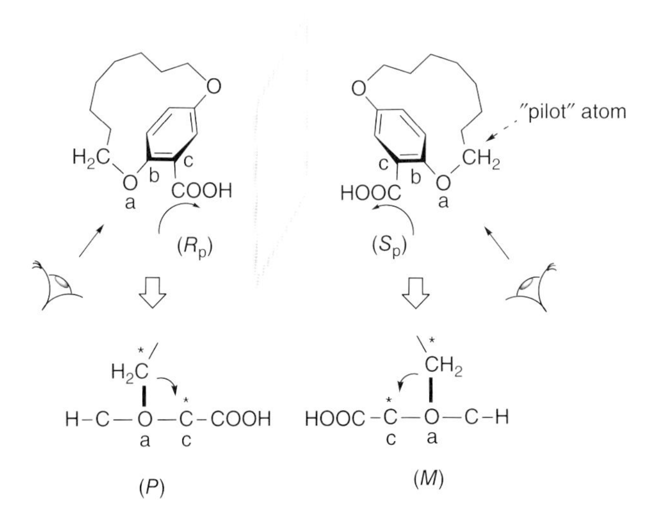


Figure 2.29 Determination of the stereodescriptors for planar chirality.

24) Note that this is at variance to axial chirality, where generally $P = S_a$ and $M = R_a$, respectively.

## 2.4 Prostereogenic Elements, Prochirality

The two most common transformations in organic chemistry are substitution and addition. Chiral (*R*)-lactic acid can be prepared in principle by means of both types of transformations starting from an achiral compound (Figure 2.30). Thus, both enantiomers of lactic acid can be prepared from the achiral propionic acid by substituting one or the other of $\alpha$-hydrogens. Similarly, both enantiomers can be prepared from the achiral 2-oxo-propionic acid by addition of hydrogen to one or the other face of the carbonyl group (reduction).

Accordingly, the $\alpha$-carbon atom (to which two heterotopic groups are attached) or the carbonyl group of 2-oxopropionic acid (which has two heterotopic faces) can be called *prochiral*.

### 2.4.1 Prostereoisomerism and Prochirality

Although the term **prochirality** is widely used, mainly by biochemists, this concept is restricted in many aspects. Stereochemical differences can arise without the molecules being chiral, as exemplified by the *cis*- and *trans*-1,3-dichlorocyclobutane or (*E*)- and (*Z*)-1,2-dichloroethenes, where none of them is chiral and do not contain centers of chirality (see Figure 2.31).

Consequently, when interpreting the origins of stereoisomerism we cannot restrict our consideration to the existence or lack of chiral elements, but we have to extend our concepts to stereogenic elements in general. A stereogenic element is, for example, one of the bridgehead atoms of a double bond (Section 2.3.3) or

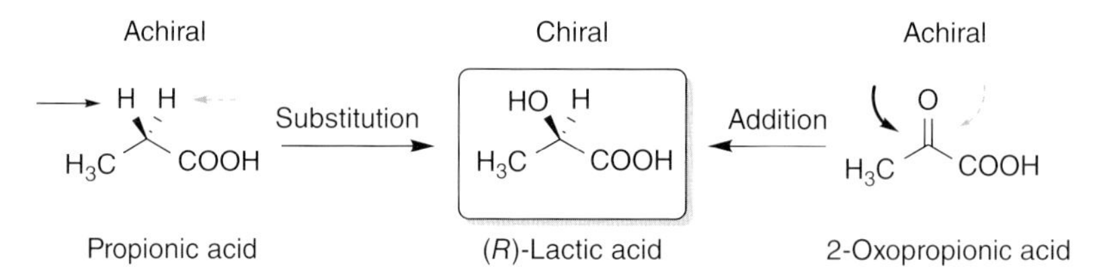


**Figure 2.30** Possible achiral precursors of (*R*)-lactic acid.

**Figure 2.31** Stereoisomers of achiral compounds and prostereogenic centers.

stereogenic centers shown in Figure 2.19. Additional stereogenic elements are discussed in Sections 2.3.4 (axial chirality, helicity, and propellenes) and 2.3.5 (planar chirality).

Evidently, the concept of **prochirality** has to be extended and generalized as **prostereoisomerism.** In this sense, the $C_2$-atom of vinyl chloride shown in Figure 2.31 along with that of (*E*)- and (*Z*)-1,2-dichloroethene is also a **prostereogenic center** since exchange of the two homomorphous hydrogens ($H_a$ and $H_b$) gives rise to different *E/Z* isomers. Similarly, the $C_3$-atom of chlorocyclobutane is a *prostereogenic center* because the exchange of the two hydrogens $H_a$ and $H_b$ generates different (*cis/trans*) stereoisomers (Figure 2.31).

In order to be able to distinguish substituents linked to the bridgehead atoms of a double bond, the descriptors *pro-E* and *pro-Z* were introduced [2–4]. If the CIP precedence of one of a pair of identical groups attached to one of the bridgehead atoms of a double bond (such as "c" in case of an abC=Ccc type compound) is increased virtually, an *E* isomer is obtained (e.g., in case of the $H_b$ of vinyl chloride in Figure 2.32) and the group is classified as *pro-E,* otherwise as *pro-Z.* The relationship of groups designated as *pro-E* and *pro-Z* is diastereotopic and their exchange for a different group results in *diastereomers.*

For distinguishing groups attached to prostereogenic centers in substituted cyclic systems, the *pro-cis* and *pro-trans* descriptors can be used in a similar way, disregarding whether substitution generates a new center of chirality or not. If virtually enhancing the CIP preference of any of the two identical groups attached to a tetrahedrally coordinated atom (most often a carbon atom) in a substituted saturated cyclic system results in a *cis* stereoisomer (e.g., in case of $H_b$ linked to carbon C-3 of chlorocyclobutane in Figure 2.33), the group is classified as *pro-cis,* otherwise as *pro-trans.* The relationship of groups distinguished by the descriptors *pro-cis* and *pro-trans* is also diastereotopic and their exchange for a different group results in diastereomers.

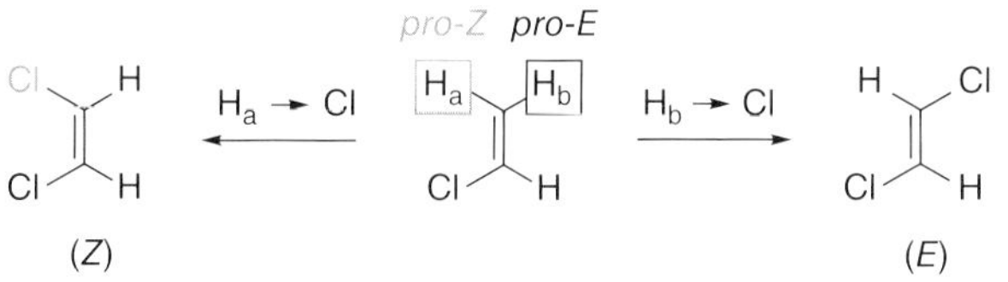

**Figure 2.32** *pro-E/pro-Z* Stereodescriptors.

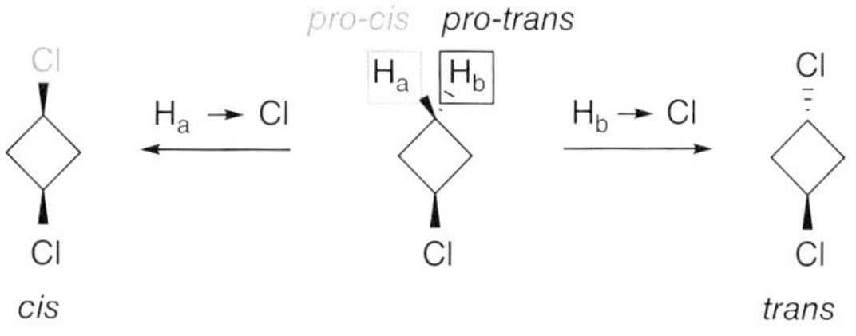

**Figure 2.33** *pro-cis/pro-trans* Stereodescriptors.

The concept of *prochirality* has several interpretations [2–4] among which some will be explained in detail in the following.

**Prochirality** can be interpreted as that particular geometrical feature of an achiral object, that it can become chiral in a single desymmetrization step. In this sense, a whole molecule can be regarded as ***prochiral*** when by exchange of an atom (or achiral group) to a different one, it becomes chiral. Achiral objects, which are transformed to a chiral one in *two desymmetrization* steps, are called ***proprochiral***. Such a molecule is, for example, $CH_3CO_2H$, which is proprochiral, $CH_2DCO_2H$, which is prochiral, and finally $CHDTCO_2H$, which is chiral.

The term **prochirality** is also used to characterize a *planar trigonal unit* of a molecule to which *addition* of a new atom (or achiral group) generates a new center of chirality. If the starting molecule is achiral, the relation of the sides of the trigonal unit is enantiotopic, while if the starting molecule contains a chiral element, the relation of the sides of the trigonal unit is diastereotopic. In this case, to characterize the sides of the trigonal unit, the descriptors *Re/Si* can be used.

One can use the concept *prochirality* also when to a tetrahedrally coordinated atom of an achiral or chiral molecule two stereoheterotopic groups are adjoined. If between the two groups only a symmetry element of second kind (generally $\sigma$) can be found, then their relationship is enantiotopic. If between them no symmetry element of any kind can be found, then their relationship is diastereotopic. The tetrahedrally coordinated atom itself can be denoted as prochiral, and the two stereoheterotopic groups can be characterized by the stereodescriptors *pro-R/pro-S*.

A concept akin to but more general than *prochirality* is that of a *prostereogenic element*. If, namely, a molecule contains enantiotopic or diastereotopic groups or faces, then exchange of one of them or addition from the direction of one of the faces generates a new stereogenic element. This newly formed stereogenic element is not necessarily a center of chirality because the product molecule can show axial or planar chirality as well.

Accordingly, since the relationship of atoms $H_a$ and $H_b$ of the allene and of the cyclophane shown in Figure 2.34 is enantiotopic and their substitution generates products with axial or planar chirality, they can be considered as prostereogenic.

### 2.4.2
### Distinction of Stereoheterotopic Groups, the *pro-R/pro-S* Notation

Similarly to stereogenic elements causing stereoisomerism, it is practical to characterize by stereodescriptors stereoheterotopic groups as well. As mentioned in

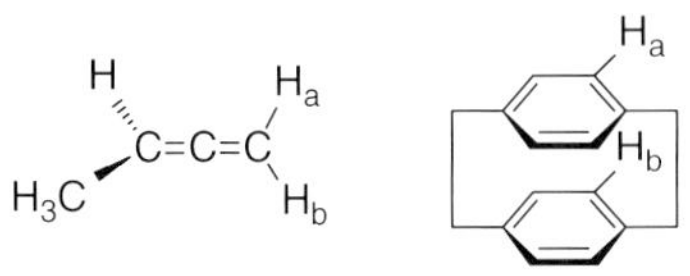

H, $H_a$, C=C=C, $H_3C$, $H_b$, $H_a$, $H_b$

**Figure 2.34** Molecules containing prostereogenic elements.

the previous section, diastereotopic and enantiotopic groups can be characterized by the descriptors *pro-R/pro-S* (Figure 2.35). A stereoheterotopic group ("c" in a tetrahedral Xabcc system) is denoted *pro-R* when enhancing its precedence virtually, the configuration of the chiral center thus generated is *R* (according to the CIP rules). The other group is consequently *pro-S*. When drawing formulas, *pro-R* groups are distinguished from *pro-S* groups by *R* and *S* letters as subscripts, respectively (Figure 2.35). Use of *pro-R/pro-S* descriptors is demonstrated in Figure 2.35.

It is important to note that the descriptors *pro-R/pro-S* do not predetermine the configuration of the product formed in an actual reaction (Figure 2.36).

It is apparent from Figure 2.36 that there is no direct correlation between the stereodescriptors of a group attached to a prochiral center and the configuration of the product formed. For example, the exchange of the *pro-R* hydroxyl group

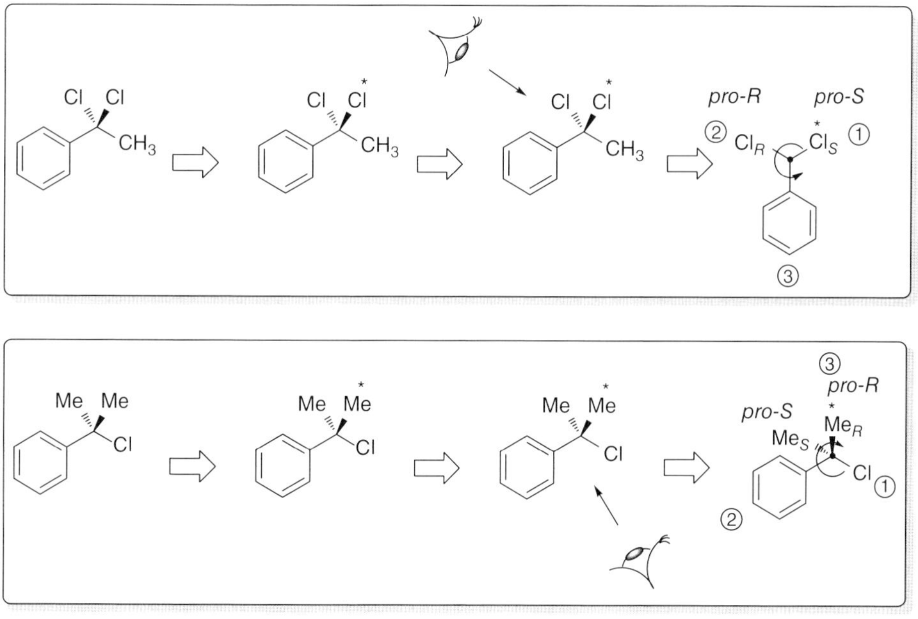

Figure 2.35 Assigning the stereodescriptors *pro-R/pro-S*.

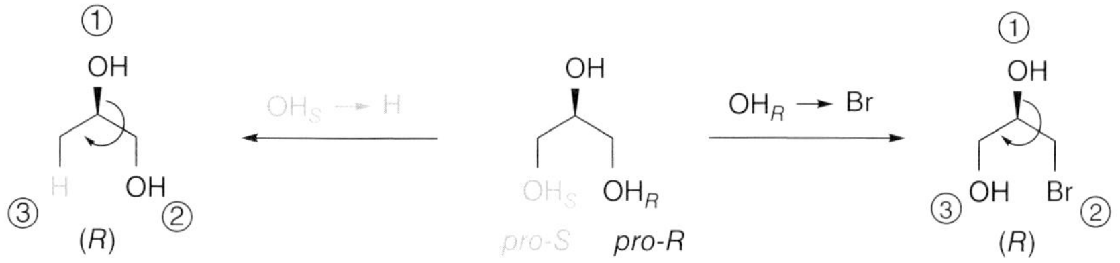

Figure 2.36 Transformations of the *pro-R/pro-S* hydroxyl groups of glycerol.

for bromine leads namely to an *R* product, similarly to the exchange of the *pro-S* hydroxyl group for hydrogen (Figure 2.36).

### 2.4.3 Prochiral and *meso* Compounds, Center of Pseudoasymmetry

Prostereoisomerism is present, when in an achiral compound replacement of diastereotopic groups results in diastereomers (see, e.g., Figures 2.31–2.34 in Section 2.4.2). We have also quoted examples for transformations of enantiotopic groups in prochiral compounds too (Figure 2.35).

A compound containing enantiotopic groups and a prochiral center can be considered as prochiral when it conforms to the first definition of prochirality (Section 2.4.1). A characteristic of a *prochiral compound* is that one can place a $\sigma$ mirror plane of symmetry across the prochiral center (Figure 2.37). When the prochiral center (encircled by a broken line) is doubled by pulling apart the molecule whereby two centers of asymmetry (encircled by a broken line), related as mirror images (*enantiomorphic*), are generated, we arrive at a *meso* compound. The *meso* compound remains symmetrical in its relationship to the mirror plane and is therefore achiral. Continuing the thought experiment, if we place an additional center between the two enantiomorphic centers of asymmetry, a *center of pseudoasymmetry* is created (encircled by a broken gray line). The compound containing the center of pseudoasymmetry is still symmetrically related to the $\sigma$ plane and is thus achiral and a *meso* compound.

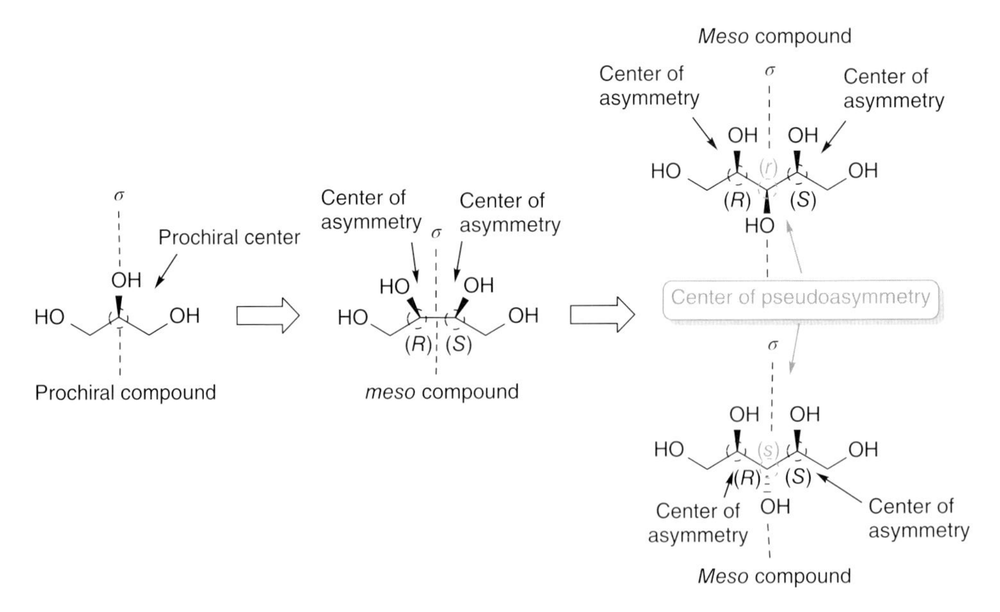


Figure 2.37 Prochiral and *meso* compounds, center of pseudoasymmetry.

**Figure 2.38** Some simple meso compounds.

Figure 2.38 shows some *meso* compounds containing no center of pseudoasymmetry. Their common feature is apparently that they contain a symmetry element of second kind.

The *center of pseudoasymmetry* is such a tetracoordinated atom, with four different ligands among which two and only two are of identical constitution but of opposite chirality (enantiomorphic). Observing the convention that an *R* configuration has precedence over an *S* one, centers of pseudoasymmetry can also be characterized by *r/s* stereodescriptors following the CIP rules.[25]

Figure 2.39 well demonstrates that centers of pseudoasymmetry are invariant to reflection (e.g., the *s* configured center of pseudoasymmetry of xylaric acid remains *s*), but changes when any two ligands are interchanged. Thus, the interchange of the ligands of the *s* configured center of pseudoasymmetry of xylaric acid leads to ribaric acid with an *r* configured center of pseudoasymmetry. Isomers having centers of pseudoasymmetry of different configuration are related to as diastereomers.

The case of hyoscyamine (Figure 2.39) demonstrates that chiral compounds can contain a center of pseudoasymmetry too, when in addition to the two enantiomorphic ligands an additional asymmetric ligand is joined to the center. In

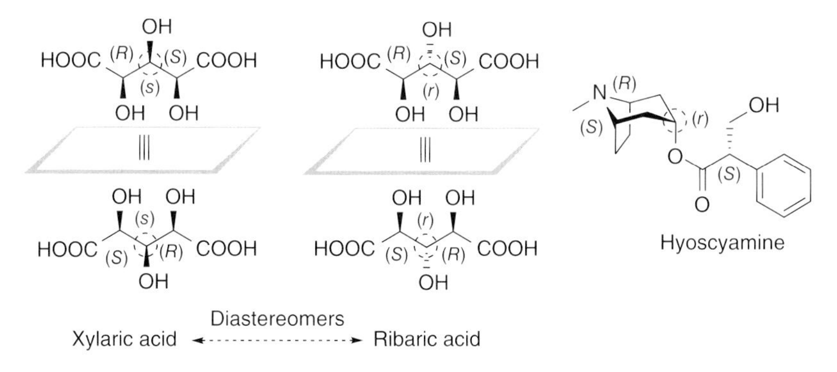


**Figure 2.39** Some compounds containing centers of pseudoasymmetry.

25) According to IUPAC recommendations *r/s*, stereodescriptors can be used to distinguish centers of pseudoasymmetry. Recent literature avoids this usage and in this case, the stereodescriptors *R/S* are applied to identify configuration.

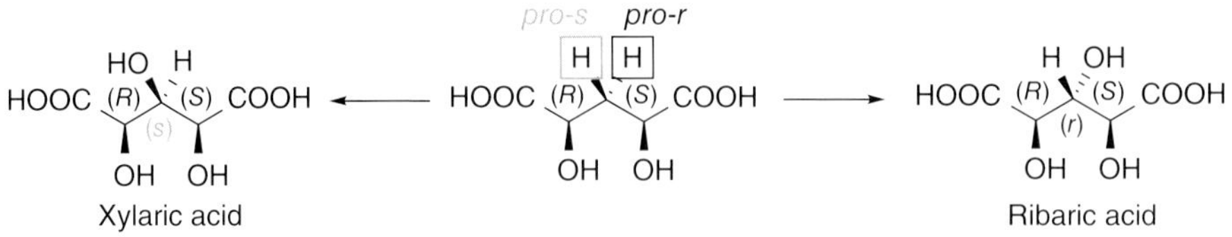

Figure 2.40 Prostereoisomerism resulting in a center of pseudoasymmetry.

Figure 2.40, it can be seen that prostereoisomerism can also be interpreted in association with centers of pseudoasymmetry.

The relationship of the two hydrogen atoms linked to the methylene carbon of the *meso* compound shown in the middle of the figure is diastereotopic. Since the carbon atom is not a center of prochirality, for the distinction of the two hydrogen atoms instead of the *pro-R*/*pro-S* descriptors used earlier for marking groups attached to centers of prochirality, the *pro-r*/*pro-s* descriptors may be applied.

### 2.4.4
### Distinguishing of Stereoheterotopic Faces, the *Re*/*Si* Notation

As it has been mentioned in Section 2.4, when discussing the significance of prostereogenic elements, among transformations of organic compounds besides

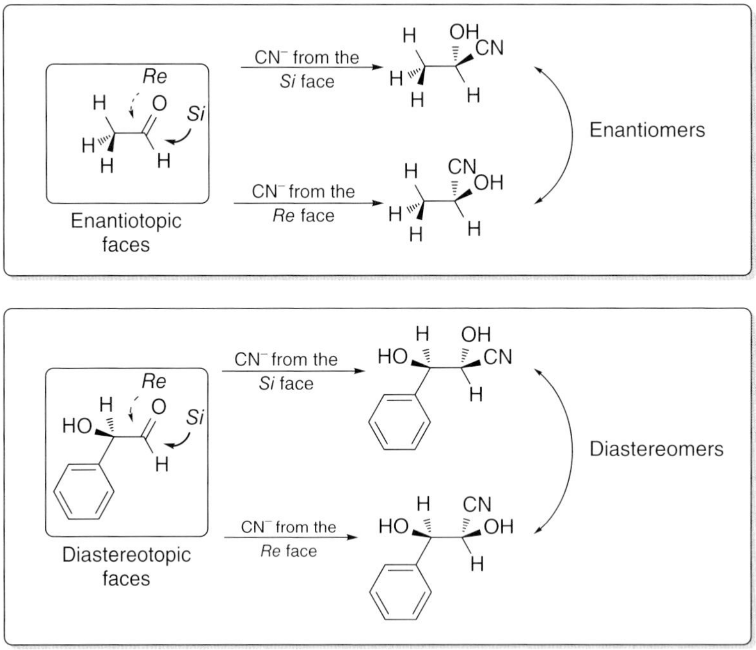

Figure 2.41 Distinguishing stereoheterotopic (enantiotopic/diastereotopic) faces.

substitution, one of the most common transformations is addition (Figure 2.30). Addition to an unsaturated trigonal system from the direction of one face or the other may lead to, depending on the relationship of the faces, a single product, with homotopic faces, or different products, with heterotopic faces (Figure 2.41).

As it is apparent from Figure 2.41 that since stereoheterotopic faces behave differently, being prostereogenic elements resulting in stereoisomerism, they have to be characterized by stereodescriptors using the symbols *Re*/*Si*. In order to be able to characterize a trigonal system, the planar moiety of the molecule in question should be placed on the plane of the paper and viewed from the direction from which we wish to classify it (Figure 2.42a). This is followed by establishing the preferences of the ligands attached to the central atom of the trigonal moiety according to the same CIP sequencing system used to determine the stereodescriptors *R*/*S* or *E*/*Z*. If the sequence of ligands ($1 \rightarrow 2 \rightarrow 3$) is clockwise, the face is classified as *Re*; otherwise, it is marked *Si*. These descriptors are valid for the characterization of faces with both in an enantiotopic (Figure 2.42a) and a diastereotopic relationship (Figure 2.42b). If both bridgehead atoms of a double bond are stereogenic centers, both of them should be analyzed separately (Figure 2.42c). It is apparent that stereodescriptors for the different prostereogenic centers at the same side may be different. A special case is a diastereotopic face addition to which a center of pseudoasymmetry is generated (Figure 2.42d). In this case, instead of the descriptors *Re*/*Si*, the stereodescriptors *re*/*si* are applied.

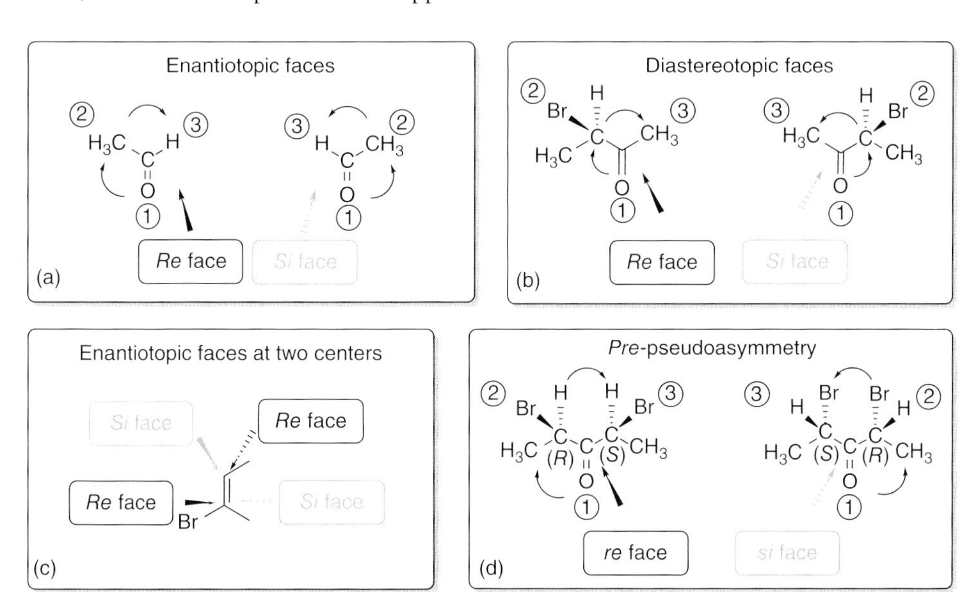


**Figure 2.42** Assigning stereodescriptors to stereoheterotopic (enantiotopic/diastereotopic) faces.

## References


1. Eliel, E.L., Wilen, S.H., and Mander, L.N. (1994) *Stereochemistry of Organic Compounds*, John Wiley & Sons, New York.
2. IUPAC (2014) *Compendium of Chemical Terminology: Gold Book*, Version 2.3.3. IUPAC, http://goldbook.iupac.org/PDF/goldbook.pdf (accessed 27 January 2016).
3. McNaught, A.D. and Wilkinson, A. (1997[ISBN 0-86542-6848]) *Compendium of Chemical Terminology*, 2nd edn, Blackwell Science, Oxford.
4. Moss, G.P. and International Union of Pure and Applied Chemistry, Organic Chemistry Division, Commission on Nomenclature of Organic Chemistry [III. 1] and Commission on Physical Organic Chemistry [III.2] (1996) *Basic Terminology of Stereochemistry* (IUPAC Recommendations 1996). *Pure Appl. Chem.*, **68** (12), 2193–2222.
5. Eliel, E.L. (1976) On the Concept of Isomerism. *Israel J. of Chem.*, **15** (1–2), 7–11.
6. Brecher, J. (2006) Graphical representation of stereochemical configuration [IUPAC Recommendations 2006]. *Pure and Appl. Chem.*, **78** (10), 1897–1970.
7. Cahn, R.S., Ingold, C.K. and Prelog, V. (1996) Specification of molecular chirality. *Angew. Chem. Int. Ed.*, **5**, 385-415.
8. Prelog, V. and Helmchen, G. (1982) Basic principles of the CIP-system and proposals for a revision. *Angew. Chem. Int. Ed.*, **21**, 567–583.

# Problems to Chapters 1 and 2

## Problem 1.1

Draw the *Lewis* and *Kekulé* formulas of borane ($BH_3$), methane ($CH_4$), and ammonia ($NH_3$)! Which of these molecules do not follow the octet rule? On the basis of the answer, explain why borane forms a complex with ammonia and why methane fails to react with the other two compounds! Draw and interpret based on the octet rule the Lewis and Kekulé formulas of the borane–ammonia complex!

$BH_3$ $CH_4$ $NH_3$

## Problem 1.2

On the basis of the linear formulas of butane and ethyl methyl ether shown below, select those particular conformations in which all heavy atoms are coplanar! When one of the methylene groups in butane is replaced by an oxygen atom, what can be found in the spatial positions that had been previously occupied by hydrogen atoms?

O

## Problem 1.3

On the basis of the space requirements of a C–H $\sigma$-bond and the lone pair of electrons on oxygen, respectively, explain why the bond angle $\alpha_1$ in butane is larger than $\alpha_2$ in ethyl methyl ether!

O

$\alpha_1$ > $\alpha_2$

*Stereochemistry and Stereoselective Synthesis: An Introduction*, First Edition.
László Poppe and Mihály Nógrádi.

Companion Website: www.wiley.com/go/poppe/stereochemistry

## Problem 1.4

Explain why are differences in physical and chemical properties between acetaldehyde and vinyl alcohol forming an equilibrium mixture larger than those between chlorocyclohexane with the chlorine atom in axial and equatorial position!

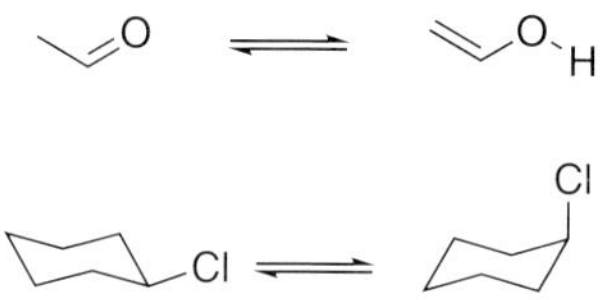

## Problem 2.1

Below, stereoformulas of two conformations of *meso*-tartaric acid and two conformations of (*R*,*R*)-tartaric acid are shown. Enumerate the symmetry elements that can be assigned to them!

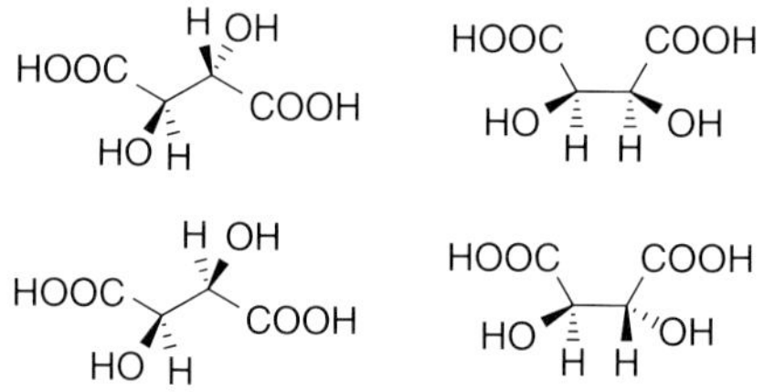

## Problem 2.2

Are acetaldehyde and vinyl alcohol, further chlorocyclohexane chiral or achiral molecules?

- What is the isomeric relationship between acetaldehyde and vinyl alcohol?
- What is the isomeric relationship between the two forms of chlorocyclohexane containing an equatorial and axial chlorine atom?
- What is the topicity relationship between the two faces of acetaldehyde?
- What is the topicity relationship between the hydrogen atoms $H_a$ and $H_b$ of chlorocyclohexane?

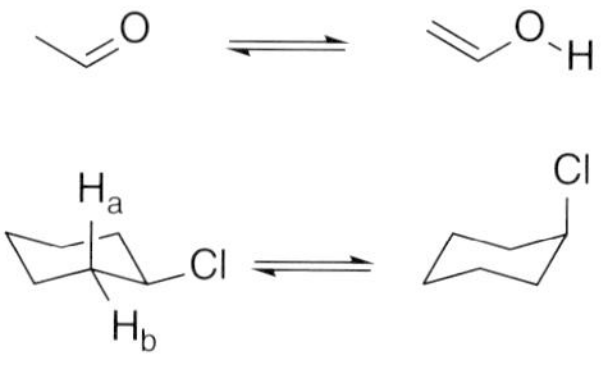

**Problem 2.2.1**

Establish whether

- the forms of chlorocyclohexane containing an axial and an equatorial chlorine atom are of different conformation
- the configuration of $C_1$ in the forms of chlorocyclohexane containing the chorine atom in axial and equatorial orientation is different or identical.

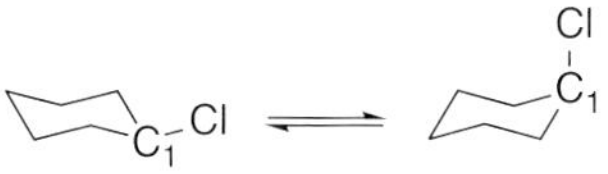

**Problem 2.2.2**

The following formulas show conformations of *meso*-tartaric acid (**1,2,5,6**), (*R,R*)-tartaric acid (**3**), and (*S,S*)-tartaric acid (**4**).

- Demonstrate that although formulas **1** and **2** are mirror images, they represent the same molecule!
- Demonstrate that formulas **1, 5,** and **6** show different conformations of the *meso* compound!
- Demonstrate that the conformations shown by formulas **5** and **6** are in enantiomeric relationship!
- Decide whether the relationship of the pairs of conformations **1** and **5** further **1** and **6** are of diastereomeric or enantiomeric!
- Select which pairs of the following compounds are enantiomers and diastereomers: *meso*-tartaric acid (**1**), (*R,R*)-tartaric acid (**3**), and (*S,S*)-tartaric acid (**4**)!

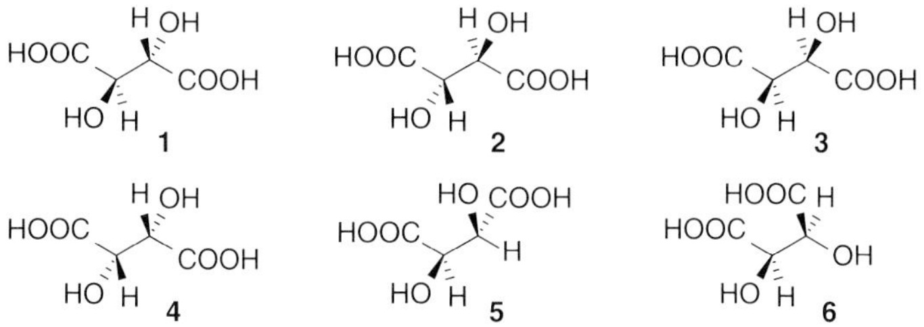

**Problem 2.2.3**

The formula below shows cyclohexyl methyl ketone. Taking into account free conformational motions, define the topicity relationships of the following pairs:

- hydrogens $H_a$, $H_b$, and $H_c$;
- hydrogens $H_{3a}$ and $H_{3b}$;

- hydrogens $H_{4a}$ and $H_{4b}$;
- hydrogens $H_{3a}$ and $H_{5a}$;
- hydrogens $H_{3a}$ and $H_{5b}$;
- two faces of the C=O bond!

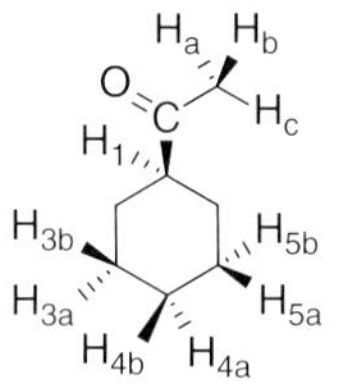

Extend your deductions with an analysis of the symmetry elements and further by analysis of the relationships between compounds obtained by substituting the individual hydrogen atoms with chlorine and by reduction of the ketone function to an alcohol!

**Problem 2.3.1**

Do the following compounds (**A**), (**B**), (**C**), (**D**), and (**E**) contain stereogenic element(s)? If yes, what are these?

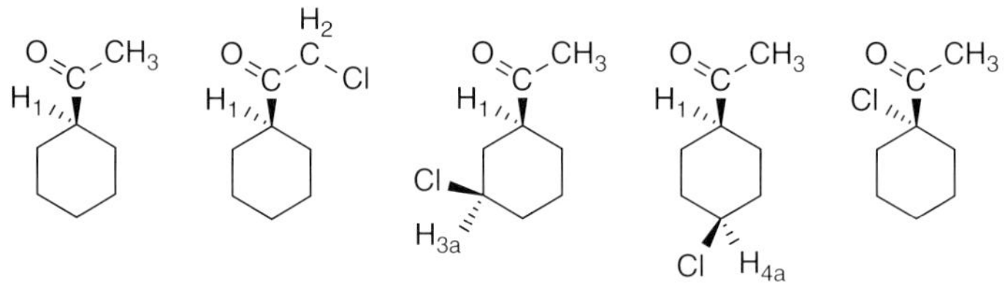

**Problem 2.3.1**

a) Draw the *Fischer* projections of (*R*,*R*)- (**I**), (*R*,*S*)- (**II**), (*S*,*R*)- (**III**), and (*S*,*S*)-2,3,4-trihydroxybutanal (**IV**) and establish which of them belong to the series of D- and L-sugars!
b) Draw the *Fischer* projections of the tartaric acids obtained by oxidation of (*R*,*R*)- (**I**), (*R*,*S*)- (**II**), (*S*,*R*)- (**III**), and (*S*,*S*)-2,3,4-trihydroxybutanal (**IV**). Which of them is the D-, L- and *meso*-tartaric acid?
c) Explain why *meso*-tartaric acid cannot be classified as a sugar acid to the D - or L-series!

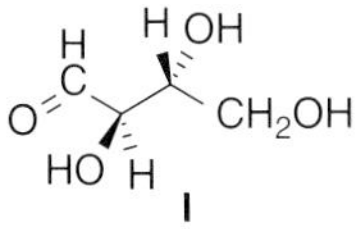

**Problem 2.3.3**

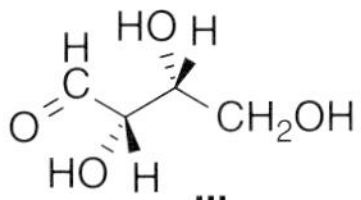

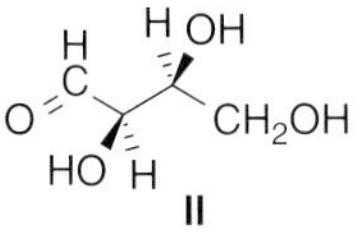

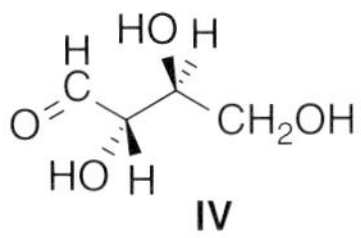

a) Draw the stereoformulas of D- (**1**) and L-erithrose (**2**), as well as of D- (**3**) and L-threose (**4**)!
b) Draw the stereoformulas of the sugar alcohols (butane-1,2,3,4-tetrols) obtained by reduction of the aldehyde group of D- (**1**) and L-erithrose (**2**), as well as of D- (**3**) and L-threose (**4**). Name them according to the C.I.P. and IUPAC conventions!
c) How many kinds of sugar alcohols may arise by reductions of the following compounds and which of them will be chiral and achiral, respectively?

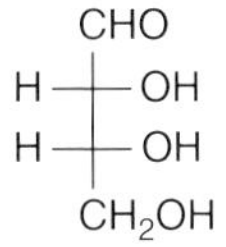

**Problem 2.3.4**

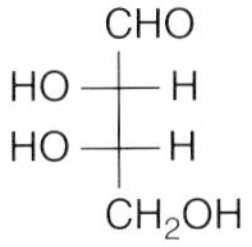

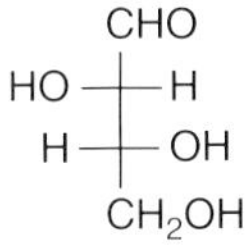

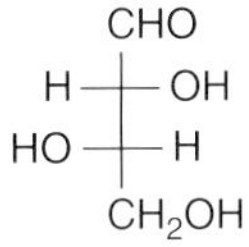

Which is the correct name of the following compound?

a) (*E*)-3-(1,3-Dioxolan-2-yl)-6-methyl-4-(propan-2-yl)hept-3-en-2-one
b) (*Z*)-3-(1,3-Dioxolan-2-yl)-6-methyl-4-(propan-2-yl)hept-3-en-2-one

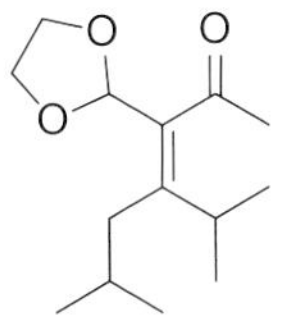

### Problem 2.3.5

Select the correct stereodescriptor(s) from among $R_a$, $S_a$, $P$, and $M$ for the binaphthol shown below!

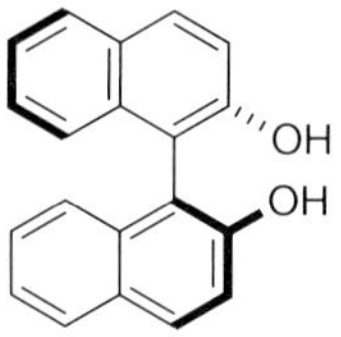

### Problem 2.3.6

Select the correct stereodescriptor from among $R_p$, $S_p$, $P$, and $M$ for the compound shown below!

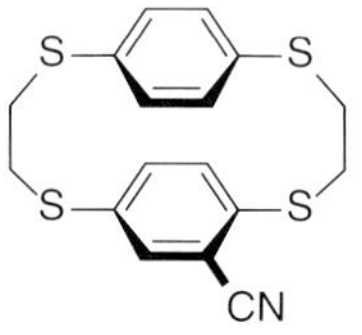

### Problem 2.4.1

Hydrogens attached to both methylene groups of 1-bromo-2-chloroethane are prochiral. Demonstrate that on monobromination of this compound, there will be among the dibromochloroethane obtained both chiral and achiral molecules!

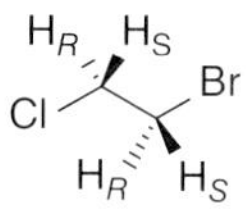

### Problem 2.4.2

a) Which of the hydrogens $H_a$ and $H_b$ of 3-(1,3-dioxolan-2-yl)but-3-en-2-one are *pro-E* and *pro-Z*?
b) Which of the hydrogens $H_a$ and $H_b$ of chlorocyclohexane are *pro-cis* and *pro-trans*?

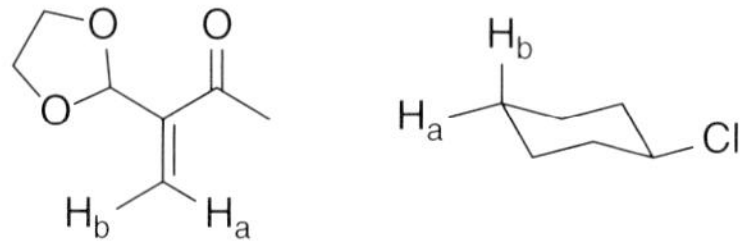

### Problem 2.4.3

Which of the indicated hydrogen atoms of chloromethyl cyclohexyl ketone are *pro-R* and *pro-S*, and which is not prochiral?

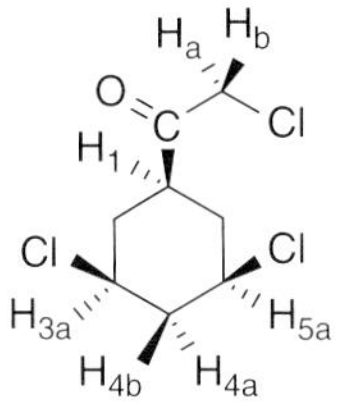

### Problem 2.4.4

a) Is 1-[(1*R*,3*R*,5*S*)-3,5-dichlorocyclohexyl]ethanone a chiral or an achiral compound?
b) Does the compound depicted below possess a pseudoasymmetric carbon atom?
c) Give stereodescriptors to $H_{4a}$ and $H_{4b}$!
d) If hydrogen atoms $H_{4a}$ and $H_{4b}$ are exchanged for a bromine atom, the product will be a chiral compound or not?

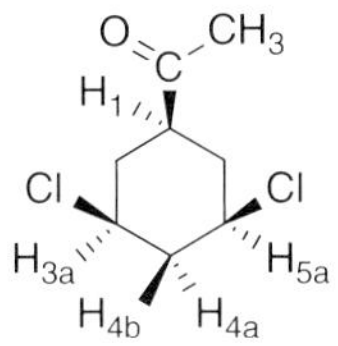

### Problem 2.4.5

Assign stereodescriptors to the faces of the two carbonyl groups in (2*R*,4*R*,6*S*)-4-acetyl-3,5-dichlorocyclohexanone! Reduction of which of the carbonyl groups gives a chiral product?

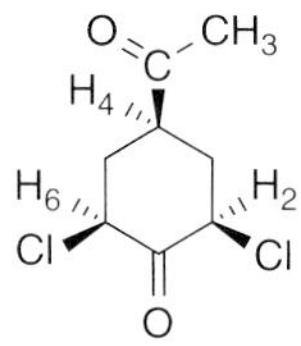

# Part II
# Properties at the Level of Material Assemblies

In Part I, we have already discussed that structure is a concept at the molecular ("*microscopic*") level, while the properties of materials are generally "*macroscopic*" ones. Difficulties in correlating molecular structure with macroscopic properties can be attributed to the quantity of material and the time required for the determination of properties. In the following chapters, concepts and their consequences treated so far mainly at the microscopic level will now be examined at the macroscopic level from the aspect of properties characteristic for material assemblies.

*Stereochemistry and Stereoselective Synthesis: An Introduction*, First Edition.
László Poppe and Mihály Nógrádi.

Companion Website: www.wiley.com/go/poppe/stereochemistry

# 3
# Timescale, Conformational Changes

It is enlightening to consider that the determination of macroscopic properties is accomplished by measurements on an assembly of at least $10^{19}$ molecules under such conditions that this assembly of innumerable molecules undergoes, during the duration of the measurement, at least $10^{12}-10^{15}$ rotational and vibrational changes of state. The significance of this fact for the determination of properties is that almost every characteristic of materials can only be stated as a conformational average of a very high number of molecules. First, let us examine some consequences of time-associated changes, which will be analyzed in detail in connection with the *stereochemistry* of conformational changes.

## 3.1
## Conformational Motion of Ethane and Its Optical Inactivity

On analyzing the conformational behavior of open-chain compounds instead of the formulas depicting their *stereostructure* (see earlier), often the s.c. *Newman* projections shown in Figure 3.1 are preferred. When drawing a *Newman* projection, the molecule is viewed along a bond axis; in the present case, along the C–C bond. The front carbon atom is presented by a dot, and the rear one by a circle.

With *ethane*, the most stable conformation is the s.c. *open* (staggered or synclinal) **conformation** [1–3],[1)] (or constellation), (Figure 3.1); all the other ones, generated by rotation around the C–C bond, are of higher energy. The maximum of potential energy is associated with the *eclipsed* conformation; further rotation involves a drop of energy until reaching the next open conformation when the molecule becomes again stabilized.

1) According to IUPAC recommendations [1–3] "a **conformer** is one among those stereoisomers, which can be characterized by a conformation associated with a defined potential energy minimum." Since the definition of conformer is not unambiguous and sometimes this very concept is used to denote potential energy maxima, in this work we avoided it and all conformational states were called ***conformations.***

*Stereochemistry and Stereoselective Synthesis: An Introduction,* First Edition.
László Poppe and Mihály Nógrádi.

Companion Website: www.wiley.com/go/poppe/stereochemistry

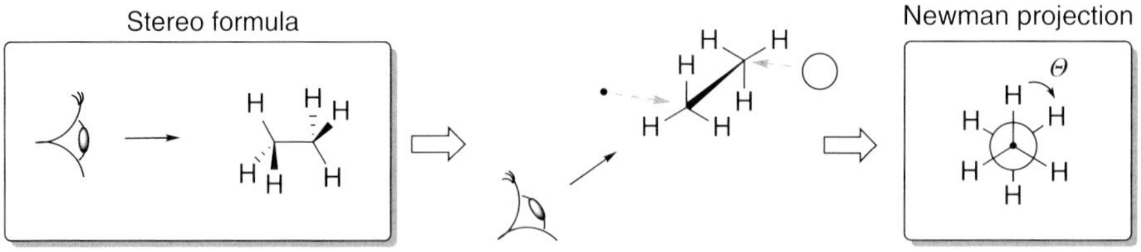

**Figure 3.1** Representation of ethane in Newman projection.

The difference between the minimum and maximum energies, the s.c. rotational energy barrier, is small ($\Delta G^{\#} = 12.5\,\mathrm{kJ\,mol^{-1}}$), and thus rotation at room temperature is virtually unimpeded. The molecule undergoes a full rotation around the C–C single bond going alternately through three identical eclipsed (0°, 120°, and 240°) and three identical open (60°, 180°, and 300°) conformations. Owing to the $C_3$ symmetry of the methyl group, the full rotation consists of three indistinguishable segments. Accordingly, it is sufficient to examine the properties of the molecular assembly involving a single segment, for example, in the range of $\Theta = 0-120°$.

In our everyday perception, ethane is regarded as an achiral molecule in agreement with IUPAC recommendations [1–3] that discourage the extension of the notions chiral/achiral to a macroscopic level. By a detailed analysis of conformational changes, it can be recognized that in the range of $\Theta = 0-120°$, only the eclipsed (**F**) and open (**N**) conformations are achiral, each of the other conformational states characterized by other torsion angles (i.e., an infinite number) are chiral, as exemplified in the figure by conformations $^{*}$**A** and A$^{*}$ derived from the eclipsed state by rotations by of +20° and −20°, respectively. Thus, it is justified to ask why ethane is regarded as achiral when in the assembly constituting the substance the majority of the conformational states are chiral.

The relative amount of two conformational states (e.g., **N** and **F**, or that of the **N** and one of the chiral conformations, e.g., A$^{*}$) depends on the difference of Gibbs' free energy ($\Delta G$) (cf. Table 3.1). Generally, in case of equilibrium A ⇆ B, the value of $\Delta G$ is calculated using the following equation:

$$\Delta G = -2.303RT\ \log K \qquad (3.1)$$

where $K$ is the equilibrium constant ($K = [\mathrm{B}]/[\mathrm{A}]$), $R$ the universal gas constant ($R = 8.314\,\mathrm{J\,mol^{-1}\,K^{-1}}$), and $T$ the absolute temperature.

It follows from this analysis that the optical inactivity of ethane on the *macroscopic level* is not a consequence of the fact that the preponderance of ethane molecules is in the energetically preferred open conformation. In fact, only a negligible proportion of ethane resides in an achiral conformation, while, as it transpires from the equilibrium data (Table 3.1 and Eq. 3.1) and the energy

**Table 3.1** Relative populations of the conformational states **A** and **B** as a function of their *Gibbs* free energy difference $\Delta G$.

| $K_{298\,K}$ | 2 | 3 | 4 | 5 | 10 | 20 | 100 | 1000 | 10 000 |
|---|---|---|---|---|---|---|---|---|---|
| $\Delta G$ (kJ mol$^{-1}$) | 1.71 | 2.72 | 3.43 | 3.97 | 5.85 | 7.52 | 11.3 | 17.1 | 23.0 |
| More stable state (%) | 67 | 75 | 80 | 83 | 91 | 95 | 99 | 99.9 | 99.99 |

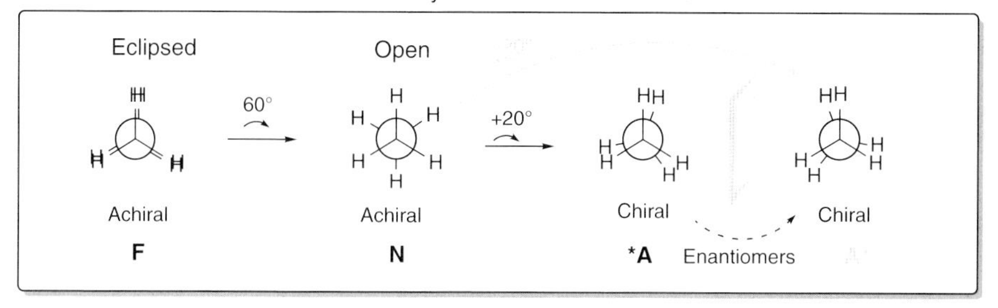


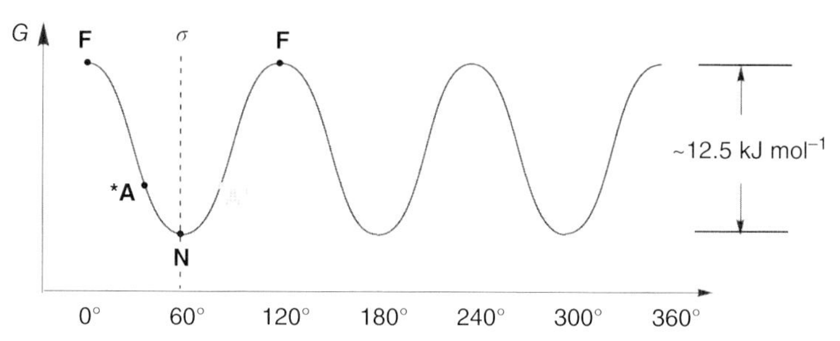


**Figure 3.2** Conformations of ethane.

diagram in Figure 3.2, an overwhelming majority of the molecules are in chiral conformational states. The real reason of the optical inactivity at the macroscopic level is the enantiomeric relationship of conformations obtained from the eclipsed state by rotations by the same angle but of opposite sign (e.g., ***A** and A* obtained by rotations of +20° and −20°, respectively). Enantiomers are known to be of equal energy, therefore, according to Eq. (3.1), the two conformations are present in exactly the same amount and behave as racemates. Thus, for example, when form ***A** rotates the plane of polarized light in a given direction, form A* rotates it in the opposite direction by exactly the same angle. Consequently, ethane as a multimolecular material assembly does not show any optical rotation.

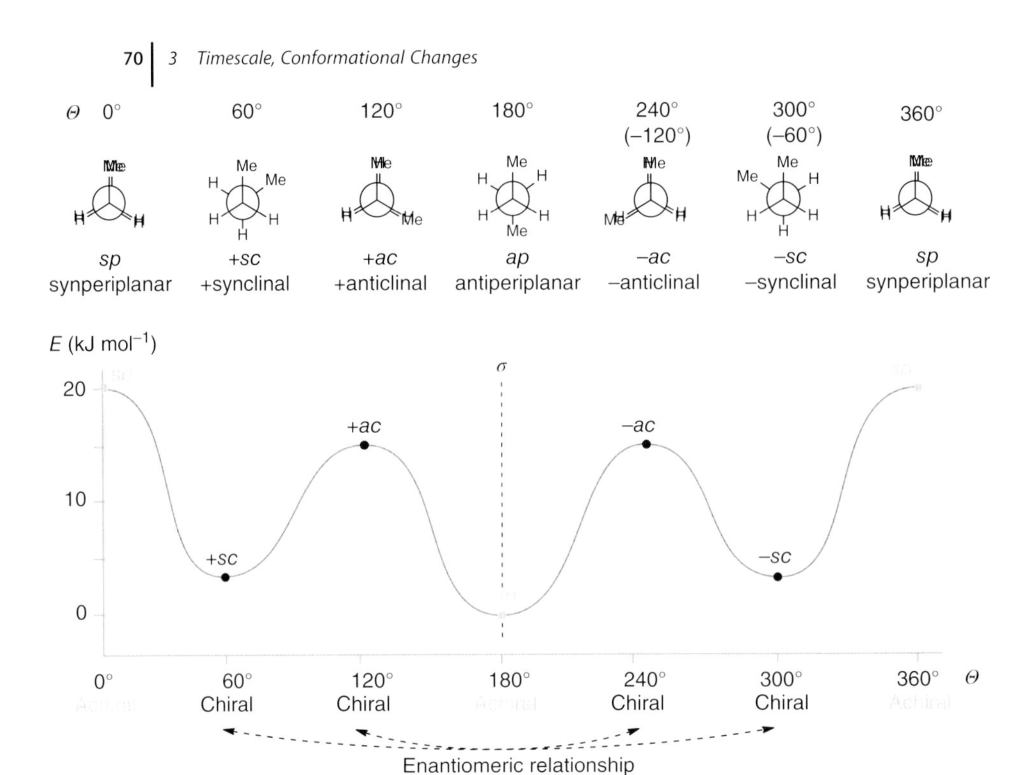


**Figure 3.3** Conformations of *n*-butane.

## 3.2 Conformations and Optical Inactivity of *n*-Butane and 1,2-Dichloroethane

*n*-Butane exists in three minimum energy conformations, associated with two different energy levels (Figure 3.3). Owing to the energy difference between them ($\Delta H = 3.7\ \text{kJ mol}^{-1}$) at room temperature, *n*-butane exists to 72% in the more stable *antiperiplanar* (*ap*) and to 28% in the less stable *synclinal* (+*sc*, −*sc*, or *gauche*) conformational state. The energy level difference between energy barriers associated with the highest energy *synperiplanar* (*sp*) conformation ($\Delta G^{\#} \sim 19\ \text{kJ mol}^{-1}$) and the *anticlinal* (+*ac*, −*ac*) ones ($\Delta G^{\#} \sim 16\ \text{kJ mol}^{-1}$) is also different, and both are higher than with ethane.[2] At room temperature, rotation around the $C_2$–$C_3$ bond is free.

Undergoing complete rotation, butane passes one *ap*, one *sp*, and two *ac* and two *sc* conformations. The two *ac* (+*ac* and −*ac*) and the two *sc* (+*sc* and −*sc*) conformations, respectively, are pairwise nonidentical mirror images and are related therefore as enantiomers. Since rotation is free, conformers interconvert

2) These values are depending on the state of the material. Energy difference between individual conformations is larger in gas phase than in liquid phase. Accordingly, in a liquid phase, the *ap* state is present in 56%, while the *sc* state in 44%.

**Table 3.2** Extended interpretation of the conformational states of *n*-butane.

| Torsion angle Θ | Conformational state |
|---|---|
| $0° \pm 30°$ | ±*synperiplanar* (±*sp*) |
| $+30°$ to $+90°$ | +*synclinal* (+*sc*) |
| $+90°$ to $+150°$ | +*anticlinal* (+*ac*) |
| $+150°$ to $+180°$ | +*antiperiplanar* (+*ap*) |
| $-30°$ to $-90°$ | −*synclinal* (−*sc*) |
| $-90°$ to $-150°$ | −*anticlinal* (−*ac*) |
| $-150°$ to $-180°$ | −*antiperiplanar* (−*ap*) |

freely, and as discussed with ethane, equal proportions of enantiomeric conformers will be present in the mixture. Since the enantiomeric conformations cannot be separated at room temperature, the compound is macroscopically optically inactive.[3)]

Note that in the case of *n*-butane, Figure 3.3 is only a simplified representation of the multitude of different possible conformational states, since methyl groups at the chain ends also rotate freely. The real conformational behavior of 1,2-dichloroethane is closer to the single conformation–energy curve represented in Figure 3.3. For *n*-butane, the interpretation of conformational states can be replaced by a more realistic concept of *conformational ranges* (Table 3.2).

## 3.3 Biphenyl and Substituted Biphenyls, Atropisomerism

Biphenyl reaches one of its maximum energy conformational state when $\Theta = 0°$ ($\Delta G^{\#} \sim 6.4 \pm 2.1\ \mathrm{kJ\ mol^{-1}}$; Figure 3.4). Although the *delocalization* between the two aromatic rings is maximal in this conformation, the energy is high due to a strong Coulomb-type repulsion because pairs of *ortho* hydrogens on the two rings are approaching each other within the range of their van der Waals radii.

When $\Theta = 90°$, the *ortho* hydrogens are at a maximum distance from each other, but the energetically favorable effect of *delocalization* between the two aromatic rings is absent and the energy barrier also becomes high ($\Delta G^{\#} \sim 6.0 \pm 2.0\ \mathrm{kJ\ mol^{-1}}$). An energy minimum can be observed at around $\Theta = 44°$ when the *ortho* hydrogens already moved away from each other, but some overlapping of the $\pi$-systems of the two rings is still present. With unsubstituted

3) At very low temperature conformations, *ap* and the two chiral conformations (+*sc* and −*sc*) behave as real isomers.

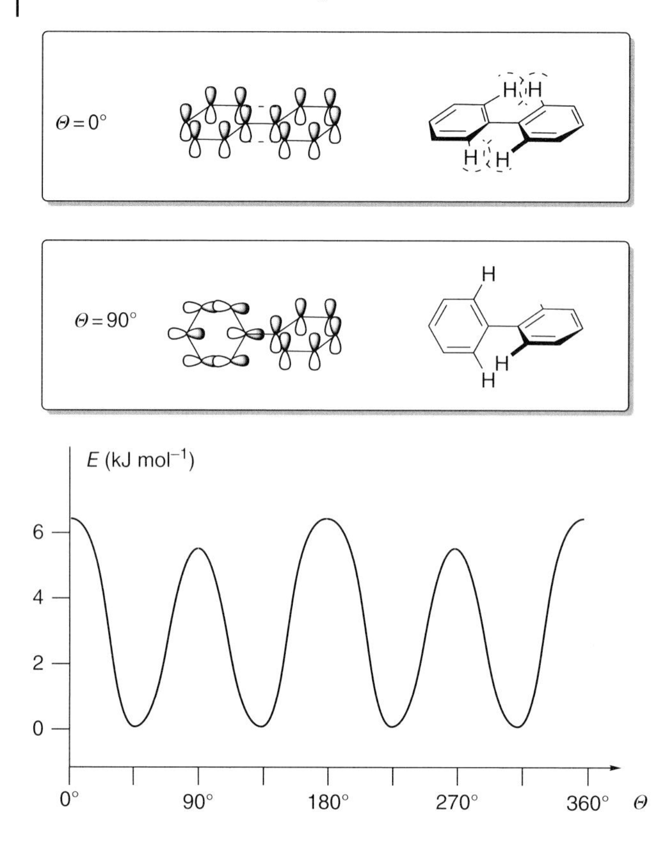


**Figure 3.4** Conformations of biphenyl.

*biphenyl*, rotation around the bond connecting the phenyl rings is possible at room temperature.

The situation becomes thoroughly changed with *ortho* substituents larger than hydrogen (Figure 3.5). Of the two energy barriers, those associated with $\Theta = 0°$ and $\Theta = 180°$, respectively (coplanar rings), increase significantly, while that at ~90° does not change much. When in substituted biphenyl derivatives, the energy barrier against free rotation cannot be overcome at room temperature; the phenomenon of *atropisomerism* can be observed.

In Section 2.3.4, dealing with axial chirality, we have discussed that the presence of four different ligands (A, B, C, and D) is not a necessary condition for axial chirality including atropisomerism; it is sufficient when coplanar pairs of ligands (A, B and C, D, respectively) are different. Because the atropisomers are in nonidentical mirror image relationship, they are enantiomers.

Since atropisomerism may be due to hindered rotation around a single bond, it is not restricted to biphenyls (Figure 3.6).

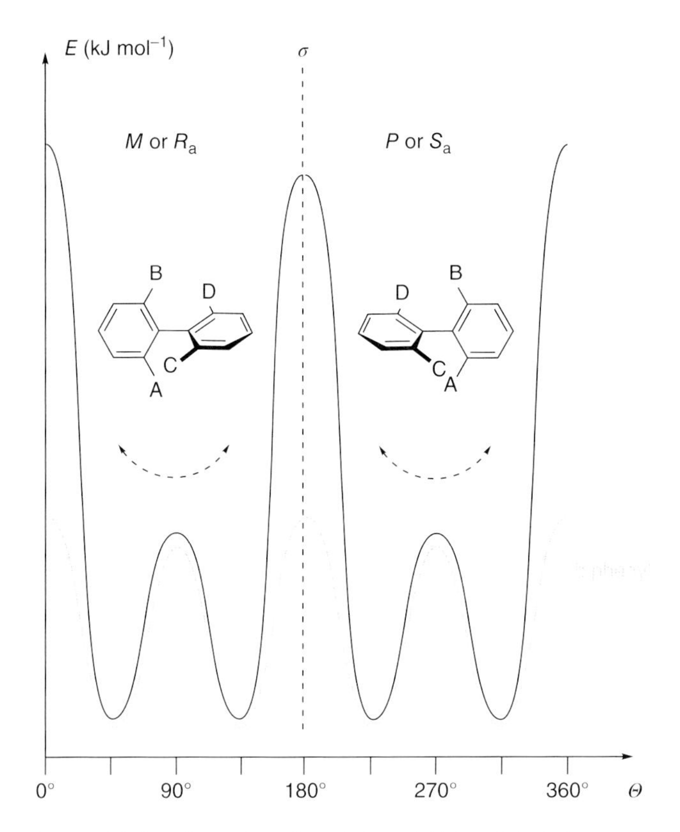


**Figure 3.5** Atropisomerism of *o*-substituted biphenyl derivatives.

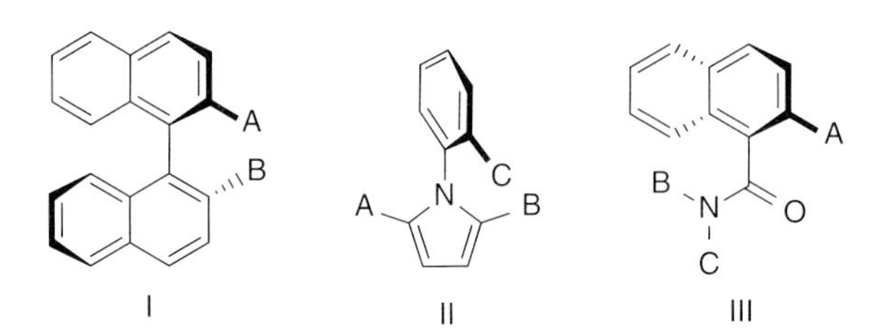


**Figure 3.6** Atropisomerism of compounds differing from biphenyls.

Hindered rotation around a single bond within various other compounds such as 2,2′-substituted-1,1′-binaphthalenes (**I**), 1-(*o*-substituted-phenyl)-2,5-disubstituted-1*H*-pyrroles (**II**), or *N*,*N*-disubstituted-2-substituted-1-naphthamides (**III**) may also result in atropisomers.

## References

1. Moss, G.P. and International Union of Pure and Applied Chemistry, Organic Chemistry Division, Commission on Nomenclature of Organic Chemistry [III. 1] and Commission on Physical Organic Chemistry [III.2] (1996) *Basic Terminology of Stereochemistry* (IUPAC Recommendations 1996). *Pure. Appl. Chem.*, **68** (12), 2193–2222.
2. McNaught, A.D. and Wilkinson, A. (1997) *Compendium of Chemical Terminology*, 2nd edn, Blackwell Science, Oxford [ISBN 0-86542-6848].
3. IUPAC (2014) *Compendium of Chemical Terminology: Gold Book*, Version 2.3.3, http://goldbook.iupac.org/PDF/goldbook.pdf (accessed 17 January 2016).

# 4
# Absolute Configuration

In Section 2.2, we had already discussed the basic terms of stereochemistry, which can be related to the structure of a molecular entity. In the context of stereochemistry, IUPAC defines the term **configuration** (stereochemical) (see p 2204 in ref. [1] and p 1099 in ref. [2]) as the arrangements of atoms of a molecular entity in space that distinguishes stereoisomers (the isomerism that is not due to conformation differences).

IUPAC also defines the so-called **absolute configuration (AC)** (see p 2197 in ref. [1]) as the spatial arrangement of the atoms of a chiral molecular entity (or group) and its stereochemical description, for example, *R* or *S*. Although this definition is clearly related to chiral molecular entities, it is rather terse and disregards the fact (discussed in Section 1.4) that the structure and chirality of a molecular entity are concepts at the molecular ("*microscopic*") level, while the properties of materials are generally "*macroscopic*" ones. In this respect, the enantiomers being chiral molecules are of special interest. In Section 2.2.2, discussing the isomerism and isomeric relationship, we have already stated that the constitutional isomers or diastereomers differ in all of their properties. On the other hand, while scalar physical properties of enantiomers are identical, they do differ in vectorial physical properties, such as optical rotation (OR). Their chemical behavior toward achiral agents is identical, while they may differ toward chiral agents. Achiral compounds can be adequately characterized invoking scalar physical properties (such as melting point, boiling point, refractive index, solubility, heat of combustion, normal spectra like UV–Vis, IR, NMR, and MS). Consequently, pairs of constitutional isomers or diastereomers can be easily distinguished by their characterization invoking scalar physical properties, but distinguishing the enantiomers being chiral compounds requires special methods. This is the reason why the term *absolute configuration* is used exclusively to characterize chiral compounds (to distinguish enantiomers), but not applied for other types of isomerism (e.g., for *E*/*Z*-isomers).

At the "*macroscopic*" level, the French scientist Louis Pasteur observed as early as in 1848 that the crystals of racemic sodium ammonium and sodium potassium tartrate were chiral and formed two asymmetric forms in equal amounts that were mirror images of one another [3, 4]. After sorting the left- and right-handed crystals, he found that solutions of right-handed crystals were dextrorotatory (rotated the linearly polarized light clockwise, similarly as the natural compound isolated

*Stereochemistry and Stereoselective Synthesis: An Introduction,* First Edition.
László Poppe and Mihály Nógrádi.

Companion Website: www.wiley.com/go/poppe/stereochemistry

from wine), while the solutions of left-handed crystals were levorotatory, with the same absolute value of specific rotation. An equal mixture of the two forms had no optical rotation. Pasteur deduced that the molecules in the crystals are asymmetric and exist as two different forms that are mirror images but nonidentical. He also found that in samples of tartaric acid isolated from nature, only one of the two forms is present.

At the "*microscopic*," that is, molecular level, Jacobus Henricus van't Hoff [5] and Joseph Achille Le Bel independently recognized, in 1874 [6], that the phenomenon of optical activity of carbon compounds can be explained by assuming that orientation of the four bonds between a central carbon and its ligands is toward the apexes of a tetrahedron. If the four ligands are all different, then there are two possible arrangements around the tetrahedron, which will be mirror images of each other.

The projective representation of configuration suggested by Hermann Emil Fischer in 1891 [7] in relation with the stereostructure of aldohexoses enabled to depict 3D structures in two dimensions, for example, the configuration of the two forms of glyceraldehyde (Figure 4.1a). It was also evident that one form of

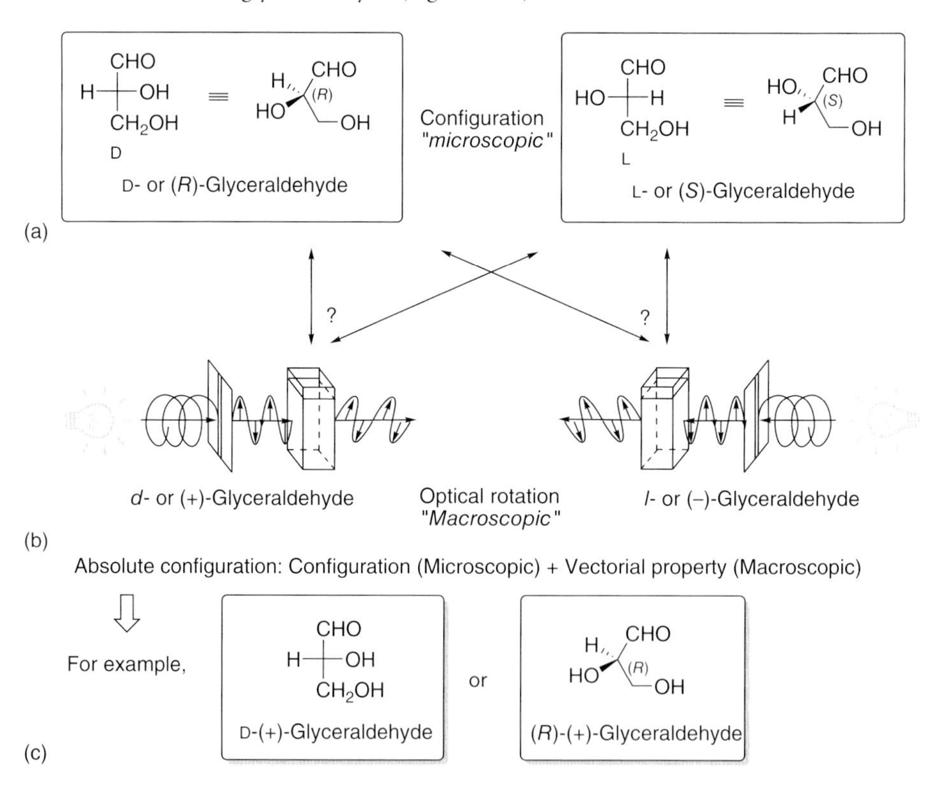


**Figure 4.1** Configuration and optical rotation of glyceraldehyde enantiomers. Due to the lack of unambiguous methods to establish the relationship between (a) the microscopic (configuration) and (b) the macroscopic vectorial (e.g., optical rotation) properties until 1951, the absolute configuration of glyceraldehyde (c) was defined arbitrarily as D-(+)[1].

the glyceraldehyde being in nonidentical mirror image relationship with the other one should be dextrorotatory (*d*), while the other one should be levorotatory (*l*) to the same extent (Figure 4.1b). It was not evident, however, which configuration belonged to which sign of optical rotation. To extend the correlation of the configuration to other classes of compounds, Martin André Rosanoff proposed in 1905 a tentative absolute configuration of D-(+)-glyceraldehyde using the *Fischer* projection (Figure 4.1c) [8].[1] The isomer of glyceraldehyde rotating the linearly polarized light to the right (*d* or +) was labeled D, while the isomer rotating the linearly polarized light to the left (*l* or −) was labeled L.

In an extended sense, the absolute configuration relates the configuration of any chiral compound to the configuration of D-(+)-glyceraldehyde as an agreed stereochemical standard.[1] Later in this chapter, we discuss the strategies of chemical correlations linking the configuration of various compounds to that of D-(+)-glyceraldehyde. If the arrangement of atoms in space in a molecule can be related to D-(+)-glyceraldehyde – or some other standard of known absolute configuration – we can state that we know that molecule's absolute configuration.

Note that the D/L nomenclature system is fundamentally different from the *R/S* system. While the latter characterizes only a single asymmetric center in the molecule, D and L descriptors relate to the entire molecule. The stereochemistry of all the other stereogenic centers of a given sugar is defined by its historic trivial name. In contrast, using the *R/S* system, stereodescriptors for each and every stereogenic center have to be defined separately.

In Section 2.2.1, we considered that chirality resulting in measurable properties is a feature of the whole molecule. Since it is also obvious that the properties (either scalar or vectorial) can only be investigated on assemblies of whole molecules, we can restrict and generalize the term of absolute configuration compared to the IUPAC recommendations [1, 9, 10]. Although the IUPAC recommendation for absolute configuration refers to spatial arrangement of the atoms of a chiral molecular entity (or group), we suggest to restrict this term only for whole molecules. Moreover, the statement of absolute configuration is restricted to a single element of chirality; it is only valid for molecules containing a single stereogenic element. In case of chiral molecules with more than one asymmetric unit, the term absolute configuration refers to the full assembly of the asymmetric centers, axes, and planes present in the molecule. Thus, when stating that the absolute configuration of glucose is D-(+), we define the configurations all of four asymmetric centers at once (Figure 4.2).

In the most general sense, we may consider the absolute configuration as a relationship between the full assembly of asymmetric elements causing the chirality of a molecular entity and a measurable vectorial property of the molecule. In this respect, a statement telling that the absolute configuration of the natural glucose is D would be satisfactory because the vectorial property here is the chiral metabolic machinery of nature leading to the natural enantiomer.

1) According to this proposition, the graphical representation of D-glyceraldehyde should correspond to the configuration of D-glyceraldehyde, that is, to the form rotating the linearly polarized light into (+)-direction.

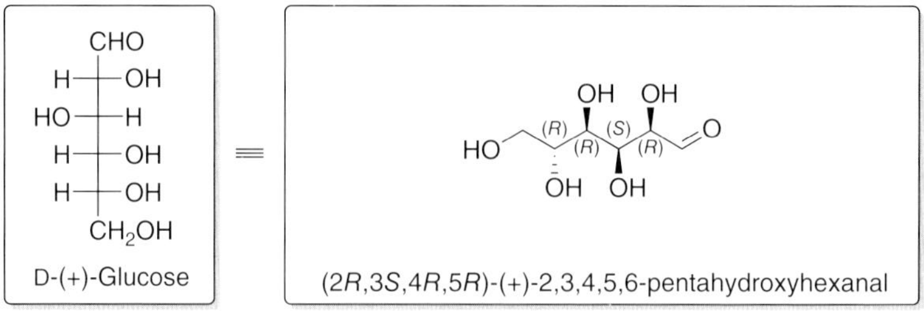

D-(+)-Glucose

(2*R*,3*S*,4*R*,5*R*)-(+)-2,3,4,5,6-pentahydroxyhexanal

**Figure 4.2** Absolute configuration of natural glucose. Note that an extended sense of configuration (i.e., the full assembly of four asymmetric units responsible for the chirality of the molecule) is related to the optical rotation to denote the absolute configuration.

## 4.1 Methods to Determine Absolute Configuration[2)]

The question of absolute configuration (i.e., the exact orientation of the atoms in molecules with asymmetry) has been a serious problem to be solved since the suggestion of van't Hoff and Le Bel on the reason of molecular dissymmetry in 1874 [5, 6].

The most direct and obvious physical method to determine the absolute configuration (AC) is the theoretical calculation of the optical rotatory power. In 1938, Kuhn suggested a method to determine the absolute configuration based on the different interaction of the two enantiomers of 2-butanol with right and left circularly polarized light [12].

The first unambiguous direct assignment of AC succeeded in 1951 when Bijvoet *et al.* [13] solved the structure of sodium rubidium salt of (+)-tartaric acid by X-ray crystallography exploiting the anomalous dispersion effect caused by a heavy atom in the crystal of a pure enantiomer. The (*R,R*) absolute configuration determined for (+)-tartaric acid confirmed that the previous subjective assignments by Fischer [7], Rosanoff [8], and Kuhn [12] were correct. Thus, all absolute configurations that had been based on correlations to (−)-L-tartaric acid or (+)-D-glyceraldehyde turned out to be correct as well. Nowadays, given good quality crystals of an enantiopure compound, X-ray structure analysis can be performed from a crystal as small as 1 μg in mass. The measurement generally yields the total connectivity and relative configuration of all nonhydrogen atoms, and most of the hydrogen atoms in small (up to 100 atoms) organic molecules in less than 1–3 h [14]. The major bottleneck for X-ray analysis is the production of suitable crystals from enantiopure sample.

In our days, predictive calculations of the chiroptical data [optical rotation (OR), electronic circular dichroism (ECD), and vibrational circular dichroism (VCD) spectra as well as vibrational optical activity (VOA) and Raman optical

2) Allenmark and Gawronsky [11], and the further reviews in the corresponding special issue on absolute configuration determination.

activity (ROA)] have become reliable methods for direct AC determination due to the development of *ab initio* quantum mechanical calculations, particularly the density functional theory (DFT) [11, 15]. An advantage of the chiroptical AC assignment methods is that these methods do not require crystalline compounds. While an ECD spectrum can be obtained from sub-microgram amounts, the VOA methods require much larger samples. The most critical requirement of AC determination by chiroptical methods is the necessary knowledge of the conformer distribution of the compound. The ECD method requires a suitable chromophore in the compound, while this is not compulsory with VCD and ROA techniques.

To cover the field completely, besides the above *direct* methods for AC determination, we have to mention several *indirect* methods. These require a chiral auxiliary such as a chiral-derivatizing agent (CDA) or a reference compound of known AC [11]. Strictly speaking, these *indirect* methods provide *relative configuration* (see Section 4.2.2), since the absolute configuration of the CDA or catalyst (enzyme) is *a priori* known.

Such indirect methods include X-ray structure analysis after cocrystallization with a compound of known absolute configuration, NMR anisotropy methods with the use of an internal source of chirality (CDAs, chiral shift reagents, and chiral solvents), enzymatic methods [computing the stereoselective reaction(s) within an enzyme's active site], and chemical correlation. The latter technique, due to its outstanding importance for the establishment of configurational relationships since the days of Fischer, is discussed in detail in the following chapter.

## 4.2 Strategies to Determine Absolute Configuration by Chemical Correlation

In course of chemical transformations aiming the determination of *absolute configuration* related to the configuration of D-(+)-glyceraldehyde (or some other standard of known absolute configuration), the relative configurations of different molecules are compared.

According to IUPAC, **relative configuration** (see p. 2217 in ref. [1]) has two different meanings.

1) The configuration of any stereogenic (asymmetric) center with respect to any other stereogenic center contained within the same molecular entity. Unlike absolute configuration, relative configuration is reflection-invariant. Relative configuration, distinguishing diastereomers, may be denoted by the configurational descriptors *R*,R** (or *l*, like, Figure 4.3a) and *R*,S** (or *u*, unlike; Figure 4.3b) meaning, respectively, that the two centers have identical or opposite configurations. For molecules with more than two asymmetric centers, the prefix *rel-* is placed before the name of one enantiomer where *R* and *S* have been used. If any of the centers has a known absolute configuration, then only *R** and *S** can be used for the relative configuration.

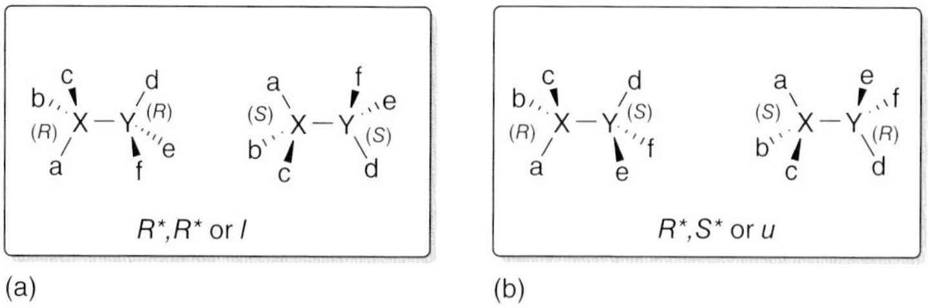

Figure 4.3 Relative configuration within diastereomers.

2) Two different molecules **Xabcd** and **Xabce** may be said to have the same relative configurations if **e** takes the position of **d** in the tetrahedral arrangement of ligands around **X** (i.e., the pyramidal fragments **Xabc** are superimposable; Figure 4.4**a**). By the same token, the enantiomer of **Xabcd** may be said to have the opposite relative configuration to **Xacbe** (Figure 4.4b). The terms may be applied to chiral molecular entities with central atoms other than carbon, but are limited to cases where the two related molecules differ in a single ligand.

Importantly, even the IUPAC Gold book [9] states that the term relative configuration can be generalized to include stereogenic units other than asymmetric centers.

On the basis of the relative configuration determinations, the following strategies can be applied to correlate absolute configuration to known stereochemical standards:

- ➢ Transformations of the ligands attached to an asymmetric unit without influencing the bonds between the asymmetric unit and the ligands.
- ➢ Degradation of multiple asymmetric units of a more complex molecule until only one remains intact.
- ➢ A stereoconstructive strategy that adds further asymmetric elements to the one of known absolute configuration without influencing the bonds between its asymmetric unit and ligands.

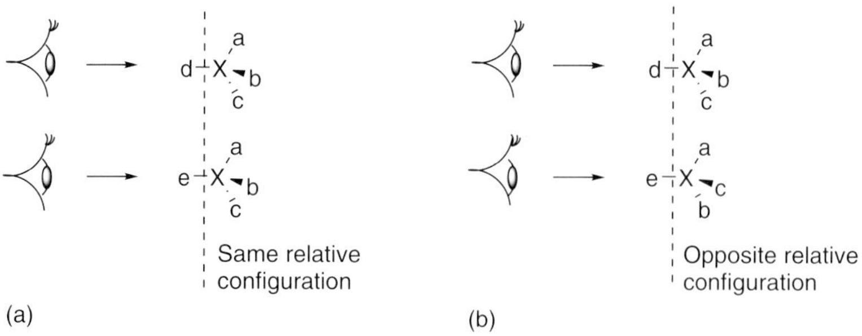

Figure 4.4 Relative configuration between molecules of different constitution.

- A comparison strategy that simplifies complex isomeric molecules containing multiple asymmetric elements by symmetrization of several asymmetric units leading to a simple compound with less (but more than one) asymmetric elements.
- Transformations involving stereospecific reactions of known stereochemistry influencing the bonds between the asymmetric unit and the ligands.

### 4.2.1 Transformations of the Ligands Attached to an Asymmetric Unit without Influencing the Bonds between the Asymmetric Unit and the Ligands

Freudenberg [16] converted (+)-glyceric acid to (+)-lactic acid by methods that did not sever any of the bonds directly attached to the asymmetric center. D-(−)-Glyceric acid was obtained by the oxidation of D-(+)-glyceraldehyde by Wohl and Schellenberg [17]. These data led to a configurational correlation between D-(+)-glyceraldehyde and (−)-lactic acid and established the designation D-(−)-lactic acid (Figure 4.5).

### 4.2.2 Degradation of Multiple Asymmetric Units of a More Complex Molecule until Only One Remains Intact

The degradation sequence named after the German chemist Alfred Wohl is a chain contraction method that can be applied in studies on stereostructures of aldoses in carbohydrate chemistry [18]. The classic example is the conversion of a hexose (glucose) to a pentose (arabinose) and with subsequent degradation sequences further to D-(+)-glyceraldehyde as shown in Figure 4.6. This degradation sequence can rationalize why the asymmetric center most distant from the most oxidized carbon atom of the carbon chain was selected to denote the configuration of sugars according to the D/L nomenclature system.

COOH
HO—H
$CH_2OH$
L-(+)-Glyceric acid

COOH
HO—H
$CH_3$
L-(+)-Lactic acid

CHO
H—OH
$CH_2OH$
D-(+)-Glyceraldehyde

COOH
H—OH
$CH_2OH$
D-(–)-Glyceric acid

COOH
H—OH
$CH_3$
D-(–)-Lactic acid

**Figure 4.5** Correlation of the absolute configuration of lactic acid to D-(+)-glyceraldehyde.

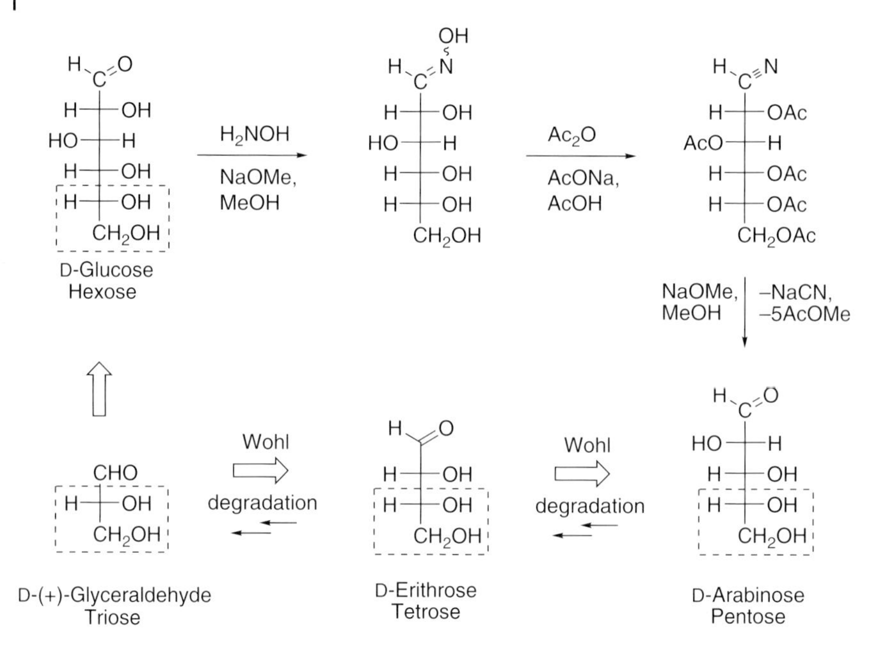


Figure 4.6 The Wohl degradation of natural glucose to D-(+)-glyceraldehyde.

### 4.2.3
### Stereoconstructive Strategy That Adds Further Asymmetric Elements to One of the Known Absolute Configuration without Influencing the Bonds between Its Asymmetric Unit and Ligands

Dutch chemists Bijvoet *et al.* [13] could determine based on anomalous dispersion of sodium rubidium salt of (+)-tartaric acid in X-ray crystallography that the salt had an (*R*,*R*) absolute configuration. Because (+)-tartaric acid could be synthesized from (−)-glyceraldehyde (−)-glyceraldehyde had to be the (*S*)-enantiomer (Figure 4.7). The fortuitous proposal, therefore, that (+)-glyceraldehyde had an (*R*)/D configuration [8] was correct.

### 4.2.4
### Symmetrization of One or More Asymmetric Units Leading to a Common Compound with Less (but More Than One) Asymmetric Elements

Three molecules of phenylhydrazine ($C_6H_5NHNH_2$) react with $C_1$ and $C_2$ of aldoses and ketoses to form derivatives called osazones [19]. The osazones being crystalline derivatives are useful for comparing the structures of sugars (Figure 4.8).

The fact that identical osazones were obtained from D-glucose, mannose, and fructose demonstrated that the configurations of fragments from $C_3$ to $C_6$ of the

L-(−)-Glyceraldehyde
HCN
L-(+)-Tartaric acid
D-(+)-Glyceraldehyde
*meso*-Tartaric acid

**Figure 4.7** Correlation of the configuration of L-(+)-tartaric acid with glyceraldehyde.

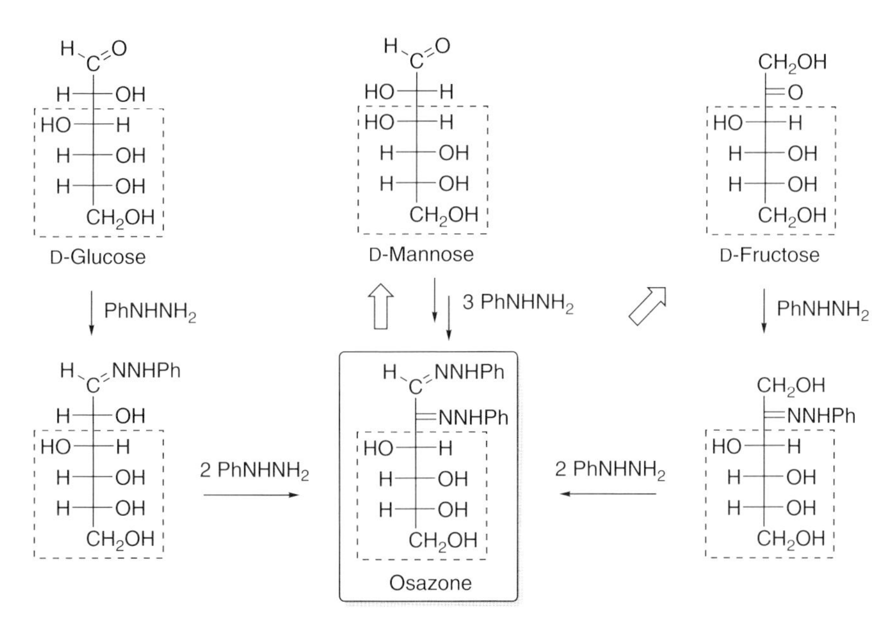


**Figure 4.8** Correlation of the configuration of glucose, mannose, and fructose by their common osazone.

D-glucose, mannose, and fructose molecules were identical. This fact could be applied to assign the configuration of natural mannose and fructose as D.

### 4.2.5 Transformations Involving Stereospecific Reactions of Known Stereochemistry Influencing the Bonds between the Asymmetric Unit and the Ligands

Emil Fischer studied the stereochemistry of alanine using reactions involving Walden inversion at the reacting centers. The reaction of (+)-alanine with NOBr

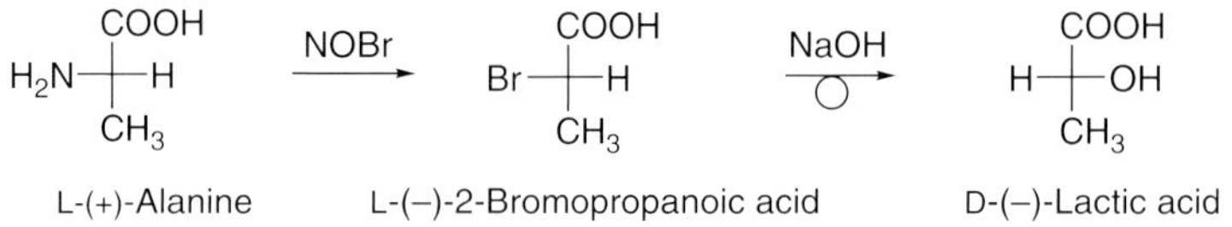

COOH / $H_2N$—C—H / $CH_3$ —NOBr→ COOH / Br—C—H / $CH_3$ —NaOH, ⊖→ COOH / H—C—OH / $CH_3$

L-(+)-Alanine L-(−)-2-Bromopropanoic acid D-(−)-Lactic acid

**Figure 4.9** Correlation of the configuration including reactions with inversion.

gave (−)-2-bromopropionic acid (Figure 4.9) [20], which upon hydrolysis with NaOH gave D-(−)-lactic acid.

Fischer concluded that a substitution process, applied to an optically active compound, may be accompanied by (a) complete inversion, (b) partial inversion or racemization, and (c) no inversion (i.e., retention) [20]. In the case of (a) and (c), the substitution product will be optically active, thus, such stereospecific reactions with known stereochemical outcome can be applied in the course of configuration correlations.

## References


1. Moss, G.P. and International Union of Pure and Applied Chemistry, Organic Chemistry Division, Commission on Nomenclature of Organic Chemistry [III. 1] and Commission on Physical Organic Chemistry [III.2] (1996) Basic Terminology of Stereochemistry (IUPAC Recommendations 1996). *Pure Appl. Chem.*, **68** (12), 2193–2222.
2. Muller, P. and International Union of Pure and Applied Chemistry, Organic Chemistry Division, Commission on Physical Organic Chemistry (1994) Glossary of terms used in physical organic chemistry (IUPAC Recommendations 1994). *Pure Appl. Chem.*, **66**, 1077–1184.
3. Pasteur, L. (1848) Mémoire sur la relation qui peut exister entre la forme cristalline et la composition chimique, et sur la cause de la polarisation rotatoire. *Compt. Rend. Acad. Sci., Paris*, **26**, 535–538.
4. Pasteur, L. (1848) Sur les relations qui peuvent exister entre la forme cristalline, la composition chimique et le sens de la polarisation rotatoire. *Ann. Chim. Phys.*, **24**, 442–459.
5. van't Hoff, J.H. (1874) Voorstel tot uitbreiding der tegen woordig in de scheikunde gebruike struktur-formules in de ruimte, benevens een daarme samehangende vermogen en chemische constitutie van organische verbindingen. *Arch. Neerland. Sci. Exact. Natur.*, **9**, 445–454.
6. Le Bel, J.A. (1874) Sur les relations qui existent entre les formules atomiques des corps organiques, et le pouvoir rotatoire de leurs dissolutions. *Bull. Soc. Chim. France*, **22**, 337–347.
7. Fischer, H.E. (1891) Über die Configuration des Traubenzuckers und seiner Isomeren. I. *Berichte der Deutschen Chemischen Gesellschaft*, **24**, 1836–1845; Über die Configuration des Traubenzuckers und seiner Isomeren. II. *Ber. Deutsch. Chem. Ges.*, **24**, 2683–2687.
8. Rosanoff, M.A. (1905) On Fischer's Classification of Stereo-Isomers. *J. Am. Chem. Soc.*, **28**, 114–121.
9. IUPAC (2014) *Compendium of Chemical Terminology: Gold Book* Version 2.3.3, http://goldbook.iupac.org/PDF/goldbook.pdf (accessed 29 January 2016).
10. McNaught, A.D. and Wilkinson, A. (1997) *Compendium of Chemical Terminology*, 2nd edn, Blackwell Science, Oxford. [ISBN 0-86542-6848]
11. Allenmark, S. and Gawronsky, J. (2008) Determination of Absolute Configuration—an Overview Related to This Special Issue. *Chirality*, **20**, 606–608.

12. Kuhn, W. (1938) Das Problem der absoluten Konfiguration optisch aktiver Stoffe. *Naturwiss.*, **26**, 305–310.
13. Bijvoet, J.M., Peerdeman, A.F., and van Bommel, A.J. (1951) Determination of the Absolute Configuration of Optically Active Compounds by Means of X-Rays. *Nature*, **168**, 271–272.
14. Thompson, A.L. and Watkin, D.J. (2009) X-ray crystallography and chirality: understanding the limitations. *Tetrahedron: Asymmetry*, **20**, 712–717.
15. Nafie, L.A. (2011) Determination of absolute configuration of chiral molecules using vibrational optical activity: a review. *App. Spectr.*, **65**, 699–718.
16. Freudenberg, K. (1914) Über die Konfiguration der Glycerinsäure und Milchsäure. *Ber. Deutsch. Chem. Ges.*, **47**, 2027–2037.
17. Wohl, A. and Schellenberg, R. (1922) Die Überführung des aktiven Glycerinaldehyds in die aktive Glycerinsäure. *Ber. Deutsch. Chem. Ges.*, **55**, 1404–1408.
18. Wohl, A. (1893) Abbau des Traubenzuckers. *Ber. Deutsch. Chem. Ges.*, **26**, 730–744.
19. Fischer, E. (1896) Synthesen in der Zuckergruppe. I. *Ber. Deutsch. Chem. Ges.*, **23**, 2114–2141.
20. Fischer, E. (1911) Waldcnsche Umkehrung und Substitutionsvorgang. *Liebigs Ann. Chem.*, **381**, 123–141.

# 5
# Methods for the Determination of Enantiomeric Composition

The most often used parameter to characterize the composition of enantiomeric mixtures is the percentage of the dominant enantiomer ($100 \times X'$, where $X'$ is the molar fraction of the enantiomer in excess).

Another very often applied method is to state the s.c. "**enantiomer(ic) excess**" (ee) defined as the absolute difference between the molar fraction of each enantiomers.

$$\text{ee} = (X' - X)/(X' + X)$$

where $X' > X$ are the molar fractions of the individual enantiomers. It is most often expressed as a *percent enantiomeric excess*.

$$\text{ee}\ (\%) = 100 \times (X' - X)/(X' + X)$$

The *enantiomeric excess* can be expressed using only the molar fraction of the major enantiomer:

$$\text{ee} = 2X' - 1 \quad \text{or} \quad \text{ee}\ (\%) = 100 \times (2X' - 1)$$

The expression "**enantiomeric purity**" (ep) is also often used. Unfortunately, its usage is inconsequent, either denoting the percentage of the major enantiomer or as a synonym for enantiomeric excess.

Earlier enantiomeric purity was often given as "**optical purity**" (op):

$$\text{op} = ([\alpha]/[\alpha]_{\text{max}}) * 100$$

where $[\alpha]$ is the measured optical rotation of the enantiomeric mixture and $[\alpha]_{\text{max}}$ the optical rotation of the pure enantiomer.

A wide variety of analytical methods is available for the determination of the composition of enantiomeric mixtures. Most often the mixtures themselves are investigated, but sometimes derivatization with a chiral or achiral reagent may be necessary. The most important methods are given in Table 5.1 [1].

Selection of the most suitable method is influenced by several factors, and not always the one requiring a minimum effort may be the best. Generally, utmost care should be exerted in the course of the procedures because the composition

*Stereochemistry and Stereoselective Synthesis: An Introduction*, First Edition.
László Poppe and Mihály Nógrádi.

Companion Website: www.wiley.com/go/poppe/stereochemistry

**Table 5.1** Methods for the determination of enantiomeric composition [1].

| | Principle of the method | The actual measurement | Handling of the sample | Material to be assayed |
|---|---|---|---|---|
| 1 | Chiroptical methods | A: $\alpha$, $\Phi$, $\Delta\varepsilon$ | $A_s$ | E or A |
| | | B: Emission circular polarization | Original mixture | E |
| 2 | Diastereotopicity (External comparison) | A: NMR spectroscopy in achiral solvents (or solid state) | $D_s$ | D |
| | | B: NMR spectroscopy in chiral solvents (chiral-solvating media) | Original mixture | E |
| | | C: NMR in the presence of a chiral shift reagent | Original mixture | E |
| 3 | Diastereomeric interaction (separation) | A: Chromatography on a diastereoselective stationary phase | $D_s$ | D |
| | | | Original mixture | E |
| | | | Original mixture | E |
| | | (i) GC, HPLC, and TLC | | |
| | | (ii) HPLC in a chiral solvent | | |
| | | (iii) B: Chromatography on an enantioselective stationary phase | | |
| | | (iv) GC, HPLC, TLC, and electrophoresis | | |
| 4 | Kinetic methods (differentiation based on enantiomer selectivity) Kinetic resolution | A: Enzymatic methods (quantitative enzyme catalyzed reactions) | Original mixture | E |
| | | B: Chemical methods | Original mixture | E |
| | | | or $D_s$ | D |
| 5 | Fusion properties | Mp, DSC | Original mixture | E |
| | | | or $D_s$ | D |
| 6 | Isotope dilution | Isotope analysis | $A_s$ | E |
| 7 | Potentiometry | Electrochemical cell | Original mixture | E |

$A_s$: sample treated with an achiral-derivatizing agent (enantiomeric mixture); $D_s$: sample treated with a diastereomer-forming agent (diastereomeric mixture); E: original enantiomeric mixture; and D: derived diastereomeric mixture.

of the sample may change during the procedure giving false results. Enantiomeric composition deduced from methods using conversion to diastereomeric derivatives may suffer from serious errors due to change of the diastereomeric compositions during chemical purification (workup, washing, recrystallization, sublimation, etc.).

## 5.1 Chiroptical Methods

These methods are simple, fast, and noninvasive, based essentially on the measurement of optical rotation. Therefore, they are very sensitive to impurities having optical rotation also, and sometimes not fully dependable.

For the determination of enantiomeric purity by optical rotation, a sample of the pure enantiomer is required ($[\alpha]_{max}$). Optical rotation depends on wavelength, temperature, concentration, and eventual contaminants. Often, earlier literature values prove to be inaccurate, mainly because the inaccuracy of $[\alpha]_{max}$ value used or the sample contained traces of solvent or contaminants of high specific optical rotation. The concentration effect is especially pronounced when optical rotations of polar molecules (e.g., alcohols and carboxylic acids) are measured in apolar solvents. Polyfunctional alcohols or hydroxy acids are also prone to show a concentration effect, since in more concentrated solutions, intermolecular interactions become significant.

Consequently, experimental values of optical purity may not be equal to enantiomeric purity. This was first demonstrated in 1969 by Horeau experiencing a discrepancy between op values of 2-methylsuccinic acid measured in apolar and polar solvents, the latter giving the true values [2]. Deviation of optical and enantiomeric purity (op and ep, respectively) being especially pronounced in the medium range of enantiomeric excess is called the *Horeau* effect.

Measuring op of compounds with very low specific rotation is prone to errors. In this case, derivatization with an achiral agent of good chromophore properties is useful. Acetylation of chiral alcohols ($[\alpha]_D \sim 0.7$) can enhance specific rotation from 0.7 to about 12. The effect of esterification with 3,5-dinitrobenzoic acid can be quite dramatic ($[\alpha]_D \sim 150$).

## 5.2 NMR Methods Based on Diastereotopicity

### 5.2.1 Methods Based on Forming Diastereomeric Derivatives

The method is based on the analysis of the composition of diastereomeric mixtures by NMR. Chiral alcohols or amines are reacted with a chiral derivatizing agent, most often an acid chloride to produce a mixture of diastereomeric esters or amides, respectively, which is then analyzed (Figure 5.1). As an example, determination of the composition of chiral aldehydes with $\alpha$-substituent by *in situ* diastereomeric Schiff base formation can be quoted. Results are in good agreement with those obtained by traditional methods. Since diastereomers differ even in an achiral environment, in fortunate cases there will be an easily distinguishable and measurable signal in the spectrum, the intensity ratio of which reflects the enantiomeric ratio of the original mixture. Provided that the

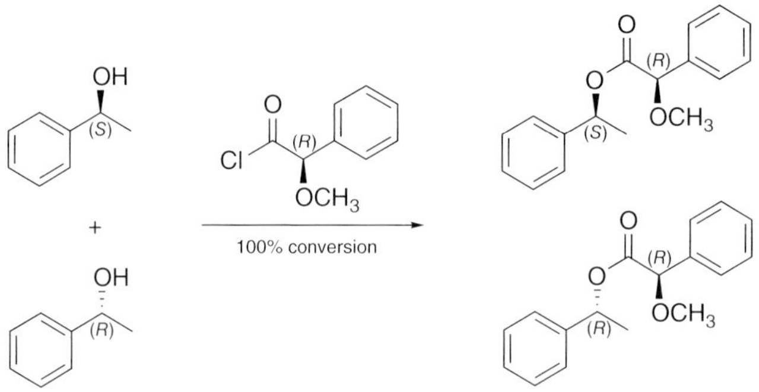

**Figure 5.1** Example of the application of a chiral derivatizing agent (e.g., using the well-separated methoxy signals in the $^{1}H$ NMR spectra).

reaction is quantitative, the ratio of diastereomeric products is exactly the same as that of the original enantiomers.

An advantage of the method is that it does not require a pure enantiomer as standard, but the derivatizing agent must be, however, 100% enantiopure and conversion must also be 100%; otherwise, an eventual rate difference between enantiomers might influence the results. Another difficulty is that the original sample cannot be recovered, and for each substrate, the best derivatizing agent has to be found. If a separation process is needed before analyzing, we have to determine (calibrate) the possible change of diastereomeric ratio.

Some often used derivatizing agents are shown in Figure 5.2.

An interesting case of diastereotopicity is when the derivatizing agent is a **bidentate achiral reagent.** In this case, three products are formed: two optically

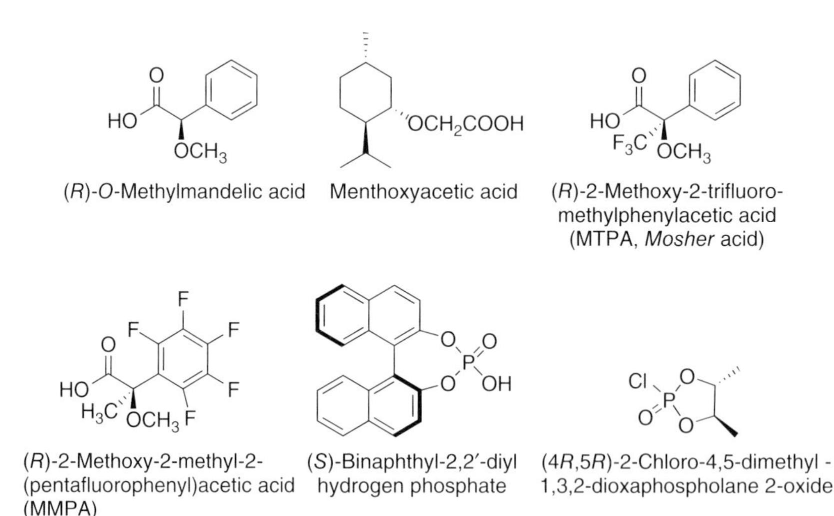


**Figure 5.2** Some chiral-derivatizing agents.

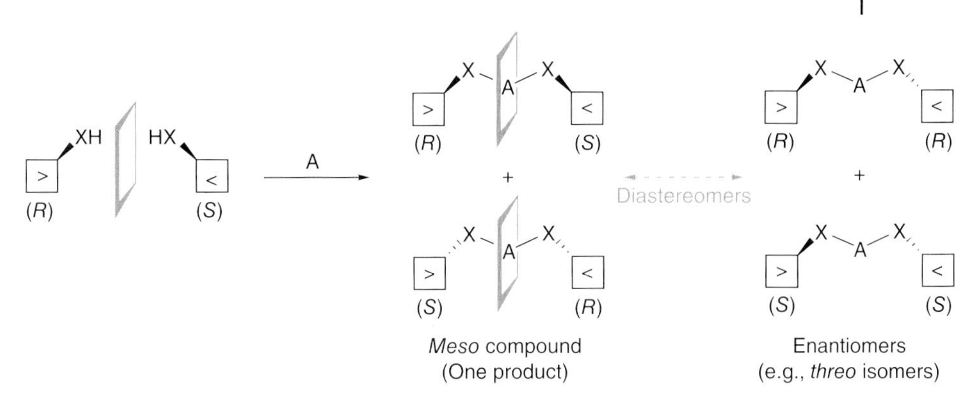


**Figure 5.3** Application of a bidentate-derivatizing agent.

**Table 5.2** Composition of a mixture obtained by applying a bidentate achiral-derivatizing agent.

| Composition of the starting mixture | *threo* Isomers formed | *meso* Isomers formed |
|---|---|---|
| Racemic mixture (100 mol) | 12.5 mol *R,R* + 12.5 *S,S* | 25 mol *R,S* (*S,R*) |
| (50% *R*–50% *S*) | Total: 50% *threo* | Total: 50% *meso* |
| Enantiomer mixture (100 mol) | 31.25 mol *R,R* + 6.25 mol *S,S* | 12.5% *R,S* (*S,R*) |
| (75% *R*–25% *S*) | Total: 75% *threo* | Total: 25% *meso* |
| Enantiopure compound (100 mol) | 50 mol% *R,R* + 50 mol%/ *S,S* | 0 mol% *R,S* (*S,R*) |
| (100% *R*) | Total 100% *threo* | Total: 0% *meso* |

active enantiomers (*R,R* and *S,S*) and a meso compound (*R,S*) (Figure 5.3). The *threo* isomers are mirror images and therefore produce a single set of signals in the NMR spectrum, whereas the *meso* derivative gives another set of signals. The ratio of signals corresponds to the ratio of enantiomers (cf. Table 5.2). (We suppose that rate of formation of diastereomers is equal. This is the limit of method's applicability.)

As a practical example, the reaction of racemic 1-phenylethanol and phosphorus trichloride is quoted giving four stereoisomeric diphenylphosphonates (Figure 5.4). In the $^{31}$P NMR spectrum of the product, three signals in a ratio of 1 : 1 : 2 can be observed. Owing to the tetrahedral geometry of the phosphorus atom (pseudoasymmetric center), two *meso*-isomers are formed.

Distinguishing diastereomers by NMR spectroscopy is also possible with ionic salts. When mixing racemic acids and bases, a mixture of diastereomeric salts is formed in which salt formation takes place between both enantiomeric acids and bases. Exchange between the ionic components of the salts exceeds the NMR timescale, and therefore only averaged signals can be recorded and the mixture behaves as a uniform substance. If, however, one of the counterions is a pure enantiomer, there is seemingly no exchange and the signals of the pairs

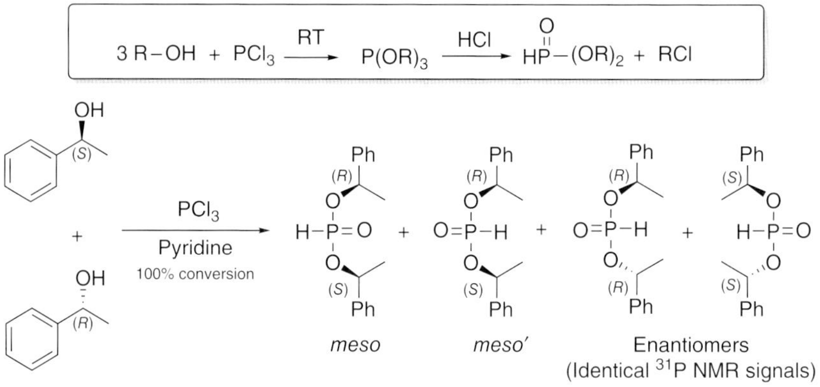

**Figure 5.4** Reaction of 1-phenylethanol with $PCl_3$.

of diastereomeric salts appear in the spectrum separately. This phenomenon can be foremost observed in aprotic solvents ($CDCl_3$, $C_6D_6$, DMSO-$d_6$, and eventually pyridine-$d_5$) in which the ion pairs are present as aggregates. Protic solvents (e.g., $CH_3OD$) may destroy the aggregates diminishing this effect considerably.

### 5.2.2
### NMR Methods Based on the Use of Chiral Solvents (Solvating Agents)

The possibility to distinguish enantiomers by NMR spectroscopy was first demonstrated in the 1960s when the $^{19}F$ NMR spectrum of 2,2,2-trifluoro-1-phenylethanol (TFPE) was recorded in (*S*)-1-phenylethylamine as solvent. A slight difference (2 Hz) in the shift of the $CF_3$ signal was observed. Later, several examples of this phenomenon were recorded and it turned out that instead of using expensive chiral solvents, adding some chiral additive (called a **chiral solvating agent**) in an amount at least equivalent to the solution of the substance to be studied in an achiral solvent suffices. This phenomenon does not differ essentially from making a diastereomeric mixture, as discussed in the previous section. The molecular complexes formed from the two enantiomers of the molecule studied and the molecules of the chiral solvent are in a diastereomeric relationship, and therefore the chemical environment of the given parts of the assembly might be different. For producing a different environment, it is sufficient when into the achiral solvent shell intrude molecules of a chiral auxiliary (Figure 5.5).

An advantage of measurements in chiral solvents is its simplicity. Principally, both the sample and the solvent can be recovered, no 100% conversion and enantiopure standard sample is required. Its drawback is that the solvent must be 100% enantiopure that makes the procedure expensive. In Figure 5.6, some chiral solvents (solvating agents) are shown.

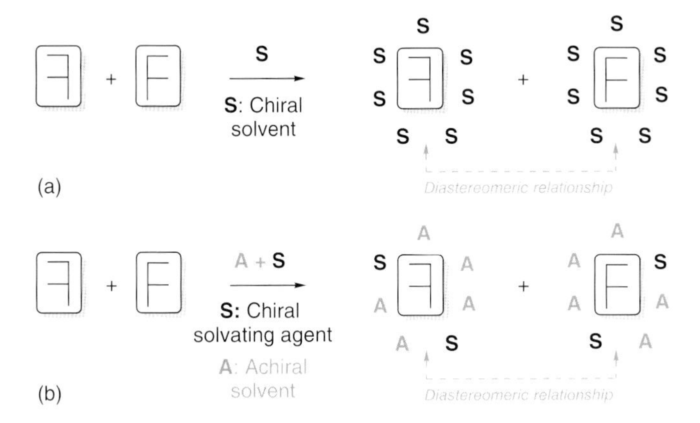


**Figure 5.5** Use of chiral solvents (solvating agents).

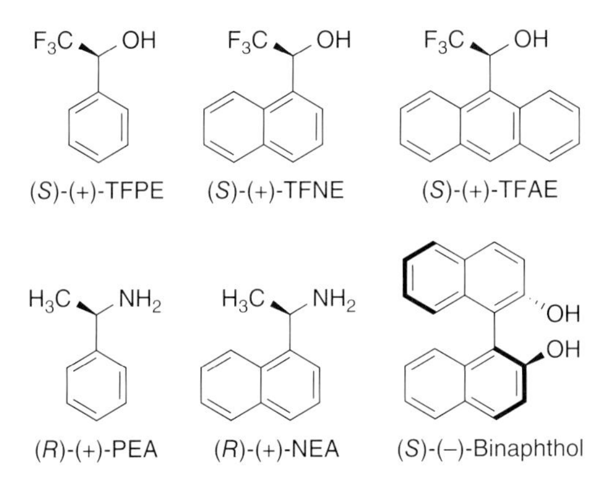


**Figure 5.6** Several chiral solvents (solvating agents).

### 5.2.3 NMR Methods Based on Chiral Shift Reagents

Complexes of lanthanides with chiral β-diketones found useful application in NMR spectroscopy. Binding of the metal to a nonbonding pair of electrons of a chiral substrate produces diastereomeric complexes that show significant chemical shift differences. Measuring the intensity ratio of analogous signals informs about the enantiomeric excess of the compound studied. Unfortunately, the complexing agents are rather expensive and not recoverable. The lanthanides most often used are lanthanum (La), europium (Eu), praseodymium (Pr), and ytterbium (Yb) (Figure 5.7). The considerations and requirements for using chiral shift reagents are the same as discussed for the chiral solvent/solvating agent-based NMR methods, for example, 100% enantiopurity of the shift reagent is a requirement (cf. Section 5.2.2 and Figure 5.5b).

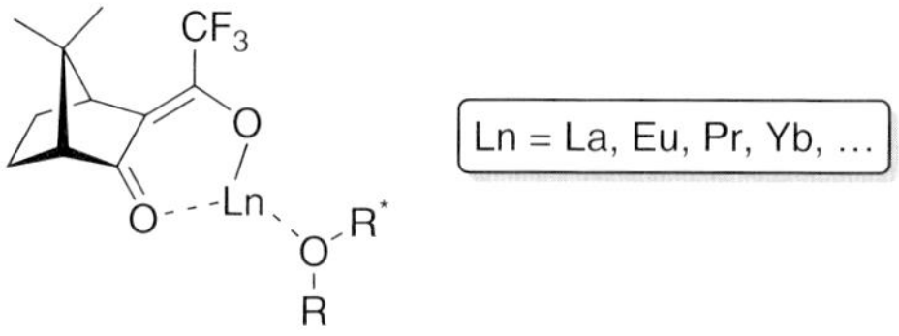

Ln = La, Eu, Pr, Yb, ...

**Figure 5.7** Complexation with a chiral shift reagent.

## 5.3 Chromatographic and Related Methods Based on Diastereomeric Interactions

For the determination of the composition of enantiomeric mixture, chromatographic methods are widely used. Earlier, it was necessary to convert the mixture of enantiomers by derivatization with a chiral reagent to a mixture of diastereomers. In this case, separation could be effected on an achiral stationary phase using gas chromatography (GC), high-performance liquid chromatography (HPLC), or other types of chromatographies.

The direct measurement (without derivatization) of enantiomers is possible using *chiral stationary phase* (*CSP*). These methods are available for both GC and HPLC. In the course of separation by HPLC, an achiral stationary phase may be combined with a chiral mobile phase.

A common feature of all these methods is the formation of stable or transient diastereomeric entities. Thus, the different solubility, stability, or adsorption is responsible for the separation of stereoisomers.

### 5.3.1 Separation on a Diastereoselective (Achiral) Stationary Phase

Diastereomers differing in all of their properties, even in an achiral environment, can be separated on an achiral stationary phase using an achiral solvent. The mixture must not be subjected to any purification operation, since this may change its composition (or we must know the influence of changes).

With mixtures of enantiomers prior to measurement, the mixture has to be transformed by means of a chiral derivatizing agent to a mixture of diastereomers. It is important that the reagent must be 100% enantiopure and the transformation has to be run to 100% completion. (Enantiomers react with chiral reactants at different rates; therefore, no starting material must remain after completion of the reaction.) An advantage is that no CSP is required and the method is highly variable.

The importance of the enantiopurity of the derivatizing agent is illustrated by the following example. The sample to be assayed (A) consists of 99.5% *R* and 0.5% *S* enantiomer (ee 99%). Using a chiral-derivatizing agent [(+)-B] of ee 99%, the composition of the product will be as follows: (*R*)-A–(+)-B 98.505%, (*S*)-A–(+)-B 0.495%, (*R*)-A–(−)-B 0.995%, and (*S*)-A–(−)-B% 0.005%. Chromatography on an achiral stationary phase would give two peaks: one for the enantiomeric pairs (*R*)-A−(+)-B and (*S*)-A−(−)-B and a second for (*S*)-A−(+)-B and (*R*)-A−(−)-B being also enantiomers of each other. Integral of the first signal will be 98.51%, that of the second 1.49%, indicating an enantiomeric excess of 97%. This example showed that ee 97% could be deduced for a sample of ee 99% due to a minimum impurity in the reagent causing a significant error in the result. The principle of chromatographic separation is shown in Figure 5.8.

Compounds mentioned in the previous section and their derivatives can be employed as chiral derivatizing agents (Figures 5.2 and 5.6).

*Gas chromatography* (*GC*): With volatile samples, the simplest method seems to be gas chromatography. Using a long capillary plate, numbers of several tens of thousands can be realized, using as carrier gas the cheap nitrogen or hydrogen. Flow rate of carrier gas and the temperature of the column can be readily controlled. A drawback is limited variability and applicability.

*High-performance liquid chromatography* (*HPLC*): Suitable also for the assay of larger molecules. Separation can be influenced by the composition of the solvent providing extensive variability. Gradient elution is also possible. It is more expensive than GC and generally of low efficiency (in terms of plate numbers achievable).

*Capillary electrophoresis:* It combines the advantages of GC and HPLC, but it is only applicable to charged system.

In the last two cases, measurements with chiral solvents or solvation agents are possible requiring neither 100% conversion nor covalent derivatization of the sample, while 100% enantiopurity of the solvent or solvation agent is a requirement.

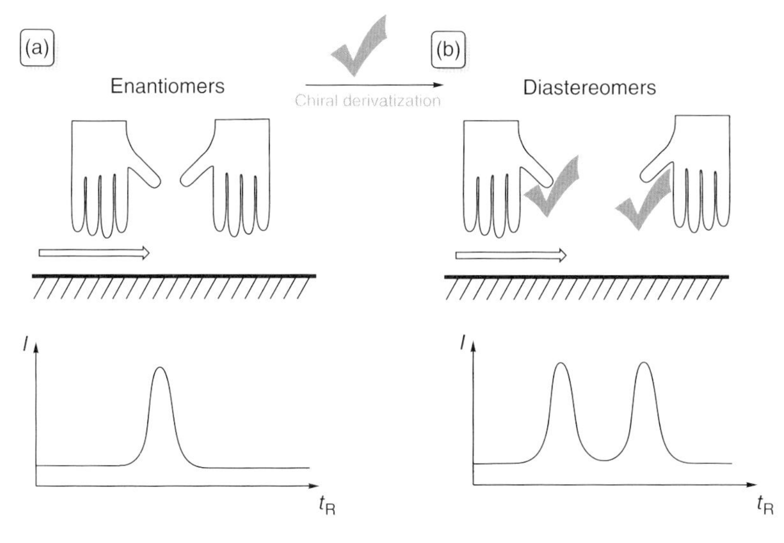


**Figure 5.8** Chromatographic separation on diastereoselective stationary phase (a: without or b: with chiral-derivatizing agent).

### 5.3.2 Use of an Enantioselective (Chiral) Stationary Phase

For the separation of enantiomers, a chiral environment that can be provided also by a CSP is required. Efficiency of separation depends on physicochemical parameters (formation energy, stability, etc.) of transient diastereomeric entities (aggregates and adsorbates). A considerable advantage of the method is that the CSP must not be completely enantiopure. The principle of the method is illustrated in Figure 5.9.

Capacity (efficiency) of separation is dependent on the number of active sites available on the stationary phase (Figure 5.10a and b). Although being not an absolute requirement, enantiopurity of the stationary phase is also important because diminished enantiomeric excess decreases the efficiency of separation (Figure 5.10c).

Except for protein-, cyclodextrin-, cellulose-, and amylose-based stationary phases, it is usually possible to prepare the phases in both configurations. Analyzing the same sample on two kinds of stationary phases, retention times are usually interchanged (Figure 5.11). It is worth mentioning that by proper derivatization of the cyclodextrin-, cellulose-, and amylose-based CSPs, the elution order may be reversed as well.

For an efficient separation, the substrate and the stationary phase must bind the favored enantiomer by at least three different kinds of interactions (principle of enantiomeric recognition) (Figure 5.12a). Obviously, in this case, the less-preferred enantiomer can interact with the CSP only by two kinds of interactions, thus binds less tightly (Figure 5.12b).

In the following, some examples for CSPs will be presented.

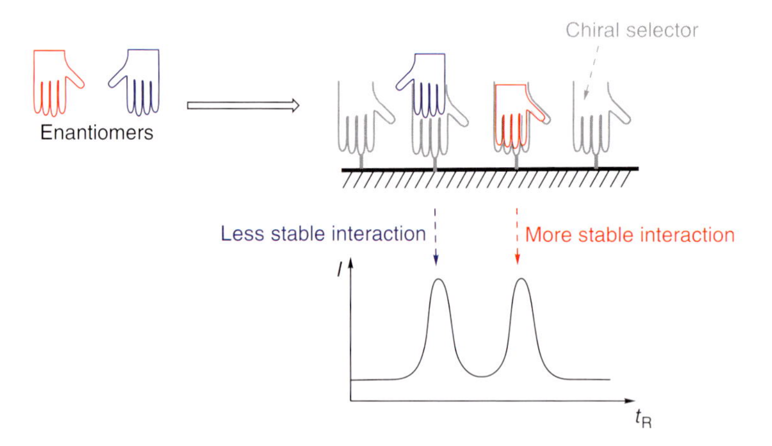


**Figure 5.9** Principle of chromatographic separation on enantioselective (chiral) stationary phase.

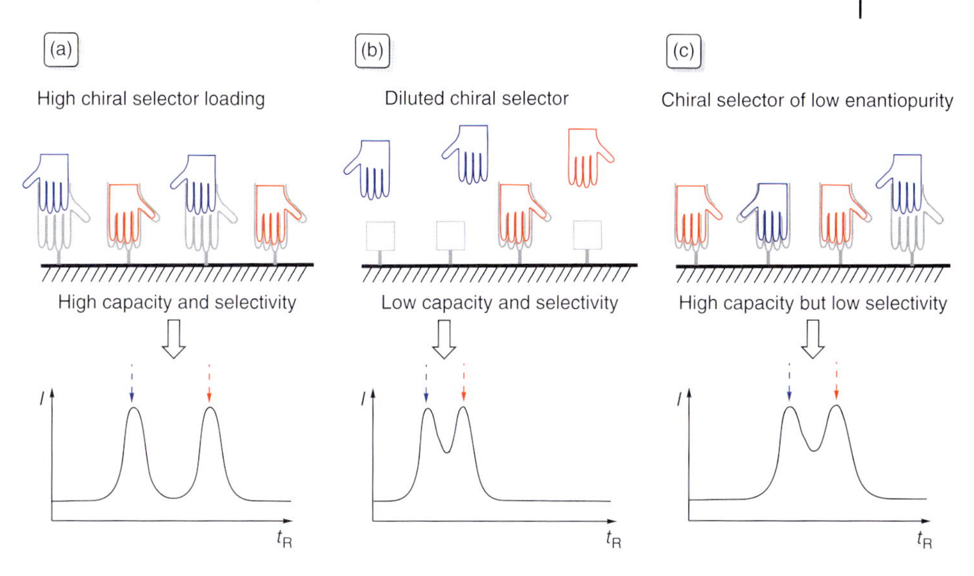


**Figure 5.10** Capacity and selectivity of chromatographic separation on CSPs.

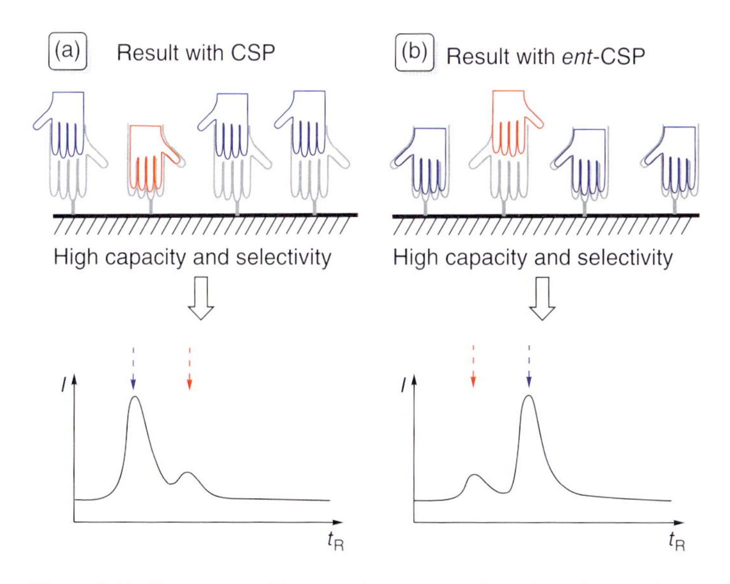


**Figure 5.11** Chromatographic separation on enantiomeric stationary phases.

#### 5.3.2.1 Gas Chromatography

As stationary phase for enantioselective gas chromatography, the chiral polysiloxane derivative Chirasil-Val™ is used, the structure of which is shown in Figure 5.13. GC is characterized by a high plate number, and therefore this relatively inexpensive CSP, which is available as both enantiomers, may result in sufficient enantiomer selectivity. Its variability is, however, restricted.

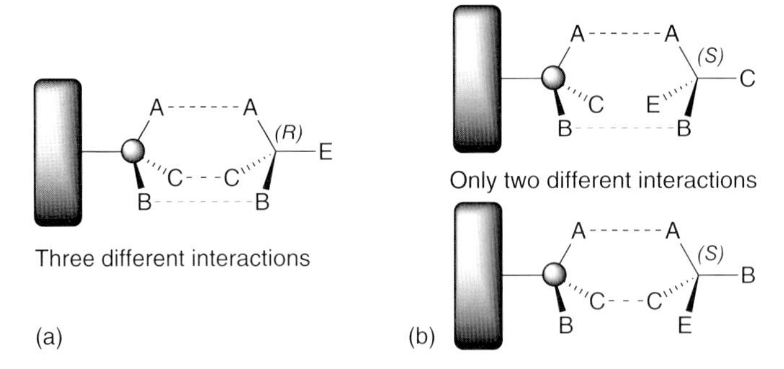


**Figure 5.12** Principle of enantiomeric recognition.

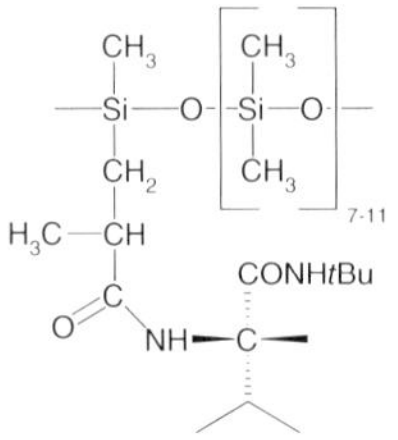

**Figure 5.13** Structure of Chirasil-Val™ coating for gas chromatographic columns.

CSPs based on cyclodextrins are also widely used. The α-, β-, and γ-cyclodextrins composed of 5, 6, and 7 glucose units are all characterized by a chiral cavity. They are very efficient systems but only available as one enantiomer. Transformation of free hydroxyl groups causes changes in the structure resulting in changes of selectivity; even the order of retention may be inverted.

#### 5.3.2.2 High-Performance Liquid Chromatography

One of the most often used approaches is the *Pirkle*-type methodology, operating both on a covalent or ionic principle. Its efficiency is dependent on hydrogen bond formation, $\pi-\pi$ interactions or dipole–dipole interactions, and steric effects. Its advantage is wide variability (solvents and additives), while the columns used are more expensive and less efficient than GC columns. In Figure 5.14, a Whelk-O1-type stationary phase is illustrated (available in the opposite enantiomeric form as well).

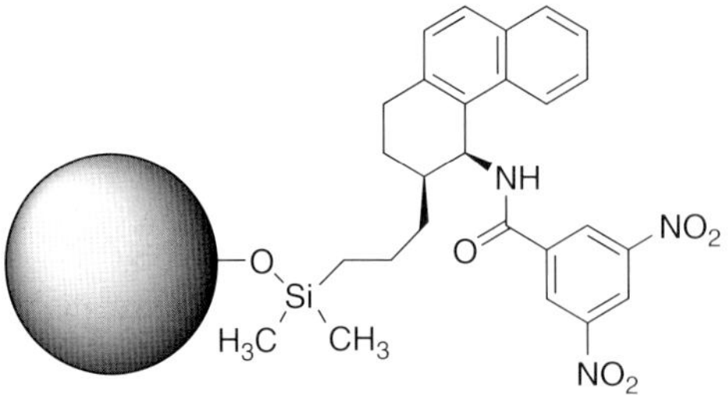

**Figure 5.14** Whelk-O1-type CSP for liquid chromatography.

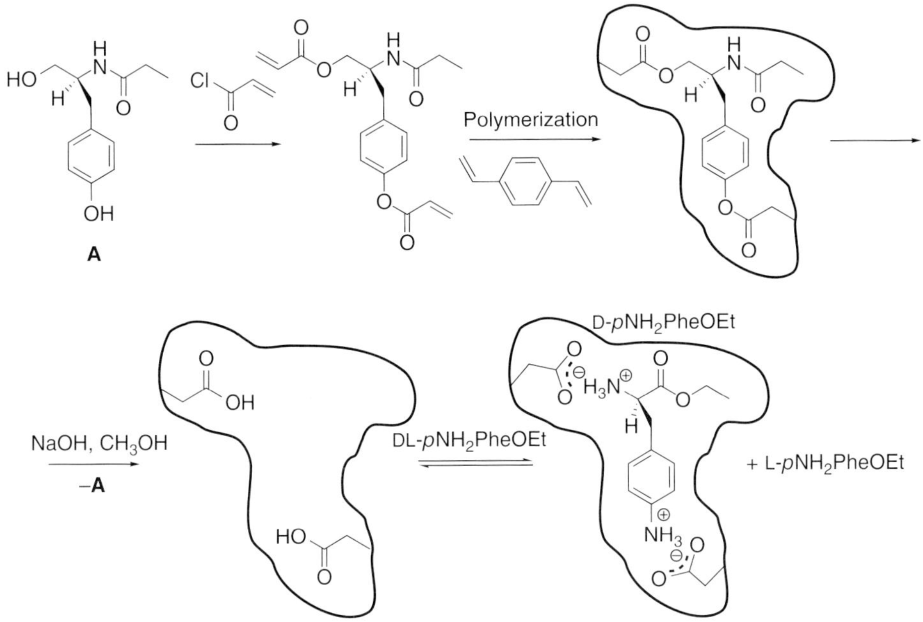

Figure 5.15 Shape-recognizing polymeric stationary phase.

For liquid chromatography, *cyclodextrin*-based columns (with free or derivatized hydroxyl groups) are also popular.

Columns with wide range and high efficiency can be prepared from *cellulose* and *amylose* derivatives. The adsorbent is of helical structure and only available as one enantiomer. The free hydroxyl groups can also be derivatized.

Highly efficient columns can be prepared using immobilized *proteins* as CSP, but such columns are of rather narrow range, sometimes unstable and only available as a single enantiomer.

Special columns are working on stationary phases based on **shape-recognizing polymers.** The chiral cavities are prepared by polymerization in the presence of a small chiral molecule used as a template. After polymerization, the template molecule is removed from the polymer by hydrolysis, leaving behind a cavity into which only one enantiomer of the material to be separated fits, providing highly efficient separation of substrates of a certain type (Figure 5.15).

## 5.4 Kinetic Methods Based on Enantiomer Selectivity

Several analytical methods are based on the fact that enantiomers react at different rate with chiral reagents. The corresponding preparative methods are called **kinetic resolution**. The source of rate difference is that the transient complexes formed in the reactions are related as diastereomers and therefore their free energy of activation is different. Kinetic methods based on enantiomer selectivity can be divided into two groups – enzymatic and nonenzymatic (chemical) ones.

### 5.4.1 Enzymatic Methods

A general property of enzymes is that they react in a highly selective manner with one enantiomer of a substrate, while the other one remains unchanged. For assaying the composition of a mixture of enantiomers, the best practice is to transform the minor "*contaminating*" enantiomer. As an example, determination of the enantiomeric excess of (*R*)-lactic acid may serve. By treating the sample with (*S*)-lactate dehydrogenase in the presence of a stoichiometric amount of nicotinamide–dinucleotide ($NAD^+$) cofactor, the (*S*)-lactic acid content of the sample is oxidized to pyruvic acid, while the cofactor is reduced to NADH. The latter can be assayed by UV spectroscopy at 341 nm (Figure 5.16). With kinetic methods, it is also important to secure that the reaction goes to completion, otherwise, the results have to be corrected with the equilibrium constant of the transformation.

There exist pairs of enzymes that are separately selective to one enantiomer of the same substrate. When, as a control, the determination of enantiomeric composition is carried out with both enzymes of such a pair, the combined results have to add up to ee 100%. For example, L-amino acid oxidase is transforming L-amino acids, while D-amino acid oxidase is acting exclusively on D-amino acids (Figure 5.17).

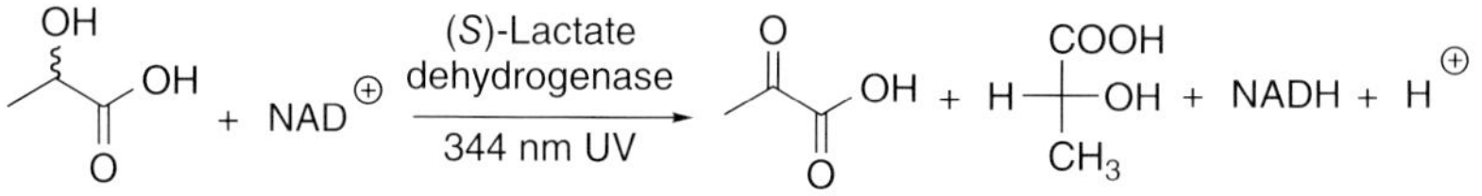

OH, OH, O + $NAD^{\oplus}$ —(*S*)-Lactate dehydrogenase, 344 nm UV→ O, OH, O + H–C(COOH)(CH$_3$)–OH + NADH + $H^{\oplus}$

Figure 5.16 Determination of the enantiomeric purity of (*R*)-lactic acid.

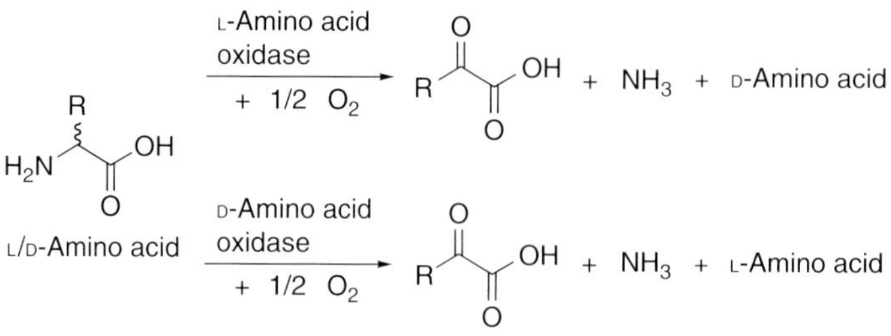

R, $H_2N$, OH, O (L/D-Amino acid) —L-Amino acid oxidase, + 1/2 $O_2$→ R, O, OH, O + $NH_3$ + D-Amino acid

—D-Amino acid oxidase, + 1/2 $O_2$→ R, O, OH, O + $NH_3$ + L-Amino acid

Figure 5.17 Determination of the enantiomer composition of amino acids by enzymatic method.

### 5.4.2 Nonenzymatic (Chemical) Methods

A condition for the applicability of nonenzymatic methods is that detection of the product should be feasible. In contrast to enzymatic systems, it is much more difficult to find systems with sufficient rate difference for the individual enantiomers. A relatively novel method is forming s.c. mass labeled "*quasi enantiomers*," which can be then subjected to mass spectral analysis (Figure 5.18).

In the course of this procedure, the mixture of enantiomeric alcohols is reacted with two homologous carboxylic acids differing in one $CH_2$ group and having centers of chirality of opposite configuration (Figure 5.19). (Sometimes a H → D change may be sufficient.) The procedure has been successfully applied for the alcohols depicted in Figure 5.19.

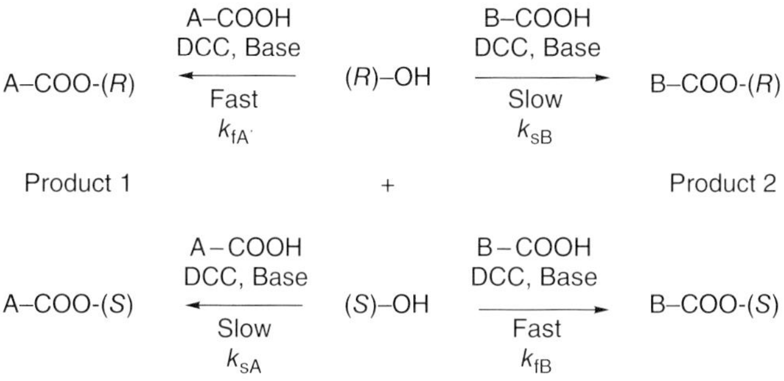

| | | | | |
|---|---|---|---|---|
| A–COO-(*R*) | A–COOH<br>DCC, Base<br>←<br>Fast<br>$k_{fA}$ | (*R*)–OH | B–COOH<br>DCC, Base<br>→<br>Slow<br>$k_{sB}$ | B–COO-(*R*) |
| Product 1 | | + | | Product 2 |
| A–COO-(*S*) | A–COOH<br>DCC, Base<br>←<br>Slow<br>$k_{sA}$ | (*S*)–OH | B–COOH<br>DCC, Base<br>→<br>Fast<br>$k_{fB}$ | B–COO-(*S*) |

Figure 5.18 Kinetic resolution with mass-labeled carboxylic acids.

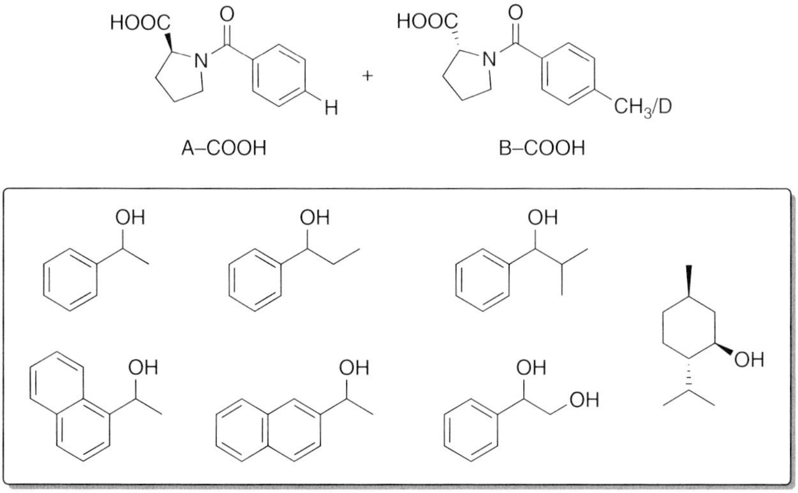

Figure 5.19 Examples of molecules used in kinetic resolutions.

Esterification of the (*R*)-alcohol is faster with one of the carboxylic acids, conversely the (*S*)-alcohol reacts faster with the other carboxylic acid. From the mass spectrum of the resulting mixture, in the knowledge of intensities ($I_{tproduct1}$, $I_{product2}$) and kinetic parameters ($k_f$, $k_s$, $q$), the composition of the alcohols can be calculated using the following formulas:

$$I_{product1}/I_{product2} = y \times q \quad \text{and} \quad \text{ee}\ (\%) = [(y-1)(s+1)/(y+1)(s-1)] \times 100$$

where $s = k_f/k_s$, $I_{product1}$, $I_{product2}$ are the measured intensities, $q$ the ionization correction factor, $k_f$, $k_s$ the rate constants, and $y$ the corrected intensity ratio.

## 5.5 Fusion Methods

Measuring the melting point of mixtures of enantiomers may give, in principle, sufficient information about the composition of the mixture since this value can be directly read from the phase diagram of binary mixtures. In case of compounds forming molecular conglomerates or when enantiomeric excess is poor (30–90%), enantiomeric composition can be directly measured by differential scanning calorimetry (DSC). Calorimetric methods can be used in a wide range and would deserve more practical attention, but the suitable equipment is rarely present in laboratories for organic chemistry.

## 5.6 Methods Based on Isotope Dilution

Isotope dilution is an analytical procedure that might be useful when the quantity of a compound (let it be "A") has to be determined in a mixture containing compounds of similar structure ("A", "B," and "C"), but the component cannot be separated quantitatively. For this approach, a minimal amount of pure component "A" is necessary, and even considerable losses during purification are admissible. The basic principle of the method is that the mixture of unknown composition is mixed with a mixture containing a known quantity of the compound labeled with an isotope, and after reisolation of either the pure enantiomer or of its racemate, the ratio of labeled and unlabeled molecules is determined. From the value obtained, the enantiomeric composition of the original sample can be calculated.

## 5.7 Potentiometric Methods

Composition of chiral ions can be determined by potentiometry in an electrochemical cell consisting of two liquid polyvinyl chloride (PVC) membranes containing the mirror image components of an electrically neutral chiral ionophore [e.g., (*R,R*)- and (*S,S*)-5-nonyl tartrate]. Each of the two membranes is selectively permeable to one enantiomer of the sample. (One is permeable to one and the other to the other.) As a consequence, a potential difference develops between the solution of the sample and the two reference solution. When properly calibrated, the data provide the enantiomeric composition of the sample.

## References

1. Eliel, E.L., Wilen, S.H., and Mander, L.N. (1994) *Stereochemistry of Organic Compounds*, John Wiley & Sons, New York. (Chapter 6.5).
2. Horeau, A. (1969) Interactions d'enantiomeres en solution; influence sur le pouvoir rotatoire: Purete optique et purete enantiomerique. *Tetrahedron Lett.*, 3121–3124.

# 6
# Tautomerism

**Tautomerism** is a special case of isomerism. *Tautomerism* or *tautomeric transformation* is called a special case of interconversion of isomers: it is a spontaneous (i.e., occurring without external interference – heating, catalysis, etc.), mutual, that is, reversible isomerization, leading to equilibrium. Thus, tautomers are defined as isomers, continuously interconverting spontaneously and mutually. The equilibrium reached in this way is called a *tautomeric equilibrium*; the relationship between such isomers is called a tautomeric relationship or tautomerism.[1] Tautomerism plays an important role in stereoselective synthetic processes (see Parts III and IV) and therefore, in our opinion, a moderately detailed discussion of this phenomenon may be useful. Apart from tautomerization, several other forms of isomerization are known: spontaneous isomerization and irreversible (i.e., unidirectional or practically unidirectional) isomerizations triggered by an extraneous effect, or reversible isomerizations taking place only in the presence of a catalyst and/or at high temperature. Examples are shown in Figure 6.1.

It has to be noted that in the last two examples in Figure 6.1, there are no fundamental differences between isomerizations taking both with and without a catalyst and reversible isomerizations proceeding only in the presence of a catalyst. The situation is the same with isomerizations taking place at room temperature and those requiring higher temperatures. In the latter case, it is pure chance that conditions on earth are such as they are and how high is the energy barrier of the transformation. For catalytic processes, it has to be stressed that in fact catalysts – provided they are applied but in catalytic quantities – only influence the kinetics of the process. Thus, also with tautomerism, only the rate of the establishment of an equilibrium, but not the thermodynamics of the process, is changed. In our example of base-catalyzed isomerization of diarylmethylene azomethines, the mechanism is completely analogous with that of oxo–enol tautomeric transformations. However, isomerization of butanes is usually not considered a tautomeric transformation.

1) In the present chapter, we use many terms as proposed by Károly Lempert in his lecture notes (1993).

*Stereochemistry and Stereoselective Synthesis: An Introduction*, First Edition.
László Poppe and Mihály Nógrádi.

Companion Website: www.wiley.com/go/poppe/stereochemistry

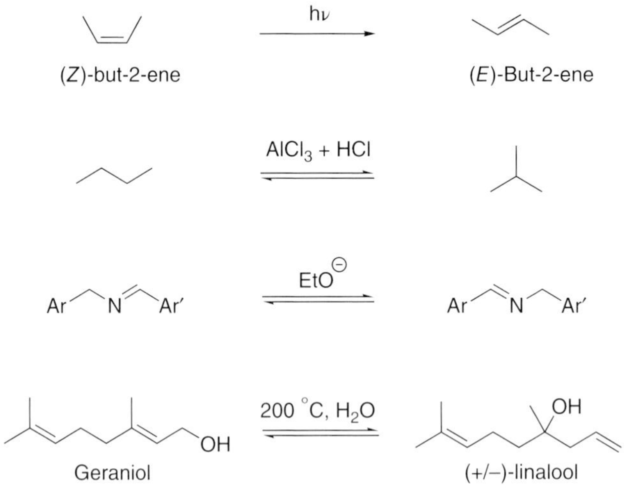

(*Z*)-but-2-ene $\xrightarrow{h\nu}$ (*E*)-But-2-ene

$AlCl_3$ + HCl

Ar–N=–Ar′ $\underset{}{\overset{EtO^{\ominus}}{\rightleftharpoons}}$ Ar=–N–Ar′

Geraniol $\underset{}{\overset{200\ ^{\circ}C,\ H_2O}{\rightleftharpoons}}$ (+/−)-linalool

**Figure 6.1** Examples of isomerizations of various types.

## 6.1 Types of Tautomerism

In the course of tautomeric transformations, a mobile ligand is migrating from one site of the molecule to another. According to the nature of the mobile ligand, two main types of tautomerism can be distinguished.

### 6.1.1 Valence Tautomerism

*Valence tautomerism* are fast, reversible processes in the course of which the bond distribution of a compound changes (certain single or double bonds disappear or are formed, respectively) without displacement of any atom or group of the molecule. Spatial dispositions of some atoms or groups may, however, change. Examples are shown in Figure 6.2.

During the interconversion of species connected by valence tautomerism, it may happen that the mobile ligand migrates from its original site of attachment to its new position, while its connection with the main molecular frame is preserved. A common feature of both types of valence tautomerism shown above is that no exchange of ligands is possible.

At room temperature, the interconversion of valence tautomers is usually very fast, and can therefore only be detected using special methods such as the study of the temperature dependence of NMR spectra by which, for example, the dynamic equilibrium of the tautomeric forms of the β-dioxo compounds shown in Figure 6.3 can be demonstrated.

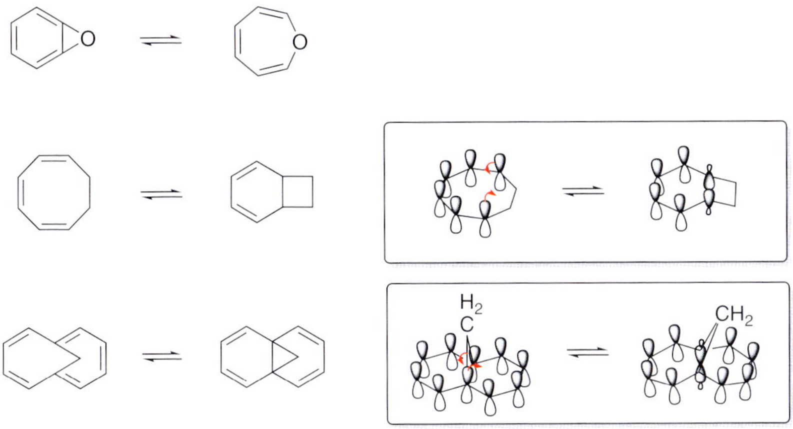

Figure 6.2 Examples of valence tautomerism.

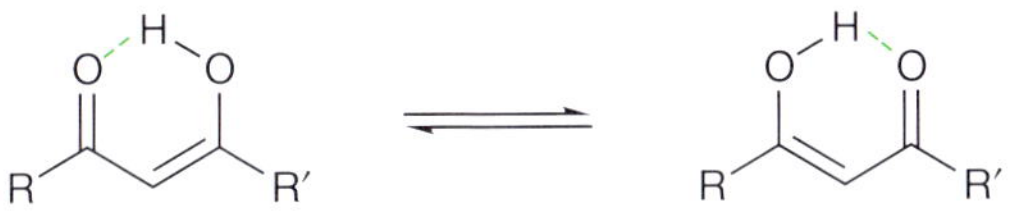

Figure 6.3 Valence tautomerism between two enol forms.

## 6.1.2
## Classic Tautomerism

Characteristic for classic tautomerism is that in the course of the interconversion of tautomeric forms, a labile ligand is separated from the molecule followed by its reattachment or that of another identical ligand to another position of the molecule. It follows that independently of the time sequence of the processes after the tautomeric transformation, it is not necessarily the same ligand that joins another point of the molecule. Intermolecular ligand exchange, as well as that with the solvent or with some other species in the solution, cannot be excluded either.

In classic tautomeric transformations, the mobile ligand is most often a proton, and therefore such transformations are denoted as **prototropy.**

Change in the position of a mobile hydrogen atom may involve other structural changes as well like change in the position of the double bond or of the character of unsaturation or even the number of double bonds.

Tautomeric processes are not restricted to proton the migration, other atoms or groups may also be transposed. In such cases, it is accompanied by the abovementioned structural changes. Some examples of such transformations are shown in Figure 6.4.

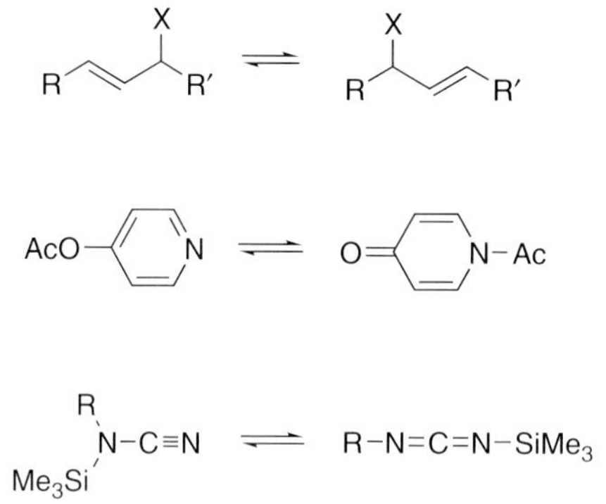

X R R′ ⇌ X R R′

AcO–(pyridine N) ⇌ O=(ring)N–Ac

R, Me$_3$Si N–C≡N ⇌ R–N=C=N–SiMe$_3$

Figure 6.4 Examples of tautomeric equilibria involving migrating ligands other than protons.

## 6.2
## Prototropy

In the earliest examples of tautomeric transformations, the mobile ligand was a proton, and therefore such transformations are called "classic." Since tautomeric transformations mostly belong to this type, we are dealing with them in more detail.

### 6.2.1
### Oxo–Enol Tautomerism

The most common example of oxo–enol tautomerism is that of aldehydes and ketones (Figure 6.5). Enolization is restricted to oxo compounds that have at least one hydrogen atom in $\alpha$-position to the carbonyl group.

In the case of ketones, when the carbonyl group is linked to two different groups and there is a hydrogen atom attached to both, enolization can give rise to two different enols (Figure 6.6). The two enols are tautomers of each other, since they can interconvert spontaneously *via* the keto form (enol–enol tautomerism). (In some cases, formation of additional isomers can also occur.)

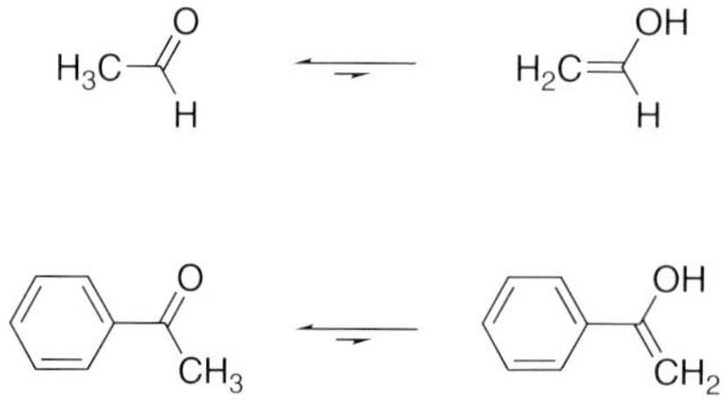

$H_3C$–C(=O)H ⇌ $H_2C$=C(OH)H

Ph–C(=O)$CH_3$ ⇌ Ph–C(OH)=$CH_2$

Figure 6.5 Examples of simple oxo–enol tautomerism.

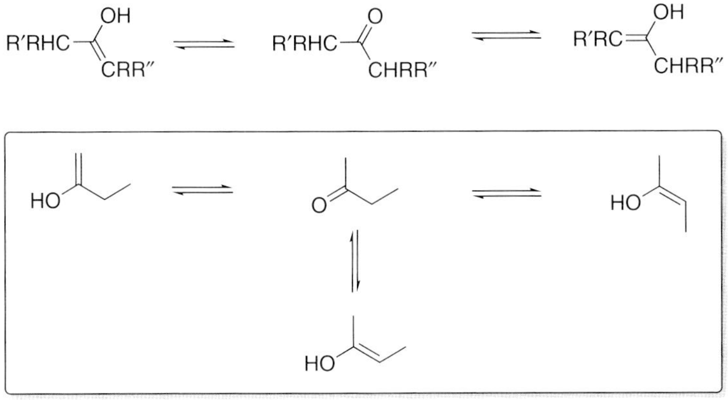

Figure 6.6 Examples of the formation of multiple enol forms from ketones.

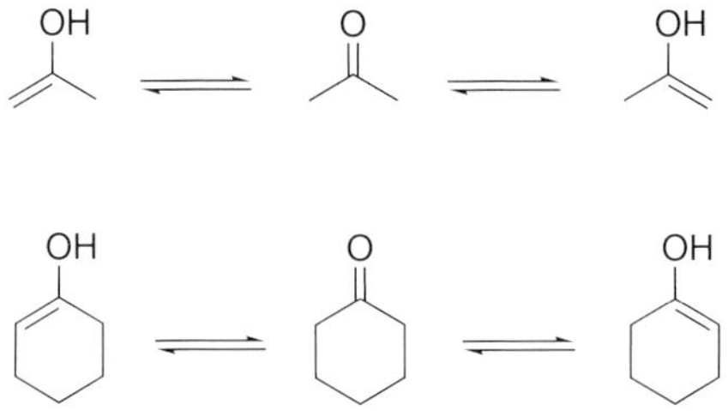

Figure 6.7 Examples of the formation of degenerate tautomers from ketones.

If the two substituents of the carbonyl group are of the same constitution and to both a hydrogen atom is attached, the two enol forms will be structurally identical (Figure 6.7). Tautomers of identical structure are called **degenerate tautomers** (*homomers*).

In oxo–enol tautomerisms so far presented, the equilibrium is predominantly on the side of the oxo form. Energy of simple enols is by about 59–75 kJ $mol^{-1}$ higher compared to the corresponding oxo form. The magnitude of this difference is in a range that cannot be influenced significantly by secondary effects, for example, by formation of hydrogen bonds with the solvent.

The rate of interconversion of oxo–enol tautomers, and therefore the rate of the establishment of equilibrium, is independent of the relative thermodynamic stability of the tautomers, but depends on the free energy of activation of the transformation. Therefore, while catalysts (acids or bases) can significantly accelerate the interconversions of tautomers, they do not affect the oxo/enol ratio. Exceptionally, interconversions may be so slow that in a kinetic sense, the thermodynamically unstable form can be stable. As an example, vinyl alcohol that is very unstable in solution is quite stable in the interstellar space as demonstrated by spectroscopy. In a high vacuum, it is relatively stable also on earth having a half-life of about 30 min. From this fact, it can be inferred that vinyl alcohol needs another

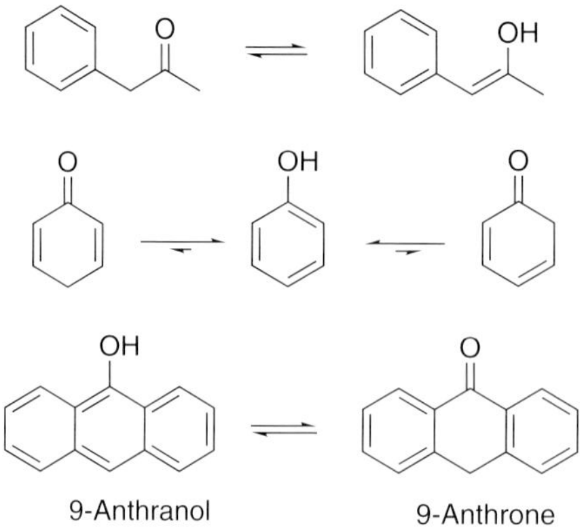

**Figure 6.8** Effect of structure on the equilibrium of oxo-enol tautomers.

molecule to be able to convert to acetaldehyde. It can be stated that oxo–enol reactions are, in general, not intramolecular.

For structural reasons, the position of the tautomeric equilibrium may change. The possibility of conjugation or formation of an aromatic system may shift the tautomeric equilibrium in favor of one of the tautomers (Figure 6.8). In the case of phenylacetone, stabilization by conjugation in the enol form (delocalization) permits its coexistence with the keto form (Figure 6.8). With phenol, the stabilization effect of mesomeric energy is so high that the equilibrium is completely on the side of the enol form. An interesting case is that of 9-anthranol, which is aromatic. In its oxo form, that is, 9-anthrone, two independent six $\pi$-electron-containing aromatic rings are present and therefore both forms can exist.

When the energy difference between the two tautomeric forms is small, environmental effects, such as secondary bonds with solvent molecules, can significantly influence the position of oxo–enol equilibria. Examples of such a situation can be met in the case of β-dioxo compounds and β-oxo-esters (Figure 6.9). With both types of compounds, there exists an enol tautomer containing both a conjugated bond system and an intramolecular hydrogen bridge. Since these effects are absent in the oxo form, the energy difference between the two forms decreases.

In the case of non-symmetrical β-diketones, the situation is more complex since two stable enol forms exist (Figure 6.10)

### 6.2.2
### Imine–Enamine Tautomerism

The imine–enamine tautomerism is the nitrogen analog of oxo–enol tautomerism. The imine form corresponds to the oxo form, while the enamine form to the enol form. Similarly to the oxo forms, the imine forms are generally the more stable tautomers (Figure 6.11). With oximes (R=OH), the imine form is exclusive and its stereoisomers are denoted as *syn*- and *anti*-forms (cf. the stereoisomers of acetophenone oxime in Figure 6.11).

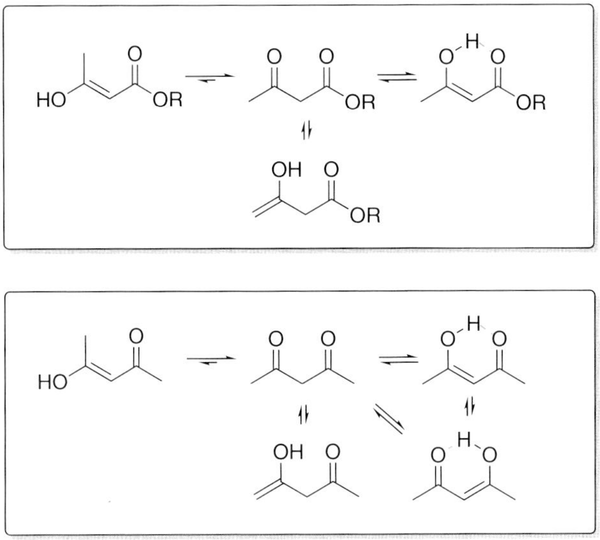

Figure 6.9 Oxo–enol tautomers with β-dioxo compounds and β-oxo-esters.

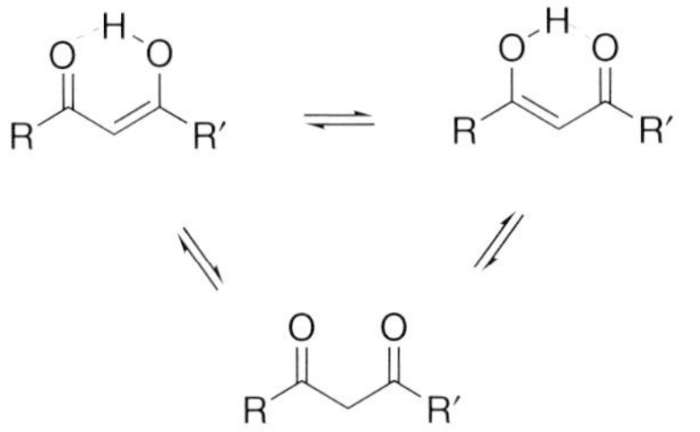

Figure 6.10 Oxo–enol tautomers of nonsymmetrically substituted 1,3-diketones.

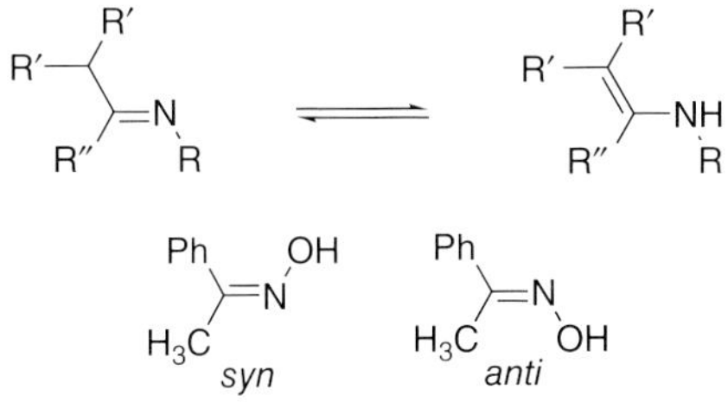

Figure 6.11 Imine–enamine tautomerism.

Stabilization of the enamine form becomes possible when the substituents R′,R″ are electron-withdrawing groups or can enter into conjugations with the double bond. (It is apparent that in this case no conjugation can take place in the imine form.)

### 6.2.3 Amide–Imido Acid and Related Tautomerisms

Amide–imido acid-, thioamide–imido thioacid-, urea–isourea-, and thiourea–isothiourea-type tautomerism are only possible when at least one hydrogen is attached to the nitrogen atom (Figure 6.12).

Owing to the high mesomeric energy of the moieties –NH–C=Y (Y = O, S), the amide, thioamide, urea, and thiourea structures are much more stable than their tautomeric forms, and therefore the equilibrium is shifted completely to the left. This situation does not change even when the abovementioned functional groups are part of a six-membered ring. An interesting situation arises when the ring also contains a conjugated double bond (Figure 6.13).

Pyridin-2-ol and pyridin-2-thiol are both aromatic compounds and it can be therefore expected that, similarly to phenol, the tautomeric equilibrium should be significantly shifted toward the enol form. In fact, however, in polar solvents and in the crystal, nonaromatic forms dominate. The aromatic tautomers are only predominant in apolar solvents or in the vapor phase.

This, at first sight, surprising fact can be properly interpreted when looking at the structure of the tautomeric form on the left, which, as can be seen in Figure 6.13, can be characterized by two mesomeric (resonance) structures. The structure on the left represents an aromatic system in which the opposite charges are separated. This results in higher energy, but the magnitude of charge separation is modest. Consequently, the energetic preference for the fully aromatic form on the right is also not very pronounced. Therefore, in such compounds, environmental effects are not negligible and can significantly influence the position of the equilibrium.

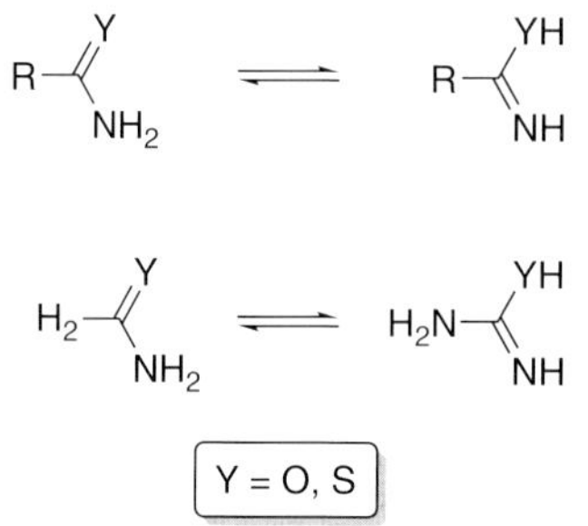

Y = O, S

Figure 6.12 Amide–imido acid-, thioamide–imido thioacid-, urea–isourea-, and thiourea–isothiourea-type tautomerisms.

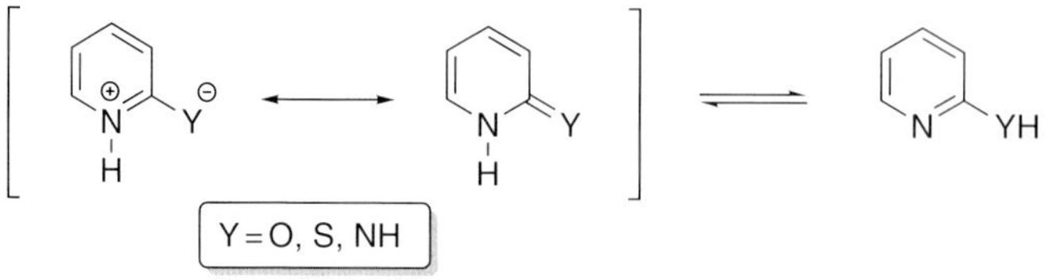

Y = O, S, NH

Figure 6.13 Amide–imido acid (Y = O), thioamide–imido thioacid (Y = S), and urea–isourea (Y = NH) type tautomerisms in cyclic compounds.

### 6.2.4
### Tautomerism of Aliphatic Nitro Compounds

All nitro compounds in which a hydrogen atom is attached to a carbon atom in α-position are capable of tautomeric transformations (Figure 6.14). Generally, the equilibrium is shifted completely toward the nitro form, but there are methods suitable for the preparation of two degenerate tautomers of the s.c. aci form.

For this purpose, first the alkali salt of the weakly acidic nitro compounds is prepared followed by acidification of the mixture, when the proton becomes attached to the "harder" oxygen atom. The aci forms are, however, unstable and revert gradually to the nitro compounds.

### 6.2.5
### Tautomerism of Carbonic Acid Derivatives Containing Cumulated Double Bonds

Tautomeric equilibria of cyanic and isocyanic acid, as well of thiocyanic acid and isothiocyanic acid, are practically completely shifted in favor of the form containing a cumulated double bond (Figure 6.15). On the other hand, the tautomeric equilibria of cyanamides and carbodiimides, as well their *N*-substituted derivatives, are shifted to the opposite side, that is, toward the form containing a triple bond. Therefore, carbodiimide itself and their *N*-monosubstituted derivatives are unknown.

Since N,N′-disubstituted carbodiimides and *N,N*-disubstituted cyanamides do not contain a mobile hydrogen atom, they do not form tautomers (Figure 6.16). Modifications with a fixed structure are stable and several derivatives are known.

**Figure 6.14** Tautomerism of nitro compounds.

| H−N=C=Y | ⇌ | N≡C−Y−H |
|---|---|---|
| Isocyanic acid | Y = O | Cyanic acid |
| Isothiocyanic acid | Y = S | Thiocyanic acid |

| H−N=C=N−H | ⇌ | N≡C−$NH_2$ |
|---|---|---|
| Carbodiimide | | Cyanamide |

**Figure 6.15** Tautomerism of cyanic acid/isocyanic acid, thiocyanic acid/isothiocyanic acid.

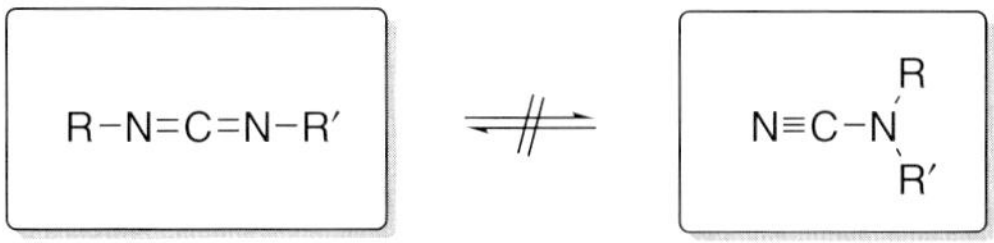

$R{-}N{=}C{=}N{-}R'$ ⇸ $N{\equiv}C{-}N(R)R'$

Figure 6.16 Unsubstituted, *N*-monosubstituted, and *N,N′*-disubstituted carbodiimides and cyanamides.

### 6.2.6 Ring–Chain Tautomerism

With alcohols, oxo compounds form in an equilibrium reaction hemiacetals. When both functional groups are in the same molecule, the reaction can also take place in an intramolecular fashion (Figure 6.17), and the resulting products are related as equilibrating tautomers.

Since the oxo group is planar, ring closure is possible in two ways and may therefore give rise to two different products. When the oxo compound was achiral ($R_1 = R_2$), the two products will be related as enantiomers in a 1 : 1 ratio. Otherwise ($R_1 \neq R_2$), the products will be diastereomers and their ratio other than 1 : 1.

Because the cyclic products can interconvert via the open-chain form, they also form a tautomeric equilibrium. A special feature of such tautomers is that they are merely stereoisomers and not, as usual with tautomers, constitutional isomers.

Figure 6.17 presents the general scheme of ring–chain tautomerism. If the distance of the reacting groups is such that in the course of ring closure a strainless ring is produced, the equilibrium is shifted toward cyclization. Several other examples exist for ring–chain tautomerism. Of them, we mention two: tautomerism of 2-formylbenzoic acid and of thiosemicarbazones (Figure 6.18).

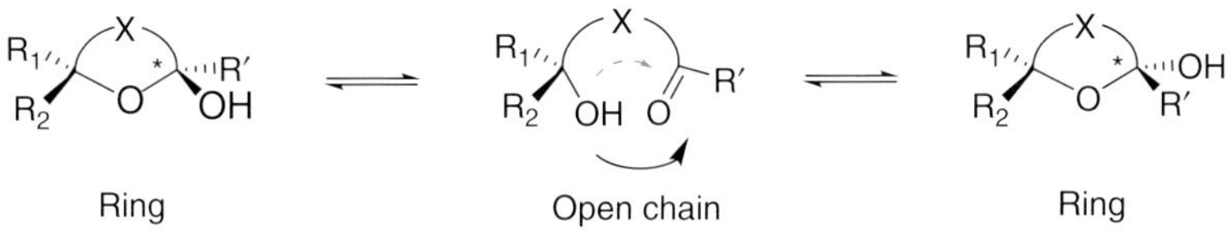

Ring ⇌ Open chain ⇌ Ring

Figure 6.17 Ring–chain tautomerism of hydroxyl–oxo compounds.

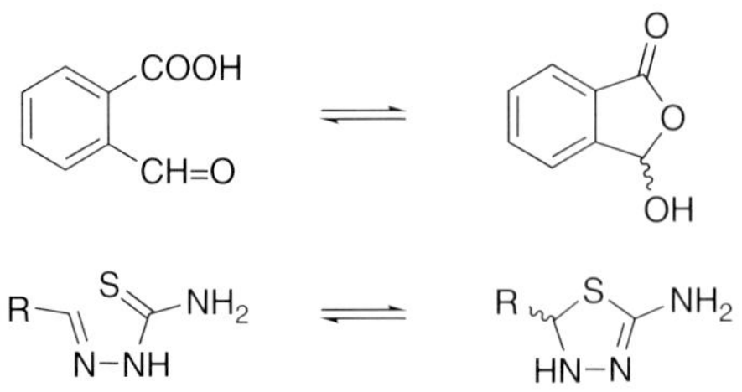

Figure 6.18 Ring–chain tautomerism of 2-formylbenzoic acid and thiosemicarbazones.

## 6.3 Methods for Studying Tautomerism

Elaboration of methods for studying the existence of tautomerism, the structure of tautomers, as well as the ratio of tautomers is one of the most remarkable achievements of classical organic chemistry. On the basis of the different reactivity of the individual tautomer, ingenious methods have been devised to assess tautomeric ratios. However, by now, most of them have been superseded by nuclear magnetic resonance spectroscopy (mainly $^1$H and $^{13}$C NMR).

In contrast to the chemical approach, physicochemical methods do not interfere with the tautomeric equilibrium. Crystalline samples of compounds presumed to form tautomeric equilibria can be studied by infrared spectroscopy, X-ray diffraction, and more recently by solid phase NMR spectroscopy. While X-ray methods are only applicable to crystalline samples, infrared spectroscopy can be applied in any state.

UV and NMR spectroscopy is used foremost in the liquid phase for the study of tautomerism (solid phase NMR, mentioned earlier is an exception). Both methods require for comparison model compounds, but nowadays a large body of literature data is available in which the necessary data can be found. Assignment of NMR signals to the individual tautomeric forms means at the same time the elucidation of their structures. Provided that one or more signals can be assigned to the individual tautomeric forms, the ratio of the tautomers can be determined. Occasionally, the interconversion of the tautomers is so fast (fast on the NMR timescale) that signals of the groups of the individual tautomers are not resolved, but produce a common signal.

## 6.4 Nitrogen Inversion

In its tricoordinated compounds, such as ammonia or amines, the nitrogen atom is undergoing fast **nitrogen inversion** (Figure 6.19). In this process – completely analogous to the inversion accompanying $S_N$ reactions – the trigonal pyramidal structure inverts owing to the oscillation of the lone pair of electrons.

Nitrogen inversion proceeds fast even at room temperature because its energy barrier is rather low (24.2 kJ mol$^{-1}$ for ammonia). This is in contrast to the phosphorus atom, which in analogous compounds, due to a much higher energy barrier (132 kJ mol$^{-1}$ for phosphine) fails to invert at room temperature. Amines with three different substituents are in principle chiral, but fast nitrogen inversion prevents to observe this and the molecule behaves as a racemic mixture. The process

R R′ N R″ ⇌ [R′ R N R″]# ⇌ R′ R N R″

**Figure 6.19** Inversion of amines containing a tetrahedral nitrogen atom.

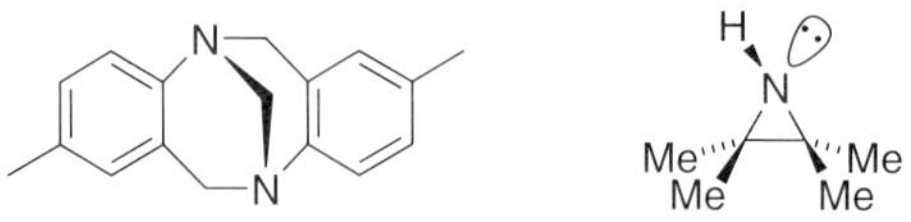

Figure 6.20 Examples of restricted nitrogen inversion.

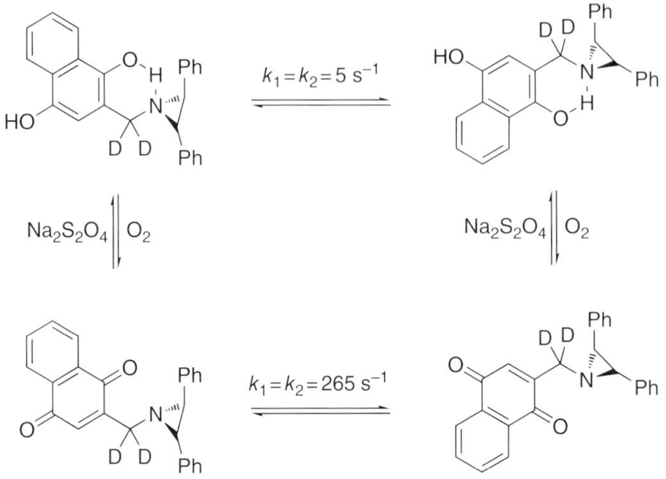

Figure 6.21 Stabilization of the configuration of the nitrogen atom by partial protonation.

can also be interpreted as a valence tautomerism or is very similar to conformational transformations. Under certain structural circumstances, for example, in certain cyclic compounds or with quaternary ammonium salts, nitrogen inversion can be restricted and the nitrogen atom becomes a stereocenter. An example is the chiral Tröger's base (Figure 6.20). Ring strain, as in aziridines, may also slow down nitrogen inversion as in 2,2,3,3-tetramethyl aziridine for which separate $^{1}$H NMR signals could be observed for the diastereotopic methyl groups at room temperature [1].

Configuration of the nitrogen atom may be fixed by hydrogen bonding, as for example, in the aziridine derivative containing a phenolic hydroxyl group (Figure 6.21), where nitrogen inversion is 50 times slower than in its oxidized derivative where formation of a hydrogen bridge and stabilization of the nitrogen atom thereby become impossible [2].

## References

1. Szántay, C., Chmielewicz, Z.F., and Bárdos, T.J. (1967) New Alkylating Agents Derived from. Diaziridine. *J. Med. Chem.*, **10**, 101–104.
2. Davies, M.W., Shipman, M., Tucker, J.H.R., and Walsh, T.R. (2006) Control of Pyramidal Inversion Rates by Redox Switching. *J. Am. Chem. Soc.*, **128**, 14260–14261.

# Problems to Part II

## Problem 3.1

Processes faster than $10^{-3}$ s cannot be followed by NMR spectroscopy because signals pertaining to the equilibrating species become averaged. Time resolution of IR spectra is about $10^{-12}$ s. If the rate of conformational motions of methylcyclohexane is about $10^{-5}$ s, is it IR or NMR spectroscopy by which it is possible to follow these motions?

## Problem 3.1.1

Explain why ethane remains optically inactive, even when liquefied and cooled to a temperature at which internal rotation is hindered.

## Problem 3.2.1

a) Will the equilibration of the sc and ap states of *n*-butane be hindered when it is cooled to a temperature when the energy required to reach the sp state cannot be invested?
b) Will the ratio of two sc states be different within a set of molecules when on lowering the temperature, the internal rotation of *n*-butane becomes totally hindered?

## Problem 3.3.1

Which of the two conformations of ($S_a$)-2,2′-dibromo-6,6′-biphenyl (**A** or **B**) shown below is more populated? Explain your answer!

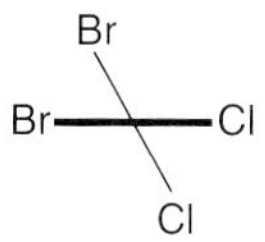

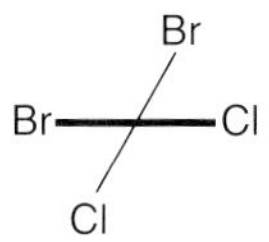

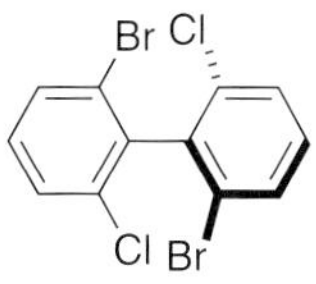

*Stereochemistry and Stereoselective Synthesis: An Introduction*, First Edition.
László Poppe and Mihály Nógrádi.

Companion Website: www.wiley.com/go/poppe/stereochemistry

**Problem 4.1**

It is known that the optical rotation of chiral materials depends on the wavelength of the light used. This phenomenon is called ORD. In the ORD spectra, away from the UV–VIS absorption range, the degree of rotation is low, then, after reaching a maximum when approaching the $\lambda_{max}$ of UV–VIS absorption, it drops dramatically. The sign of rotation changes at $\lambda_{max}$ (Cotton effect), and later in the opposite half (opposite sign) its absolute value reaches a maximum and then decreases again. Knowing this phenomenon, provide an explanation as to why it was reasonable to choose for reference wavelength in optical rotation measurements the D line of Na (589 nm)!

**Problem 4.2.1**

It is known that the decomposition of (−)-nicotine led to (−)-hygrinic acid. N-Methylation of (−)-hygrinic acid resulted in the same compound that may be obtained also by N-methylation of natural (*S*)-proline. Knowing the above, determine the absolute configuration of (−)-nicotine!

MeI MeI

(−)-nicotine (−)-hygrinic acid (*S*)-prolin

**Problem 4.2.2**

D-Glucose can be crystallized in two forms as $\alpha$- or $\beta$-D-glucopyranose. The freshly prepared solution of $\alpha$-D-glucopyranose has an optical rotation of $[\alpha]_D$ +112, while $[\alpha]_D$ +18.7 can be measured for the freshly prepared solution of $\beta$-D-glucopyranose. If the NMR spectra of the freshly prepared solutions of the two crystalline forms are taken, a coupling constant of 3.5 Hz in the $\alpha$-D-glucopyranose solution can be measured for the signal of the $C_1$–H hydrogen atom, while in the $\beta$-D-glucopyranose solution, the $C_1$–H coupling constant is 7.7 Hz. We know that the typical axial–equatorial coupling constant is 3.5 Hz, while a 7.7 Hz coupling is characteristic for hydrogens at axial–axial positions. In view of the foregoing, determine the absolute configuration of $\alpha$- and $\beta$-D-glucopyranose, and assign the (**1**) or (**2**) formula to them!

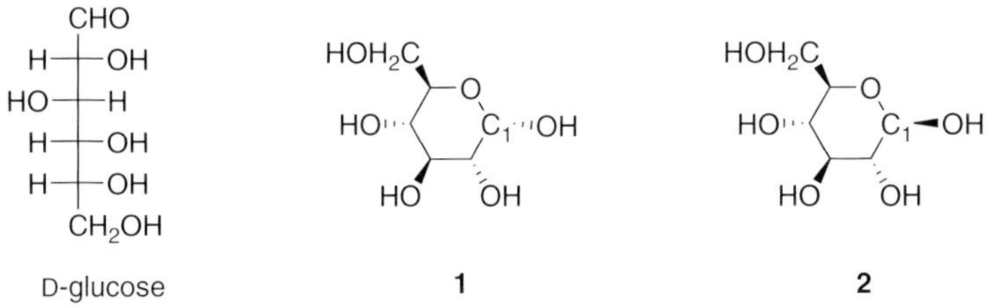

CHO
H—OH
HO—H
H—OH
H—OH
$CH_2OH$

D-glucose

$HOH_2C$ O HO $C_1$ OH HO OH

1

$HOH_2C$ O HO $C_1$ OH HO OH

2

**Problem 5.1**

The enantiomeric composition of lactic acid in a sample was determined enzymatically. On treatment of the sample (100 μmol) with (*S*)-lactate dehydrogenase and equimolar $NAD^+$, formation of 0.01 mmol NADH was detected. Determine the composition and ee value of the sample!

**Problem 5.2**

The lactic acid sample of the previous problem was used to determine the unknown enantiomeric composition of a 1-phenylethanol sample. The composition of ester formed from complete conversion of the 1-phenylethanol sample with equimolar amounts of lactic acid was measured by HPLC on a nonchiral stationary phase, and the relative stereochemistry was determined by NMR spectroscopy.

- On the basis of the HPLC analysis, a diastereomeric mixture of 0.598 : 0.402 ratio was formed.
- According to NMR investigation, the *like* diastereomer was the major component, and the *unlike* was the minor one.

a) Determine which of the **1**, **2**, **3**, and **4** compounds are the *like* and *unlike* diastereomers!
b) Determine the proportion of compounds **1**, **2**, **3**, and **4** in the product mixture!
c) Determine the enantiomeric composition and ee value of the 1-phenylethanol sample!

OH + OH

OH COOH + OH COOH

OH O O 1

OH O O 2

OH O O 3

OH O O 4

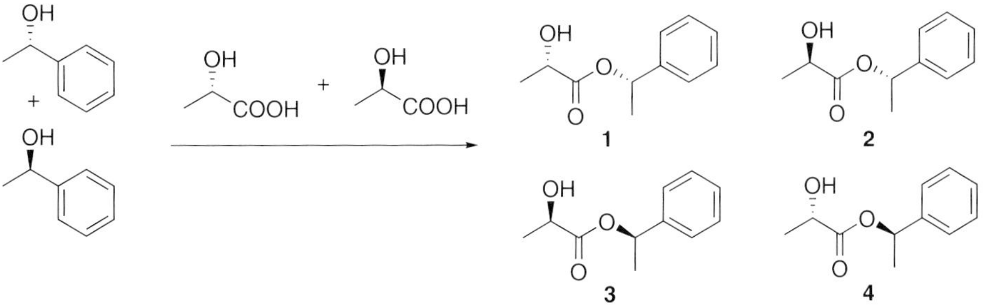

### Problem 5.3

The result of the previous measurement is also checked by reacting the 1-phenylethanol sample with $PCl_3$ (over three equivalents) and assaying the product mixture by NMR spectroscopy.

- How many diastereomers are expected to form?
- What is the proportion of the diastereomers forming?
- Which ratio of compounds **5**, **6**, **7**, and **8** is expected in the product mixture?

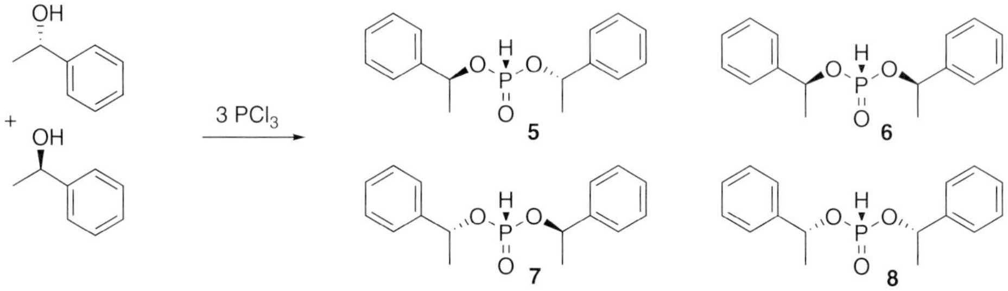

OH + OH —3 $PCl_3$→ **5**, **6**, **7**, **8**

### Problem 6.1

Of (*E*)- and (*Z*)-1-phenyl-1-chloropropene, the latter is more stable ($\Delta = -6.6\,\text{kJ mol}^{-1}$). In the presence of a catalytic amount of HCl, within 2 h an *E*/*Z* equilibrium composition of 1 : 35 is established. Give a possible mechanism for the reaction and explain the stability relationship of the two stereoisomers!

(*E*) Cl, H, Ph, Me ⇌ (HCl) (*Z*) Cl, Me, Ph, H

### Problem 6.1.1

In the course of Cope rearrangement, an equilibrium between the two dienes shown below is established.

- What kind of transition state is involved in this equilibration?
- Which is the major component of the equilibrium and why?

**Problem 6.1.2**

Which one of the two rearrangements shown below is a valence isomerization and which is a classic tautomeric rearrangement? How can you prove experimentally the existence of valence tautomerism?

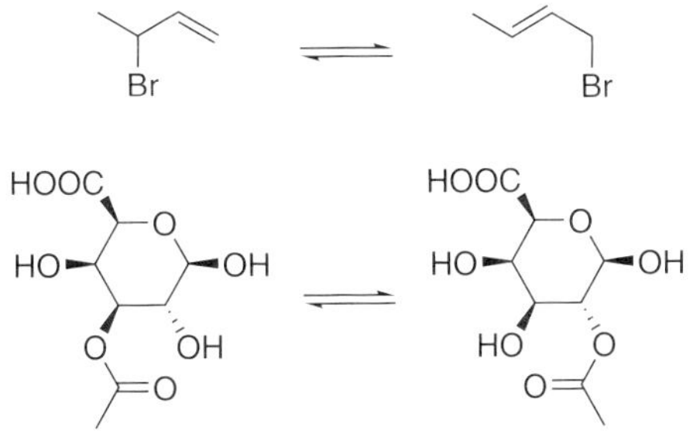

**Problem 6.1.3**

Is the following sigmatropic rearrangement a prototropic transformation?

**Problem 6.2.1**

What are the stereochemical consequences of a base-catalyzed oxo–enol tautomeric equilibrium involving the following ketone?

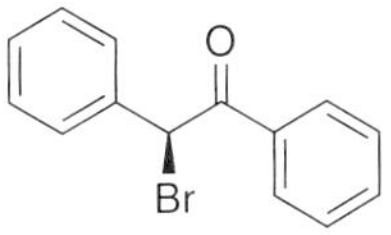

**Problem 6.2.2**

Explain the fact that while cyanuric acid is more stable as an amide, melamine is preferring the aromatic form!

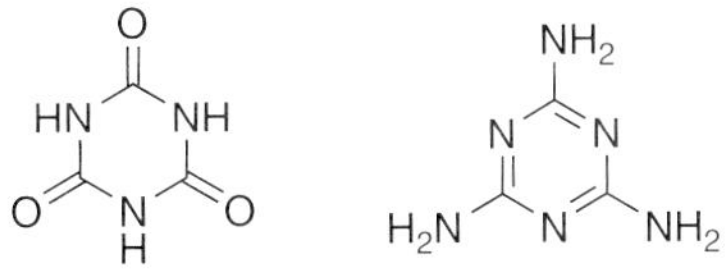

**Problem 6.2.3**

Draw up the conceivable tautomeric forms of guanine and explain why in aqueous media the form shown below is practically exclusive?

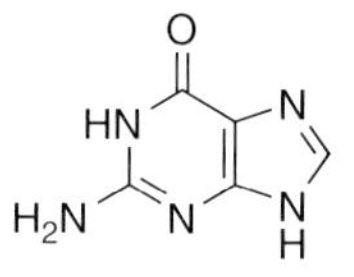

**Problem 6.2.4**

What is the stereochemical consequence of a base-catalyzed tautomeric equilibrium involving (1*R*,2*S*)-1,2-diphenyl-1-nitropropane?

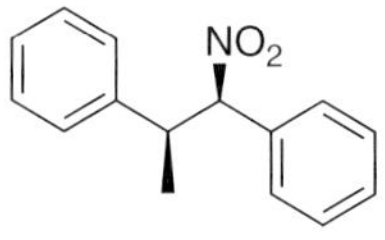

**Problem 6.2.5**

The tautomeric equilibrium between cyanamide and carbodiimide forms as well as of their monosubstituted derivatives is totally shifted toward the cyanamide form. On the basis of this fact, decide which is the more reactive compound: *N,N'*-dicylohexylcabodiimide or *N,N*-dicyclohexylcyanamide?

**Problem 6.2.6**

Draw up the possible cyclic tautomers of D-fructose!

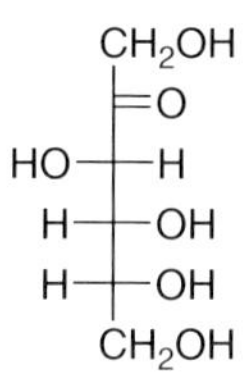

**Problem 6.3.1**

$^1$H-NMR and $^{13}$C-NMR spectra of pentane-2,4-dione recorded in $CDCl_3$ at room temperature provide the following data:

$^1$H-NMR ($CDCl_3$): $\delta = 2.00$ (s, 5.16 H, $CH_3$, enol), 2.17 (s, 0.84 H, $CH_3$ oxo), 3.62 (s, 0.28 H, $CH_2$ oxo), and 5.57 (s, 0.86 H, enol) ppm;
$^{13}$C-NMR ($CDCl_3$): $\delta = 24.3$ ($CH_3$ enol), 30.2 ($CH_3$ oxo), 58.2 ($CH_2$ oxo), 100.3 (CH enol), 191.4 (CO enol), and 201.8 (CO oxo) ppm.

- On the basis of the above data, give the composition of the oxo and enol forms in the equilibrium!
- On the basis of the above data and considering the resolution time of $10^{-3}$ s of NMR, estimate the rate of oxo–enol conversion ($\nu_1$) and the rate of the interconversion of two enol forms ($\nu_2$).

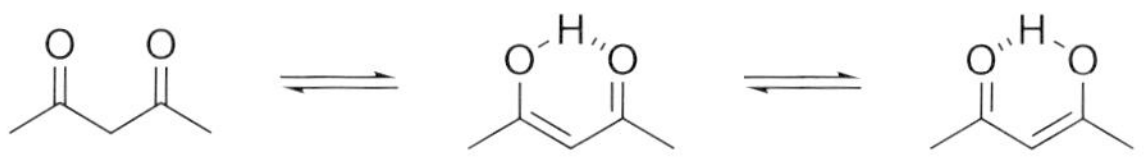

### Problem 6.3.2

Prove the validity of the following correlation formula for the equilibrium constant ($K_T$) of the tautomeric equilibrium:

$$K_T = \frac{K_{a\,oxo}}{K_{a\,enol}} = \frac{[enol]}{[oxo]}$$

where $K_{a\,oxo}$ and $K_{a\,enol}$ are the acid-dissociation constants of the oxo and enol forms, respectively.

- If the p$K_a$ value of ethyl acetoacetate is 10.2 and a tautomeric equilibrium constant of $6.2 \times 10^{-2}$ is measured as the neat liquid, calculate the p$K_a$ value of the enol form.

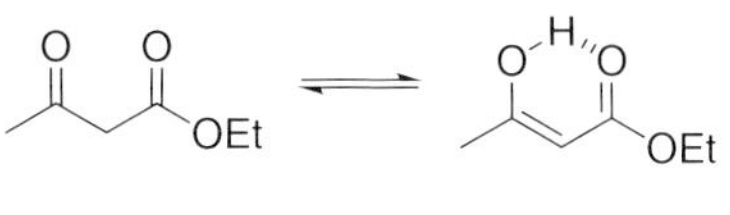

### Problem 6.4.1

What is the stereochemical consequence of transforming *N*-ethyl-*N*-methylaniline with peracetic acid to the *N*-oxide?

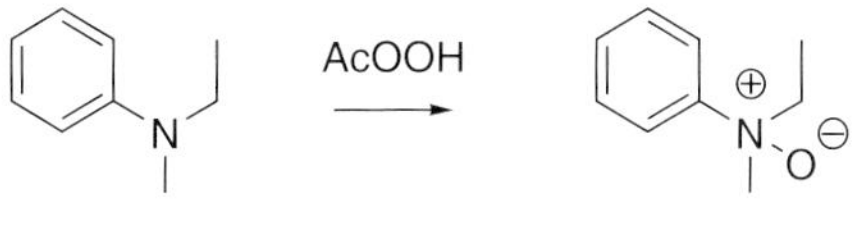

AcOOH

# Part III
# General Characteristics of Stereoselective Reactions

Selectivity is an all-important key feature of chemical reactions. Selectivity enables multistep reactions to take place economically. Moreover, without selectivity it would be impractical to analyze and purify the products of reactions. After defining stereospecificity and stereoselectivity, this part provides the reader with a systematic analysis of the various types of selectivties with special emphasis on selectivities leading to single enantiomeric products.

*Stereochemistry and Stereoselective Synthesis: An Introduction*, First Edition.
László Poppe and Mihály Nógrádi.

Companion Website: www.wiley.com/go/poppe/stereochemistry

# 7 Types and Classification of Selectivities

Selectivity in chemistry is interpreted in various ways. The terms chemoselectivity, regioselectivity, and stereoselectivity (including dia- and enantioselectivity) are widely used. However, there is no general agreement on their precise meaning. Therefore, the following chapters are dedicated to this topic and attempt to provide a uniform interpretative framework.

## 7.1 Main Types of Selectivity

In chemistry, two main types of selectivities can be defined; one depends mainly on the properties of the substrate, while the other differs in the products of a reaction [1].

These two main types of selectivities are shown in the following figures.

### 7.1.1 Substrate Selectivity

A reagent or a catalyst is **substrate selective** (Figure 7.1) when it transforms different substrates ($S_1, S_2, \ldots$) to products ($P_1, P_2, \ldots$) under identical conditions at different rates ($k_1 \neq k_2$).

### 7.1.2 Product Selectivity

A reagent or catalyst is **product selective** (Figure 7.2) when it permits the formation of more than one product at different rates ($k_1 \neq k_2$) from a single substrate ($S$) whereby the products ($P_1, P_2, \ldots$) are formed in a ratio differing from the one statistically expected.[1]

1) The statistically expected ratio depends on the number of reactive sites. On the mononitration of toluene, this ratio is, for example, 2 : 2 : 1.

*Stereochemistry and Stereoselective Synthesis: An Introduction*, First Edition.
László Poppe and Mihály Nógrádi.

Companion Website: www.wiley.com/go/poppe/stereochemistry

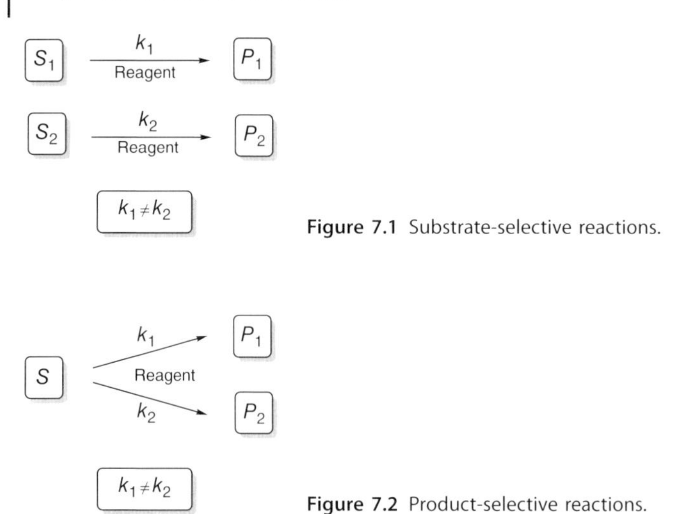


**Figure 7.1** Substrate-selective reactions.

**Figure 7.2** Product-selective reactions.

## 7.2 Classification of Selectivities

The two main types of selectivities described earlier can be used for a precise interpretation of the terms "*chemoselectivity*," "*regioselectivity*," "*diastereoselectivity*," and "*enantioselectivity*," which are often used but defined not adequately enough in the literature (Figure 7.3) [2].

Figure 7.3 shows that **chemoselectivity** and **regioselectivity** can manifest themselves in either a substrate-selective or a product-selective manner. This fact, however, should not lead to confusion, since when discussing chemical problems, it is accepted that a mixture of substances of different constitution or a mixture of constitutional isomers should be regarded as a mixture of substances. Accordingly, when discussing *chemo-* or *regioselectivity*, it will be always indicated whether the starting material was a mixture of substances or a pure compound.

In the case of stereoisomeric mixtures being less readily separable, it is often ambiguous whether the starting material in question was a mixture or a pure substance. In the chemical literature, racemic substances are often regarded as homogeneous, despite being a mixture of two enantiomers. Therefore, instead of the ambiguous terms **diastereoselectivity** and **enantioselectivity**, a more precise terminology has to be adopted (Figure 7.3) [2].

When a mixture of diastereomers is undergoing a selective transformation, the reaction is denoted **diastereomer selective**, while a reaction distinguishing between diastereotopic faces or groups of a single compound is called **diastereotope selective.** Similarly, starting from a mixture of enantiomers, the reaction may be **enantiomer selective**, while reactions distinguishing between enantiotopic groups or faces are **enantiotope selective.**

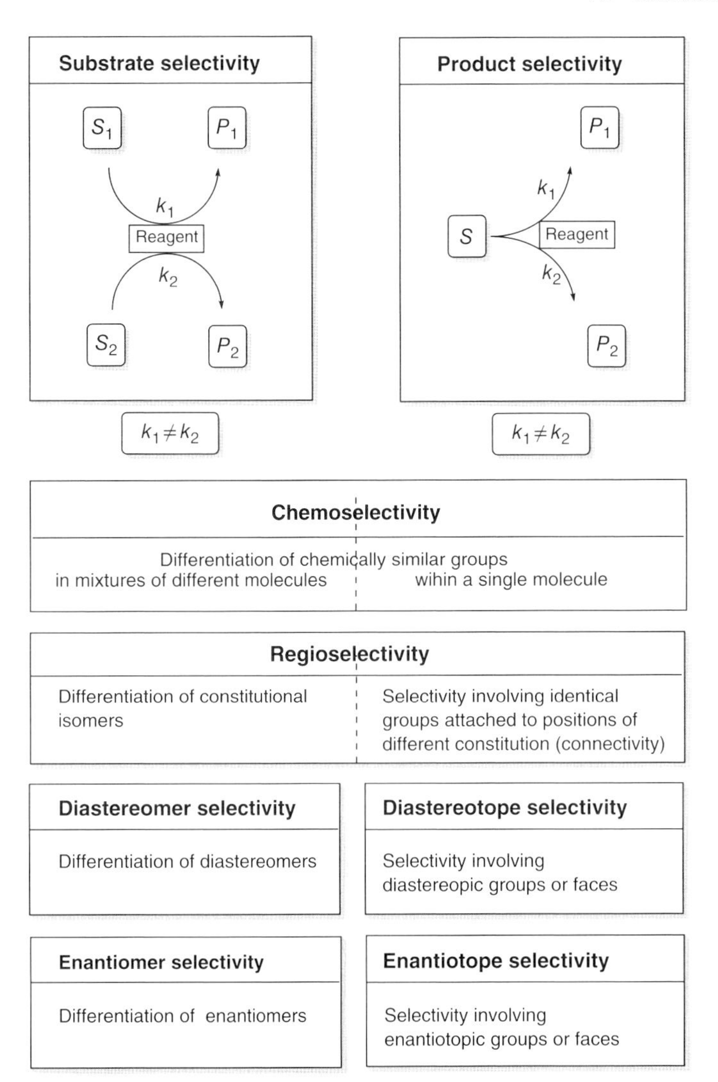


**Figure 7.3** Classification of selective reactions [2].

It is important to note that the types of selectivities discussed in the present chapter are referring to the selectivity of a single reaction and cannot be applied to a consecutive series of reactions. In the latter case, each individual steps constituting the multistep process may be characterized by different types of selectivities. Therefore, each partial reaction of the multistep process and its consequences should be analyzed separately. Pure manifestations of the type of selectivity in one-step reactions will be discussed next, while later chapters will present some more complex processes. The latter will contain an analysis of the types of selectivities and their consequences for each partial step.

## 7.2.1 Chemoselectivity

### 7.2.1.1 Interpretation of Chemoselectivity

Chemoselectivity is a multifaceted concept, and several interpretations can be encountered in the chemical literature.

a) Differentiation between groups that are capable of reacting in a given reaction. Examples are the transformations of 2-oxocyclohexane carboxylic acid shown in Figure 7.4. Chemoselective reduction of the keto group (A) gives a hydroxy acid, while that of the carboxyl group (B) a hydroxy ketone.
b) Selective reactions, when among one or more identical functional groups only one is transformed (Figure 7.5).
   An example is the monoesterification of succinic acid avoiding the formation of a diester (A) or its reduction to 4-hydroxybutanoic acid (or γ-butyrolactone) without the formation of 1,4-butane-diol (B).
c) A process can also be regarded as chemoselective when the reaction of a given functional group gives rise to a product, which might further react with the same reagent but fails to do so. For instance, the reaction of ethyl benzoate (Figure 7.6) with a metal organic compound (e.g., methyl magnesium iodide) in a way that acetophenone should be obtained preferentially without giving rise to a tertiary alcohol.



Figure 7.4 Chemoselective reductions.



Figure 7.5 Chemoselective reactions favoring the transformation of a single functional group among two or more available.



Figure 7.6 Avoiding further reaction of the product in a chemoselective reaction.

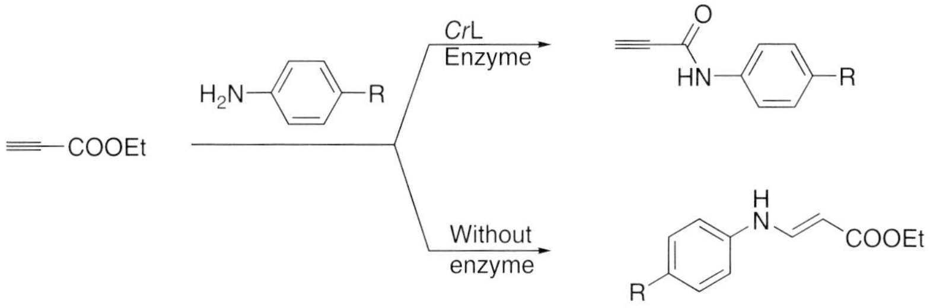

Figure 7.7 Different reactivity of ethyl propargylate in chemical- and enzyme-catalyzed reactions.

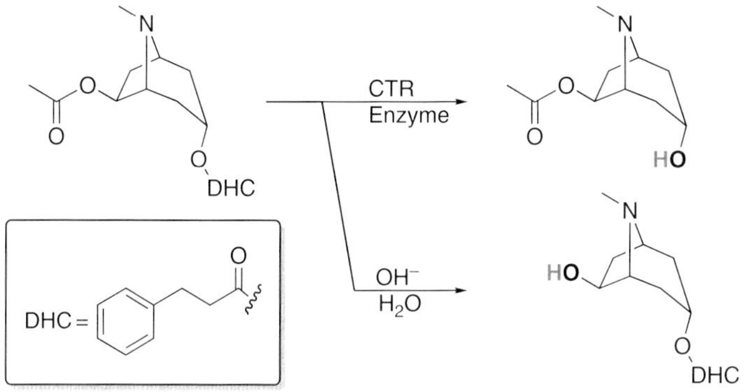

Figure 7.8 Chemoselective alternative protecting group elimination by alkaline hydrolysis and chymotrypsin, respectively.

#### 7.2.1.2 Chemoselectivity in Biocatalysis

Biocatalytic processes can provide different chemoselectivity than their chemical counterparts (Figure 7.7). For example, in a nonbiocatalytic reaction, ethyl propargylate with aromatic amines gives *Michael*-type adducts, while with *Candida rugosa* lipase (CrL) in organic solvents, the amide can be obtained.

Special chemoselectivity of biocatalysts can be exploited for manipulation of protecting groups. While chemical hydrolysis of the diester in Figure 7.8 permits the chemoselective elimination of the more reactive acetyl group, with chymotrypsin (CTR), chemoselective hydrolysis of the 3-phenylpropyl group can be realized.

### 7.2.2 Regioselectivity

**Regioselectivity** is a concept somewhat akin to chemoselectivity. In the present context, "**region**" means a site of a molecule at which a reaction can take place. If there is more than one such site and the rates involving the different sites are unequal, the reaction is regarded as **regioselective.**

$OSO_3^-$ Sulfatase (*Helix pomatia*) → OH + $OSO_3^-$

α- and β-Ester β-Naphthol α-Ester

**Figure 7.9** Regioselective hydrolysis of sulfate esters with sulfatase enzyme.

Regioselectivity is generally interpreted in the context of product selectivity. A classic example is the manifestation of *Markovnikov's* rule for the addition of hydrochloric acid onto olefins, when the chloride ion is preferentially attacking at the more substituted carbon atom. Of course, regioselectivity of a substrate-selective type can also be possible.

#### 7.2.2.1 Substrate Regioselectivity

When analyzing the types of selectivities, substrate regioselectivity is rarely mentioned because chemists regard regioisomers automatically as different compounds and their separation can be generally readily accomplished. There are, however, some processes, when it is problematic to separate the mixture of regioisomeric products. In such instances, regioselectivity may be of help, as, for example, in the enzyme-catalyzed reaction shown in Figure 7.9. Separation of the sulfate esters of 1- and 2-naphthol can be accomplished by regioselective hydrolysis using a sulfatase enzyme. A similar procedure can be carried out with sulfate esters of *o*- and *m*-, as well as with *o*- and *p*-substituted phenols.

#### 7.2.2.2 Product Regioselectivity

By regioselectivity in a chemical reaction, usually product selectivity taken in a narrower sense is understood. Even within this category, more than one type can be distinguished, and regioselectivity is often mixed up with the concept of chemoselectivity.

According to one interpretation, **regioselectivity** means competition between different regions of the same functional group. For instance, selectivities in the direct 1,2-addition, as well as in *Michael*- and 1,4-additions, are put by some authors into this category (as quoted in association with chemoselectivity shown in Figure 7.7).

According to an alternative interpretation, **regioselectivity** means selectivity manifested between groups of the same constitution but of different connectedness. The examples in the following two sections demonstrate this case.

**Regioselectivity in Enol and Enolate Formation** A classic example for regioselectivity associated with functional groups is the dependence of enol/enolate formation on reaction conditions (Figure 7.10).

Variations of *regioselectivity* in enolization can be traced back fundamentally to the difference of kinetically and thermodynamically controlled enolization

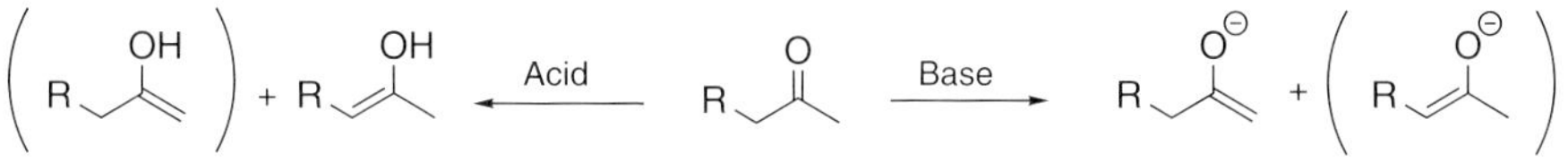

Figure 7.10 Regioselectivity in enolate formation.

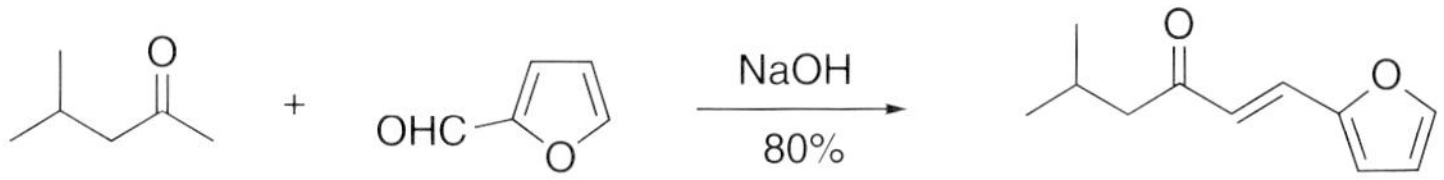

Figure 7.11 Kinetically controlled regioselectivity in a base-catalyzed enolization.

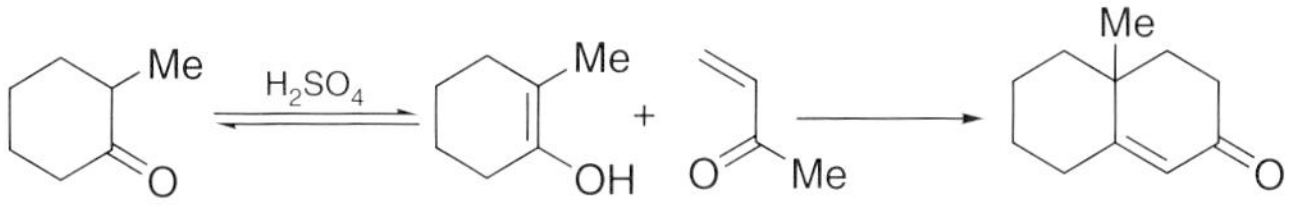

Figure 7.12 Thermodynamically controlled regioselectivity in acid-catalyzed *Robinson* anellation.

mechanisms. Protons of the ketones in Figure 7.10 (where R is a simple electron donating alkyl group) are somewhat more acidic at the methyl group. Accordingly, in kinetically controlled base catalysis, regioselectivity prefers the methyl side.

Regioselectivity of base-catalyzed enolization can be well observed when R = *i*Pr and the reaction partner is an aldehyde lacking an α-proton such as furfural (Figure 7.11). Condensation using NaOH is regioselective and produces the (*E*)-enone in 80% yield.

Under acid catalysis, however, mainly when the reaction is an equilibrium process, the more substituted enol/enolate is formed preferentially, that is, the one that contains a thermodynamically more stable double bond. Selectivity is also more pronounced when the two sides are more different. This is exemplified by the cyclic ketone shown in Figure 7.12. *Robinson* anellation of the ketone takes place primarily on the more substituted side.

**Regioselectivity Involving Functional Groups** In the initial steps of the synthesis of Lipstatin, examples for regioselectivity between groups of identical structure but linked to sites of different constitution are depicted in Figure 7.13.

Reduction of (*S*)-malic acid dimethyl ester with borane takes place in a regioselective manner. Owing to its more polar character, the carbonyl group of the ester group adjacent to the hydroxyl group was reduced selectively. Tosylation of the resulting diol was also regioselective affording, as the main product, the tosylate of the sterically more accessible primary hydroxyl function.

Enzymatic methods are well suited to realize similar selective reactions. For example, regioselectivity between two ester groups of different constitution was exploited in the synthesis of a β-lactam by selective hydrolysis of diethyl tartrate

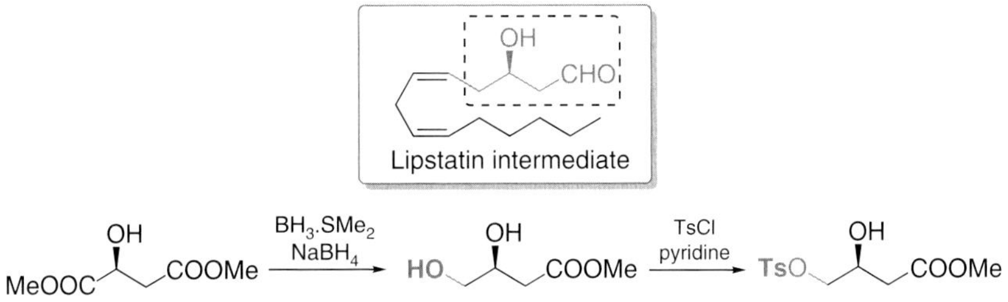

Figure 7.13 Regioselectivity between groups of identical constitution.

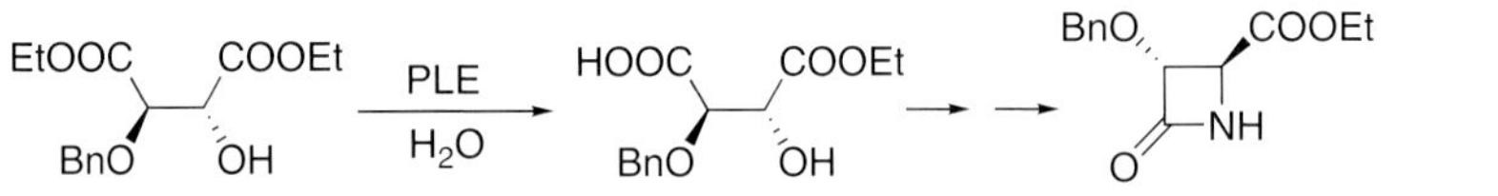

Figure 7.14 Regioselective hydrolysis with porcine liver carboxyl esterase (PLE) enzyme.

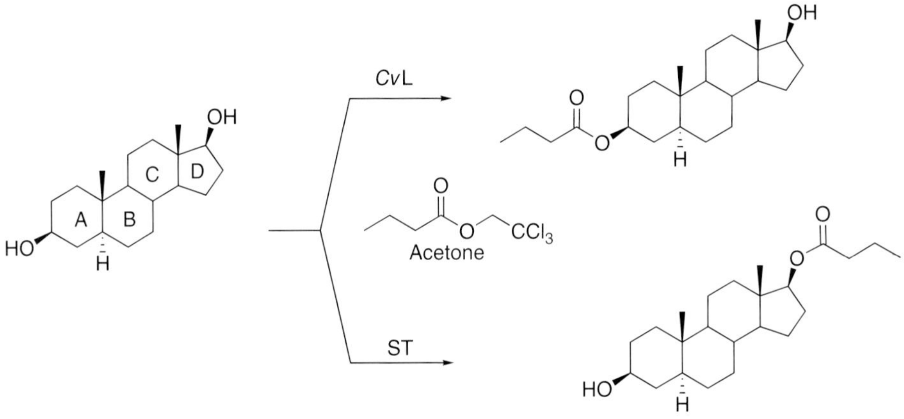

Figure 7.15 Regioselective enzymatic acylation in steroid synthesis.

protected at one of its hydroxyl groups with porcine liver carboxyl esterase (PLE) (Figure 7.14).

Using different enzymes, it is often possible to realize different kinds of regioselectivities with the same substrate. Thus, synthetic schemes without protecting groups can be elaborated, as examples from the area of steroid synthesis demonstrate (Figure 7.15).

By acylation with the same reagent, that is, trichloroethyl butyrate, regioselective acylation can be achieved with the proper selection of the enzyme. Thus, using *Chromobacterium viscosum* lipase (*Cv*L), the hydroxyl linked to ring A can be selectively transformed, while with Subtilisin (ST), it is the hydroxyl at ring D that can be regioselectively esterified (Figure 7.15).

## References

1. Nógrádi, M. (1995) *Stereoselective Synthesis: A Practical Approach*, 2nd edn, VCH, Weinheim, New York.
2. Poppe, L. and Novák, L. (1992) *Stereoselective Biocatalysis – A Synthetic Approach*, VCH, Weinheim, New York.

# 8
# Stereoselective and Stereospecific Reactions

The concepts **stereospecific** and **stereoselective** refer to the stereochemical outcome of a reaction. Unfortunately, there is no general agreement about the meaning of these concepts. In Figure 8.1, we present on hand of examples how the original meaning of these concepts can be interpreted.

## 8.1
## Stereospecificity

In a **stereospecific reaction**, one particular stereoisomer of the reactant produces exclusively one stereoisomer of the product, while the other stereoisomer gives rise, also exclusively, to the other stereoisomer. $S_N2$ reactions are deemed to be stereospecific (Figure 8.1), the reaction proceeds consistently with an inversion of the stereogenic center. This is illustrated by the reactions of (1*r*,3*r*)- and (1*s*,3*s*)-1-bromo-3-methylcyclobutane with thiophenolate anion. From the 1*r*,3*r* isomer, [(1*s*,3*s*)-3-methylcyclobutyl](phenyl)sulfane is formed, whereas from (1*s*,3*s*)-1-bromo-3-methylcyclobutane, the product is the 1*r*,3*r* stereoisomer. A similar stereospecific reaction is the epoxidation of olefins with *m*-chloroperbenzoic acid (MCPBA, Figure 8.1(b)). From (*E*)-but-2-ene, (2*R**,3*R**)-2,3-dimethyloxirane is formed, while epoxidation of (*Z*)-but-2-ene gives (2*R**,3*S**)-2,3-dimethyl-oxirane.

For the study of reaction mechanisms, the observation of *stereospecificity* is of prime importance, since in case of a stereospecific reaction, transformation of different stereoisomers cannot involve common *intermediate states.*

Taking into consideration the features of the reaction mechanism, the concept of *stereospecificity* can be extended to cases when the starting material exists in a single state only, but out of several stereoisomeric products, the mechanism of the reaction permits only the formation of but a single stereoisomer. For instance, phenylacetylene can react with the zirconium reagent only in one way; therefore, bromination can produce exclusively (*E*)-(2-bromovinyl)benzene.

*Stereochemistry and Stereoselective Synthesis: An Introduction,* First Edition.
László Poppe and Mihály Nógrádi.

Companion Website: www.wiley.com/go/poppe/stereochemistry

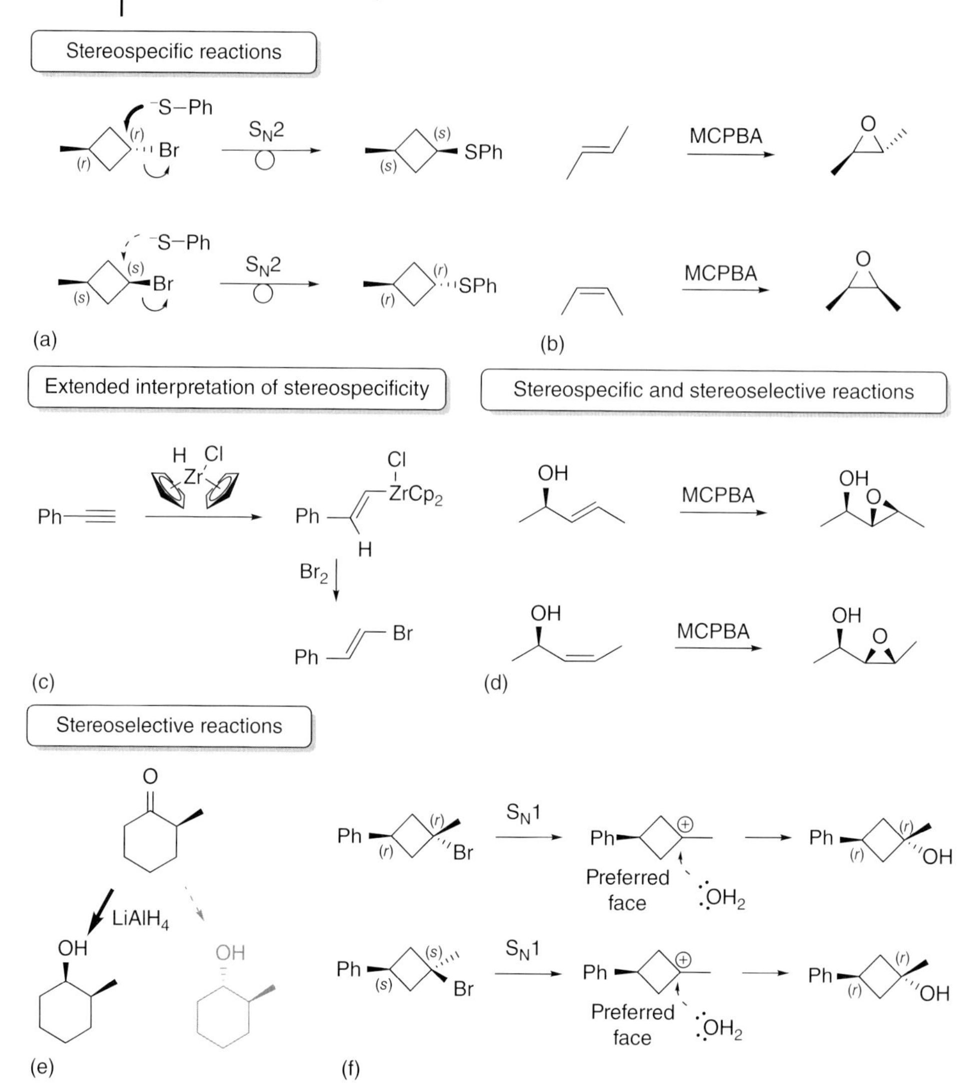


**Figure 8.1** Stereospecific and stereoselective reactions.

## 8.2 Stereoselectivity

In **stereoselective reactions**, a single starting material can be transformed into two or more stereoisomeric products, and the formation of one of them is preferred for some reason, even if only to a small degree. An example is reaction (e) in Figure 8.1, where reduction of 2-methylcyclohexanone with $LiAlH_4$ gives predominantly *cis*-2-methylcyclohexanol.

A reaction is regarded as stereoselective even when two stereoisomers of the starting material lead to the same stereoisomers of the product in the same ratio, provided that this ratio differs from 1 : 1. An example for this situation is the hydrolysis of [(1*r*,3*r*)-3-bromo-3-methylcyclobutyl]benzene and its 1*s*,3*s* diastereomer (Figure 8.1(f)), in which an $S_N1$-type reaction leads as the main product to the same (1*r*,3*r*)-1-methyl-3-phenylcyclobutanol.

A reaction can be both stereoselective and stereospecific (Figure 8.1(f)). Epoxide formation from (*R,E*)- and (*R,Z*)-pent-3-en-2-ol stereoisomers (Figure 8.1(d)) is stereospecific in a way that the relative orientation of the substituents of the double bond is conserved, but stereoselective in the respect that epoxidation is preferred from the direction of the face from which the hydroxyl group is protruding. In this sense, every stereospecific reaction is also stereoselective.

Unfortunately, a "*diluted*" interpretation of the concept of stereospecificity permitting its use for reactions that are not fully stereoselective is also often used. Thus, for example, qualifying a reaction producing an 80 : 20 mixture of stereoisomers as being "80% stereospecific" is incorrect. Since such interpretations do not permit a clear distinction between stereoselectivity and genuine stereospecificity, its use has to be discouraged.

Similarly, it is to be avoided to call a reaction stereospecific if in the reaction the formation of a single stereoisomer is observed because this is a rather relative statement and depends on the analytical method applied. A sample declared to be a pure enantiomer based on its optical rotation might prove to be a 99.8:0.2 mixture of enantiomers when examined with chiral GC techniques. In addition, this manner of interpretation disregards real differences based on reaction mechanisms.

Consequently, in our opinion, the concept of stereospecificity should be strictly confined to cases when it follows from the mechanism of the reaction that only a single one of the conceivable stereoisomers is formed. In all other cases, the concept stereoselectivity should be applied.

## 8.3
## Selective Syntheses of Enantiomers

In the earlier and also in the more recent literature, the term "*asymmetric synthesis*" is used to cover syntheses involving both enantio- and diastereoselective reactions. In addition, manipulations with pure enantiomers originating from the "*chiral pool*"[1] are also often denoted as asymmetric syntheses. Since in our opinion the term "*asymmetric synthesis*" is both restrictive and inaccurate, we recommend to use the term enantiomer or enantiotope selective, whichever applies for the transformation in question. Chemoselectivity and regioselectivity as well as diastereomer selectivity discussed in the preceding chapters are

1) In this context and in the following under "chiral," a *nonracemic* (usually highly enantiopure) chiral substance is understood.

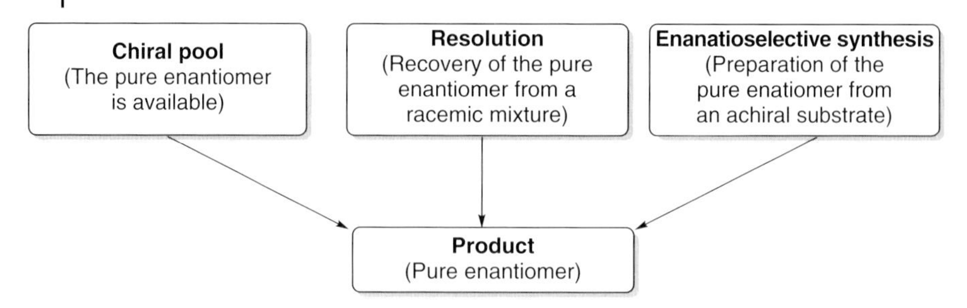


**Figure 8.2** Methods available for the preparation of pure enantiomers of a chiral compound.

"simple" cases of selectivity since constitutional isomers and diastereomers are chemically distinct entities differing in all of their scalar properties. They can be therefore separated rather easily and differ in their achiral interactions too. With enantiomers, the situation is quite different. Since all the scalar characteristics of enantiomers are identical, their separation is a complex task, and to realize enantiomer selectivities, chiral interactions are required.

If one wishes to prepare a pure enantiomer of a chiral compound, multiple means are at our disposal as summarized in Figure 8.2.

Utilization of a **chiral pool** is straightforward. If there exists a substance readily available in an enantiopure form, it can be utilized directly for the preparation of an enantiopure product. The original substance may either be incorporated into the product or only its chirality is exploited. In the latter case, the substance itself does not appear in the product but serves as a chiral aid, reagent, or catalyst. Nowadays, besides compounds isolated from natural sources in an enantiopure form, the available chiral pool is extended by an ever-increasing number of synthetic products.

If a reaction leading to the formation of a new stereogenic element is carried out under achiral conditions, the result is a racemic mixture. Thus, preparation of racemic mixtures does not involve extra difficulties. **Resolution** of such a racemic mixture, that is, recovery of one or both enantiomers in a pure form is more complex and requires the application of some sort of chiral interaction. For resolution, kinetic and thermodynamic processes can equally be utilized (e.g., kinetic resolution, "*classic*" resolution).

***Selective synthesis of enantiomers*** requires the involvement of some chiral interaction. This can be provided by a chiral environment (e.g., a chiral solvent) or some chiral aid. In kinetically controlled reactions, a good chiral aid is one that secures high asymmetry in the transition state of the rate-determining step of the process generating the new stereogenic element.[2)] In thermodynamically controlled reactions, an efficient chiral aid provides significant difference in the

2) In this case, it is important that the chiral auxiliary should only permit the formation of a single transition state. With alternative reaction paths, both selectivity and the possibility to predict it suffer.

stability of the diastereomeric intermediates. For the feasibility and efficiency of enantioselective syntheses, the following features of the process and the chiral aids are highly important:

(i) High selectivity and high yield of the reaction.
(ii) The selective reaction should be applicable for a wide range of substrates.
(iii) The chiral auxiliary reagent/catalyst should be recoverable without racemization.
(iv) The chiral auxiliary reagent/catalyst should be inexpensive and its both enantiomers should be available in an enantiopure form.

In real processes, not all of these requirements can be fulfilled completely and compromises must be considered to find an optimum (see in later chapters).

## 8.4
## Diastereomer Selectivity

Since diastereomers differ in all of their properties, in principle, they are supposed to show a different relationship toward any reagent, although this difference may be small. This means that diastereomers react with any reagent at different rates, that is, always in a diastereomer-selective manner. This selectivity is the more pronounced, the larger the differences between the given diastereomers, the closer the reaction occurs to the stereogenic element involved.

Several types of diastereomer selectivities can be distinguished:

1) Transformation of diastereomers may give the same product, but at different rates, as, for example, in the oxidation of equatorial and axial hydroxyl groups with $CrO_3$ ($k_{ax}/k_{ekv} = 3.2$) as exemplified in Figure 8.3.
2) A reaction starting from a pair of diastereomers may result in another pair of diastereomers at different rates, as in the iodide ion-catalyzed debromination of diastereomeric 2,3-dibromobutanes (Figure 8.4(a)).[3]
   Transformation of different diastereomers at different rates, that is, in a diastereomer-selective manner can also be accomplished using biocatalysts, facilitating thereby the separation of diastereomeric products that have often rather similar properties. A typical example is the separation of the *E/Z*-isomeric pair of methyl esters with the aid of diastereomer-selective hydrolysis using porcine liver esterase (PLE) (Figure 8.4(b)). This enzyme performs selective hydrolysis of the *E* methyl ester, which can be then separated by extraction with aqueous alkali.
3) Selective transformation of diastereomers may even lead to products of different constitution. A classic example is the *Beckmann* rearrangement of (*E/Z*)-isomeric oximes, which gives in a diastereomer-selective reaction constitutional isomer amides (Figure 8.5).

3) It follows from its mechanism that this is essentially a diastereomer-specific reaction.

Figure 8.3 Formation of the same product from a pair of diastereomers at different rates.

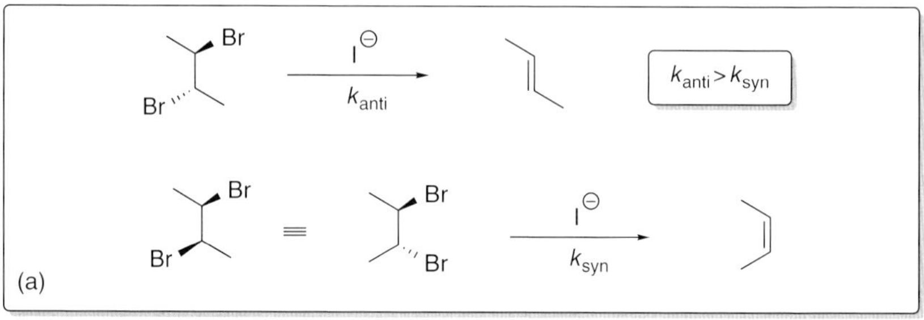

Figure 8.4 Formation of diastereomers at different rates starting from a mixture of diastereomers in a chemical (a) and a biocatalytic process (b).

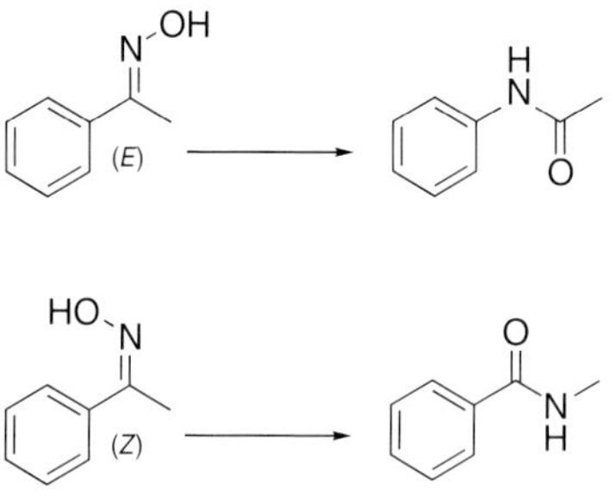

Figure 8.5 Selective transformation of diastereomers to products of different constitution.

## 8.5
## Diastereotope Selectivity

When discussing, in Section 2.2.3, the concepts concerning the relationship of specific parts of a molecule and the manner in which a molecule can be approached (topicity), we have described that exchange of one or the other diastereotopic atom or group for a new one produces different diastereomers and in principle in a ratio other than 1 : 1 (Figure 2.13). Similarly, in the case of diastereotopic faces, addition from different diastereotopic faces produces different diastereomers, also, in principle in a ratio other than 1 : 1 (Figure 2.18). In accordance with these considerations, a certain degree of diastereotope selectivity can be achieved with any reagent or catalyst. High selectivity, of course, requires special conditions.

### 8.5.1
### Diastereotope Selectivity with the Formation of a New Chiral Element

Diastereotope selectivity of a reaction can be conveniently followed, since the ratio of racemates related as diastereomers (e.g., the ratio $[RR] + [SS]/[RS] + [SR]$) can be readily analyzed by routine methods. Diastereotope selectivity assumes that there is already some kind of asymmetry within the molecule, which controls the formation of a new site of asymmetry (**asymmetric induction**).[4] If the reaction starts from a racemic mixture and is performed with an *achiral reagent (catalyst)* and there is no molecular interaction between the molecules of the substrate, enantiomers of the substrate are transformed at the same rate and the diastereomers are formed therefore in equal amounts:

$$[RR] + [SS]/[RS] + [SR] = [RR]/[RS] = [SS]/[SR] \tag{8.1}$$

In a *diastereotope selective reaction carried out with an achiral reagent (catalyst)* (**A**), the ratio of diastereomers (in other words, the magnitude of asymmetric induction[4]) is thus independent of the enantiomeric composition of the starting material (Figure 8.6(a)). It follows that the corresponding transition states are pairwise in a diastereomeric relationship, and therefore the ratio of diastereomers of the racemic products, that is, $[(R,R) - P + (S,S) - P]/[(R,S) - P + (S,R) - P]$, only depends on the ratio $k_1/k_2$ and is not influenced by the initial ratio of the enantiomers.

This case is illustrated by the diastereotope selective reduction of 2-methylcyclohexanone (Figure 8.7). Using an achiral reagent (e.g., $NaBH_4$), only diastereotope selectivity is manifested and independently of the initial enantiomeric composition, the same *cis/trans* ratio is obtained. As it can be expected, reduction proceeds preferentially from the face opposite to the methyl group and is therefore *cis*-selective.

In *diastereotope-selective reactions carried out with a chiral reagent (catalyst)* (A*, Figure 8.6(b)), the four transition states leading to the four products are all different, that is, comparing any two of them, their relationship is diastereomeric. Consequently, the four reaction rates are different and the diastereomer ratio in the product depends on the enantiomer composition of the substrate. Therefore, in such case, the *cis/trans* ratio in Figure 8.7 will be different whether one starts from one or the other enantiomer. Further, even when one starts from a racemic

4) Asymmetric induction was first recognized by H. E. Fischer in 1894 when studying the addition of HCN on aldoses.

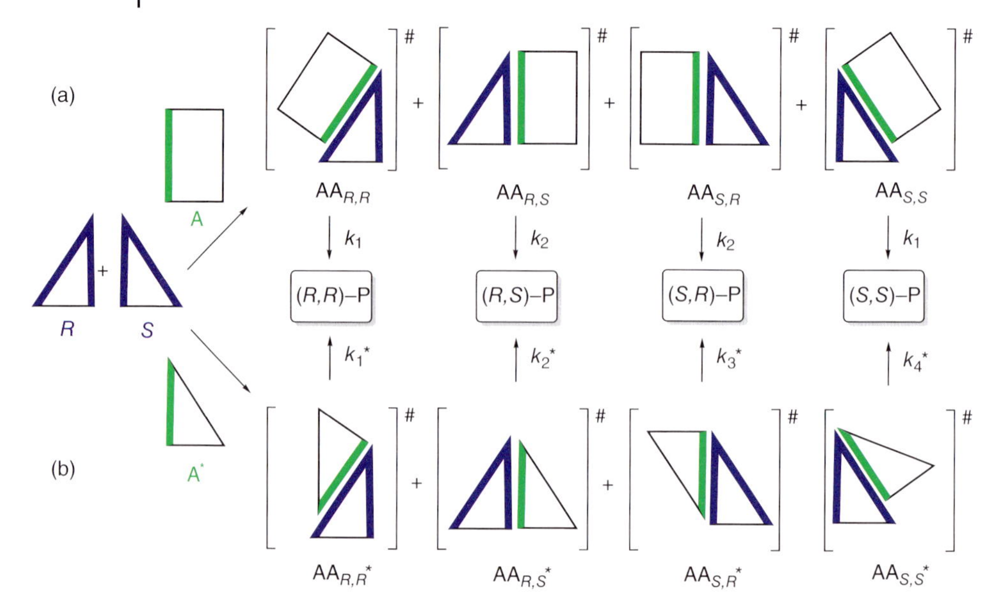


**Figure 8.6** Diastereotope-selective transformation of a racemic substrate with (a) an achiral or (b) a chiral reagent (catalyst).

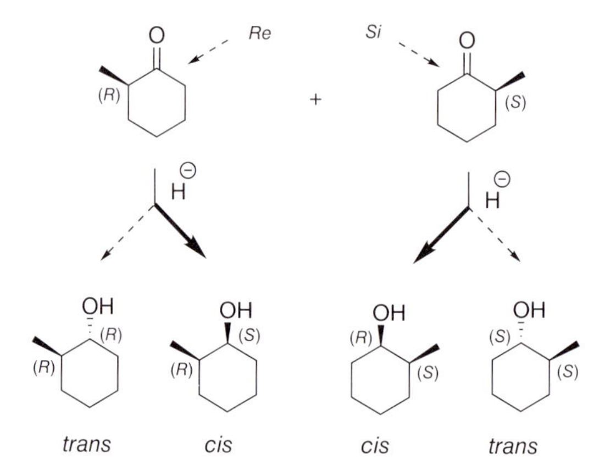


**Figure 8.7** Diastereotope-selective reduction of a racemic ketone.

mixture, the diastereomers formed will be optically active. It is understood that the attack occurs preferentially from the face opposite to the methyl group and is therefore *cis*-selective.

Diastereoselective reactions of racemic substrates having diastereotopic groups or faces yield a mixture of racemic diastereomers. Combined diastereotope and enantiomer selectivity can, however, be realized under chiral conditions, for example, using a chiral reagent. Thus, if the reductions shown in Figure 8.7 are carried out using (−)-diisopinocampheylborane, the *cis* diastereomer formed in a 92% yield contains the (1*R*,2*S*)-enantiomer in 1.8% excess, while in the *trans* isomer, obtained in 8% yield, the excess of the (1*S*,2*S*)-enantiomer was 13.3%. This phenomenon is called **double asymmetric induction** and becomes significant when in a synthetic sequence the starting material is a pure enantiomer and is reacted with a chiral reagent. In this case, the two enantiomers of the reagent will produce different diastereomer ratios. The combination giving a higher ratio of the desired diastereomer is called a *matched pair*, while the other one a *mismatched pair.*

### 8.5.2 Chiral Auxiliary Groups

In synthetic chemistry, chiral auxiliary groups are employed when starting from an achiral substrate; the aim is to achieve stereoselectivity at a newly formed stereogenic center without the use of chiral reagents or catalysts. This can be realized by linking the substrate containing enantiotopic groups or faces to a *chiral auxiliary group*. In this way, the environment of the prochiral center to be transformed is rendered diastereotopic and the diastereotopic faces or groups (Figure 8.8(a) and (b)) become, as described earlier, distinguishable without a chiral agent.

Usually, the chiral auxiliary group is removed at a later stage of the synthesis, and thus a nonracemic mixture of enantiomers, ideally even a pure enantiomer, can be obtained. Since the origin of the chirality of the product is in fact the chiral auxiliary, this strategy is often called ***chirality transfer***. Depending on the method of its removal, the chiral auxiliary group may be *recoverable* or *nonrecoverable*.

Ideally, when applying a recoverable chiral auxiliary group, it can be isolated in its original form and can be recycled directly (Figure 8.9). An important requirement for its reuse is that no racemization should accompany its introduction or

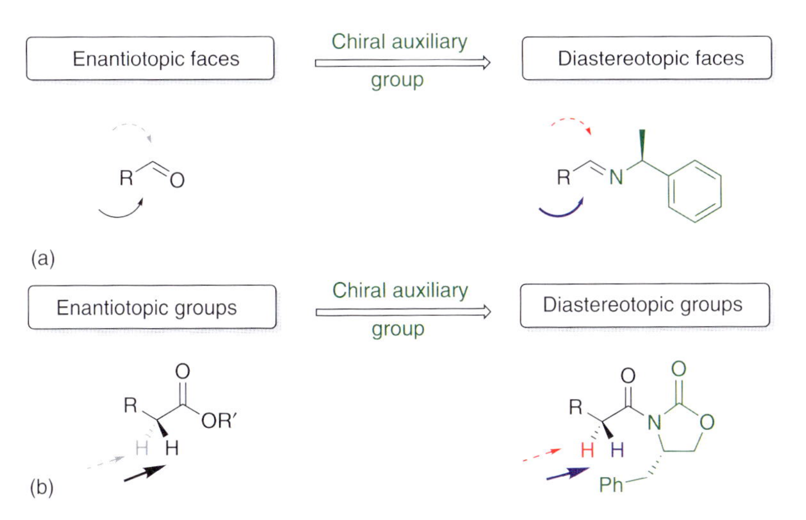


Figure 8.8 Changing enantiotopicity to diastereotopicity using a chiral auxiliary group.

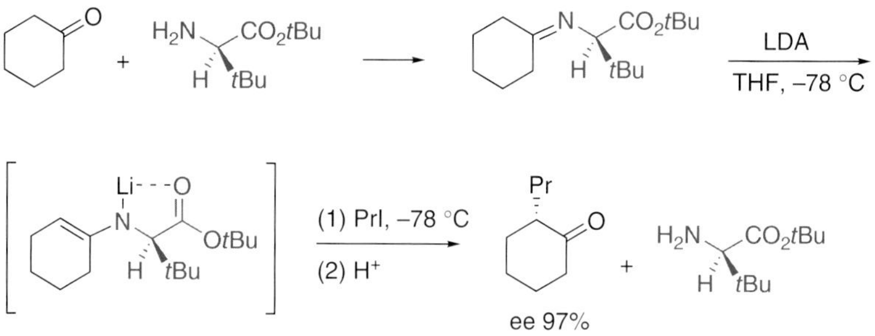

Figure 8.9 Use of a recoverable chiral auxiliary group.

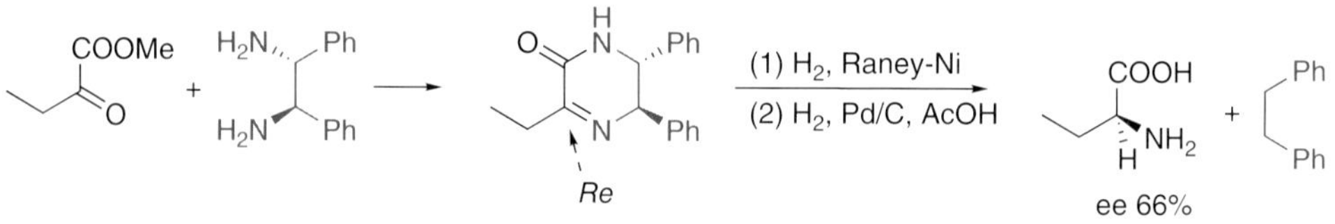

Figure 8.10 Use of a nonrecoverable chiral auxiliary group.

removal/recovery. When the chiral auxiliary group cannot be recovered (e.g., Figure 8.10), it is of prime importance that it should be inexpensive.

Note that reaction with a chiral auxiliary of compounds having enantiotopic faces gives often a mixture of *E/Z* isomers. Since there may be a significant difference between the stereoselectivity of the *E/Z* isomers, their formation should be, as far as possible, avoided. Occasionally, formation of *E/Z* mixtures is precluded as in the case of cyclohexanone (Figure 8.9) by the symmetry of the auxiliary or ring formation (Figure 8.10).

In the stereoselective step of the process involving a chiral auxiliary group, the product is necessarily a mixture of diastereomers, which becomes a mixture of enantiomers after having removed the auxiliary group. The importance of the separability of this mixture will be demonstrated in Section 8.5.4.

### 8.5.3 Enantiodivergent Synthetic Strategy Employing Chiral Auxiliary Groups

As discussed in Section 8.3, in an ideal situation, both enantiomers of a chiral auxiliary are available as the pure enantiomers. If this is not the case, or one enantiomer is prohibitively expensive, nevertheless both enantiomers of the end product are needed; the s.c. **enantiodivergent strategy** may become useful (Figure 8.11).

The essence of the strategy is that one may choose which one of the different functional groups of the end product should be elaborated first, exploiting the

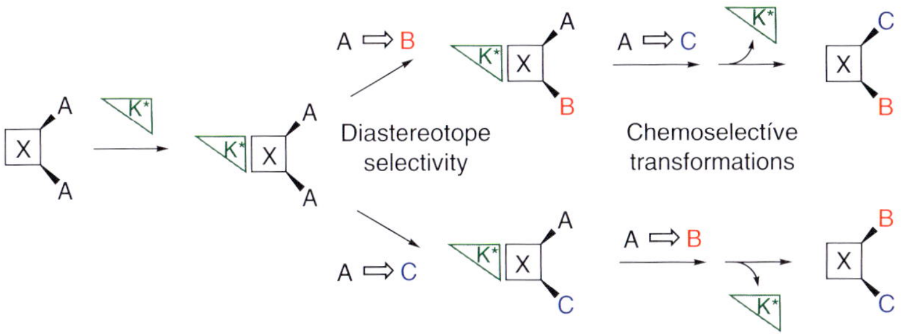

Figure 8.11 Enantiodivergent transformations using one enantiomer of the chiral auxiliary for the preparation of both enantiomers of the target compound.

diastereotope selectivity secured by the chiral auxiliary group. It can be seen in Figure 8.11 that utilizing the same type of diastereotope selectivity, group **A** can be exchanged for either group **B** or group **C**. After that, the remaining group **A** can be replaced, again in a targetable manner for either group **B**, or group **C**. Now it is sufficient to exploit the chemical difference of the groups. After removal of the chiral auxiliary group, an optional enantiomer of a given product can be prepared (cf. an example in Chapter 4).

### 8.5.4
### Significance of Separating the Diastereomers Formed

In the previous section, we emphasized that in syntheses employing a chiral auxiliary group, separation of the diastereomeric products should be feasible. This is illustrated in Figures 8.12 and 8.13.

It is not surprising that after the separation of diastereomers, the enantiomeric excess of the product (and also that of the "*unwanted*" diastereomer) is identical to the enantiomeric excess of the chiral auxiliary compound used.

Note, however, that while one of the diastereomers formed is richer in one form of the newly formed stereocenter (ee 98%), the enantiomeric excess of the

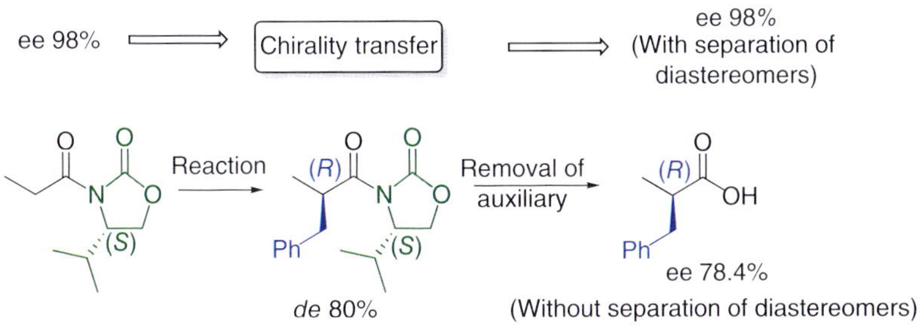

Figure 8.12 Enantiomeric composition of the end product with or without diastereomer separation.

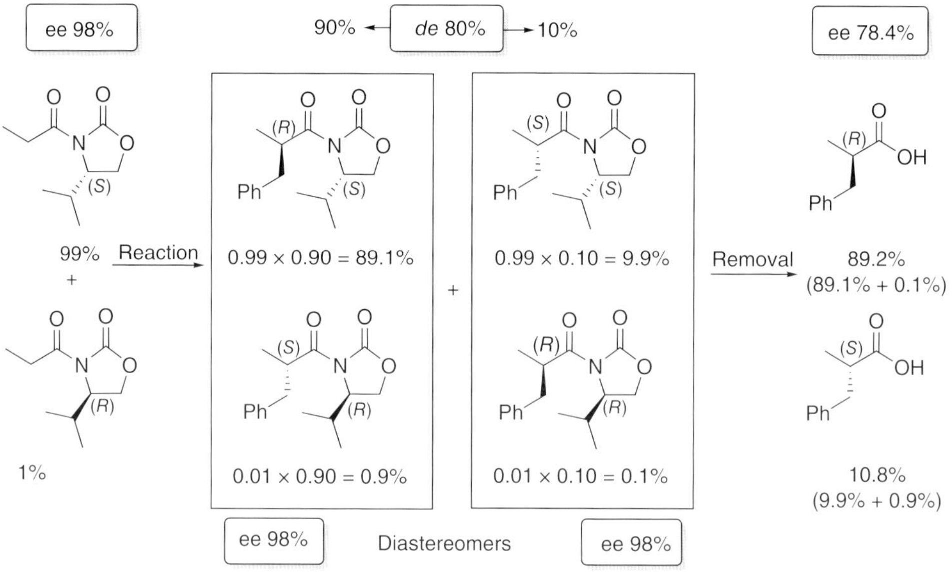

**Figure 8.13** Enantiomeric composition of the product of a process with no diastereomer separation.

oppositely configured stereocenter is the same (ee 98%). It follows that if the chiral auxiliary group is removed without separation of the primary diastereomeric products, it is not the original enantiomeric excess of the chiral auxiliary that is transferred to the end product (ee 98%), but a value diminished by the diastereotope selectivity of the reaction, that is, ee $0.98 \times 0.80 = 78.4\%$.

Significance of the separability of diastereomers in processes employing chiral auxiliaries is demonstrated in Figures 8.12 and 8.13. In processes involving a chiral auxiliary, namely, a "**transfer of chirality**" is taking place. Therefore, if the diastereomers formed are separated and the diastereomer containing more of the "bad" enantiomer is discarded, the enantiomeric excess of the auxiliary can be fully transferred to the product. Since here in a complex process a "transfer of chirality" is accomplished and not the diastereotope selectivity of the reaction but the separability of the diastereomers formed is exploited. (Note that the magnitude of diastereomer selectivity does not influence the enantiomeric excess of the product.)

As long as the diastereomeric products of the reaction are separable, the enantiomeric excess of the product conserves the enantiomeric excess of the chiral auxiliary even in the case of no diastereotope selectivity (de 0%), but without further interference it cannot be enhanced. This points out that optimally the chiral auxiliary should be enantiopure.

(R) OH — ee 92% —$(COCl)_2$→ Diastereomer separation → (R,R)-Diester [+ (R,S)-Diester + (S,S)-Diester] —Hydrolysis→ (R) OH — 78%, ee 99.6%

**Figure 8.14** Enhancing the enantiomeric excess of (*R*)-1-(naphth-2-yl)ethanol by Horeau's method.

#### 8.5.4.1 Purification of Enantiomers by the Separation of Diastereomers Formed with the Aid of an Achiral Bidentate Reagent

Separation of diastereomers may be important not solely in the abovementioned procedures, but the enantiomeric excess of a product formed along similar principles can be improved significantly as well (Figure 8.14).[5]

Figure 8.15 illustrates with the aid of an idealized procedure the principles along which the enantiomeric excess of (*R*)-1-phenylethanol can be increased from moderate to high enantiomeric excess (ee 80–97.6%) without a chiral reagent simply by forming a derivative with an achiral bidentate reagent (e.g., oxalyl chloride), followed by separating the diastereomers obtained and finally recovering 1-phenylethanol by hydrolysis.

The key step here is again the separability of diastereomers. On the expense of the *meso*–diastereomer, a racemic alcohol is formed in 18% yield, and (*R*)-1-phenylethanol can be recovered in an acceptable (82%) yield and high ee (97.6%).

### 8.5.5 Diastereotopic Version of Asymmetric Transformation by Induced Crystallization (CIAT) Involving Diastereotopic Interactions

In this part, we wish to emphasize that the selectivities of an elementary step and that of the overall process are not identical, and thus the final overall product ratio ("*selectivity*") can deviate from the ratios of the individual steps. It follows that the enantiomeric excess of a product recovered at the end of a diastereotope selective series of transformations does not correspond necessarily to the ratio in which the diastereomers are formed in the elementary diastereotope selective steps.

Reasons for this can be as follows: (i) diastereomers interconvert in a dynamic equilibrium (epimerization), (ii) during workup, one of the diastereomers of the mixture is enriched, or (iii) the product racemizes in the course of removal of the chiral auxiliary group. It is apparent that processes under (iii) diminishing the enantiomeric excess of the product should be avoided.

Case (ii) corresponds to "chirality transfer" processes accomplished using chiral auxiliary groups and presented in the previous section. These are suitable for "transmitting" the enantiomeric excess of the chiral auxiliary to the product and

5) The method is used to be called Horeau's principle [1].

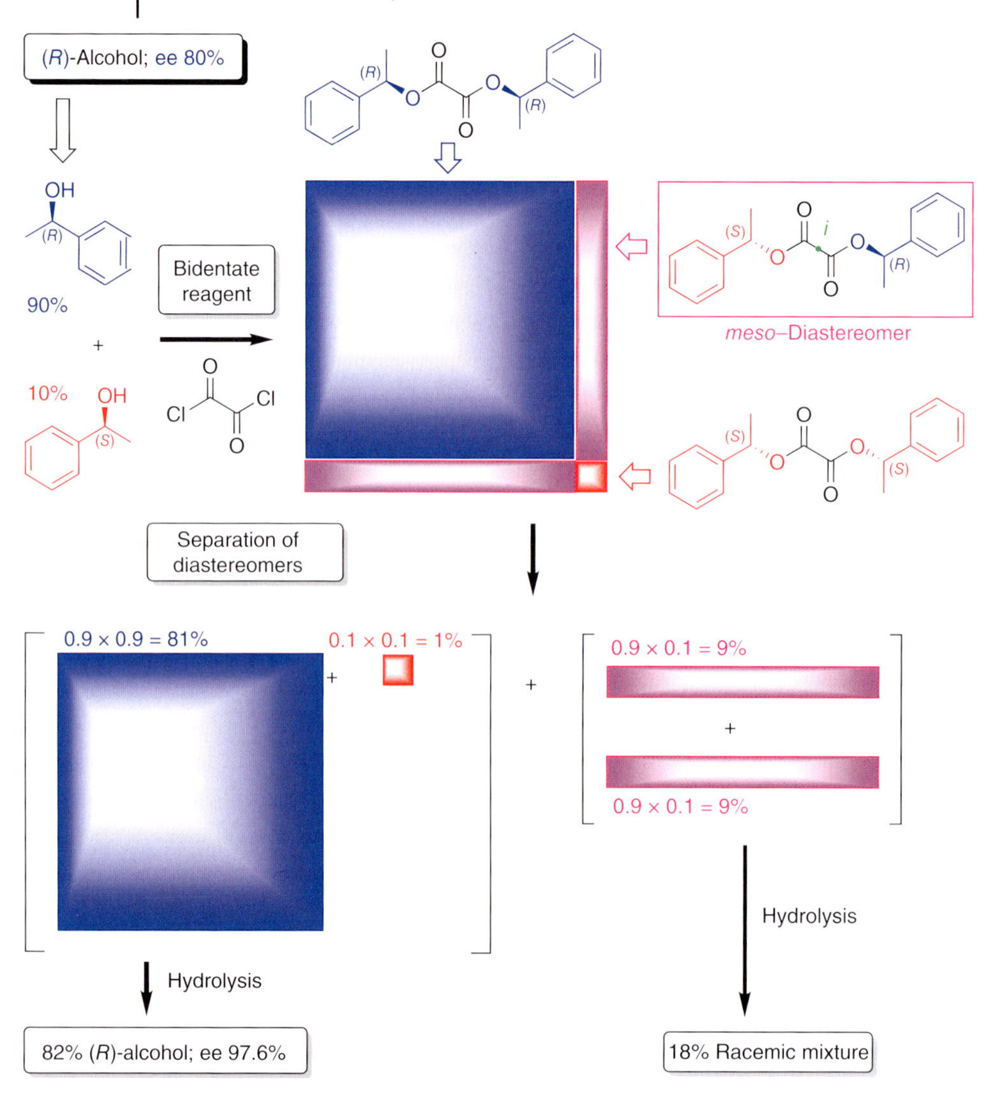


**Figure 8.15** Upgrading enantiomeric excess by separating diastereomers formed with an achiral bidentate reagent (*Horeau's* method).

preserve it even when the diastereotope selectivity of the reaction is poor. After separating the diastereomers, the one containing more of the unwanted enantiomer of the final product is discarded.

Processes corresponding to case (i) can also be useful (Figure 8.16). In *crystallization-induced asymmetric transformations* (CIAT), using a chiral auxiliary an achiral starting material (A) is transformed to a chiral product having diastereotopic groups or faces (C). In the process, diastereomeric products ($dP_R$

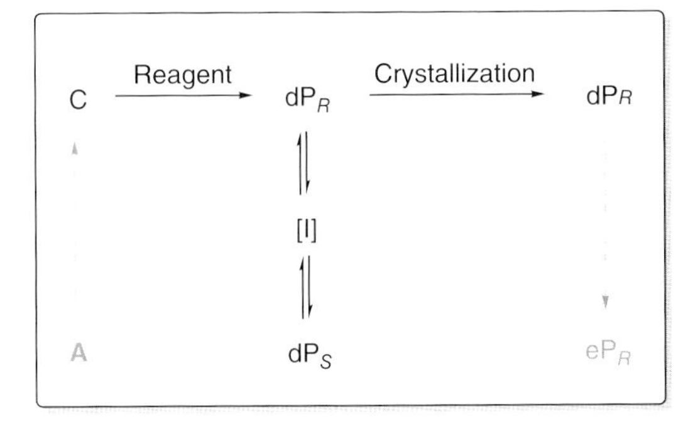


**Figure 8.16** Crystallization-induced asymmetric transformation (CIAT) with diastereotope selectivity.

and dP$_S$) may equilibrate with each other by epimerization via an intermediate (I). Precipitation of the more readily crystallizing product (in the present example dP$_R$) shifts the equilibrium in favor of the latter. In this process, it becomes possible to transfer to the product enantiomer (eP$_R$) the enantiomeric excess of the chiral auxiliary, even in quantitative yield.

To processes presented in Figure 8.16, various names are given in the chemical literature. In the older literature, diastereotope selectivity associated with concurrent epimerization and crystallization (and often also enantiotope selectivity manifested in a similar manner) were termed as **asymmetric induction of second kind** or **asymmetric transformation of second kind**. Later, the same process was also named as **CIAT** not differentiating enantio- and diastereotope selective processes. For similar dynamic stereoselective processes, several different terms were used (asymmetric transformation of the second kind, dynamic kinetic resolution (DKR), dynamic transformation, diastereoselective resolution, dynamic diastereoselective process, deracemization, and asymmetric disequilibrating transformation; [2]). In view of the rather confusing terminology of this area, it is advisable to consider not only the terms used but also the details of the partial processes and their stereoselective features.

Process(es) shown in Figure 8.16 are essentially rather similar to those already discussed in connection with chiral auxiliary groups in Section 8.5.2. In these, a chiral auxiliary group is linked to the often achiral substrate (**A**). In the derivative, the environment of the prochiral center intended to be reacted becomes diastereotopic. In descriptions of the dynamic process, this step (shown in gray in the figure) is often omitted, but they just start with the compound linked to the chiral auxiliary group (Figure 8.16, **C**). Formation of the diastereomeric products in the reaction (dP$_R$ and dP$_S$) proceeds with a certain diastereotope selectivity. In accordance with what has been discussed earlier, in this case, a "transfer of chirality" takes place and the enantiomeric excess of the chiral auxiliary group appears in the newly formed stereogenic element of the diastereomeric products. Should the process be coupled with an epimerization equilibrium of the products via an intermediate state (**I**) and crystallization of one of the diastereomers ($dP_R$) is preferred, crystals with an enantiomeric excess equal to that of the chiral auxiliary can be isolated. In favorable situations, even a quantitative yield of the end product can be realized ($dP_R$) independently of the diastereotope selectivity in the step producing the diastereomers. By removing the chiral auxiliary, the required enantiomer is obtained. Since also the removal of the auxiliary group is often not mentioned, this step is also drawn in gray.

The procedure is of course not restricted to the preparation of the *R* product. The *S* product can also be obtained provided that it is the $dP_S$ product that crystallizes preferentially.

An example for such a process is the preparation of (*S*)-α-methyl-phenylalanine (Figure 8.17). Addition of NaCN on to the Schiff's base creates a new stereocenter. Crystallization of the diastereomeric adduct containing an *S* center is favored, and therefore this is what can be exclusively isolated. The fact that from the crystallizing diastereomer practically one of the enantiomers can be obtained pure does not give a clue neither about the magnitude of the diastereotope selectivity of the addition nor about the state of the equilibrium of the diastereomers in solution.[6)]

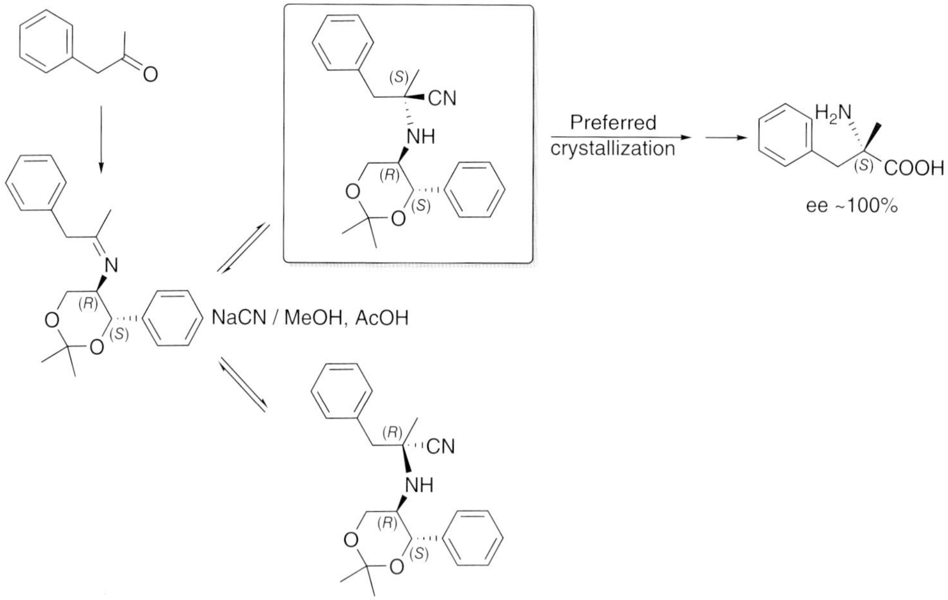

**Figure 8.17** Crystallization-induced asymmetric transformation (CIAT) in the preparation of (*S*)-α-methyl-phenylalanine.

6) The two diastereomers form in chloroform a 2 : 1 equilibrium mixture [3].

### 8.5.6
### Diastereotope Selectivity in Biotransformations

As any other tool (reagents, catalysts), diastereotope selectivity of biocatalysts can also be exploited for differentiation between diastereotopic groups or faces. Earlier, we have seen that in chemical systems, it is foremost the relative steric disposition of the reacting part of the molecule that determines diastereotope selectivity. In contrast, in biocatalytic processes, the stereochemical outcome is mainly controlled by the shape and orientation of the entire substrate molecule. This is illustrated by the following two examples.

#### 8.5.6.1 Selective Biotransformations of Diastereotopic Groups

Biocatalysts are suitable for the selective transformation of diastereotopic groups in a natural way. In Figure 8.18, a case is presented in which a biocatalyst, in this case the filamentous fungus *Beauveria sulfurescens,* is capable for the selective oxidation of the diastereotopic methyl groups in both enantiomers of a tricyclic compound. Thus, the enantiomers lead to different diastereomers (both in high enantiopurity).

#### 8.5.6.2 Selective Biotransformation of Diastereotopic Faces

It is not surprising either that biocatalysts can be used to distinguish between diastereotopic faces. In the reaction depicted in Figure 8.19, diastereotope-selective biotransformations can be efficiently implemented for the preparation of polyhydroxy compounds using fructose-1,6-diphosphate aldolase (FDPA). In this remarkable transformation, two new centers of chirality are generated simultaneously. The conspicuous feature of FDPA is that in the catalytic process,

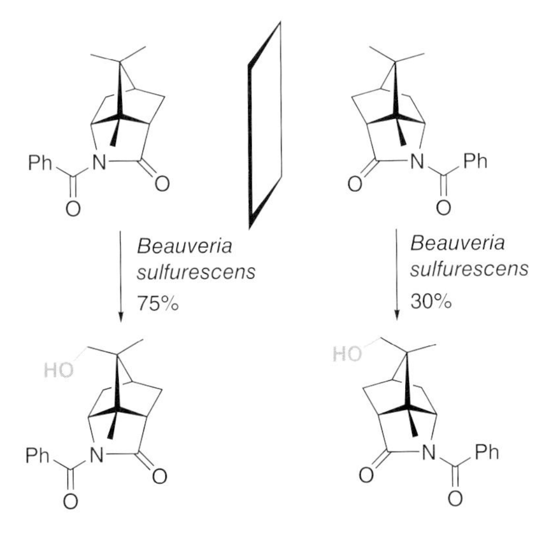


**Figure 8.18** Selective biotransformations of diastereotopic methyl groups.

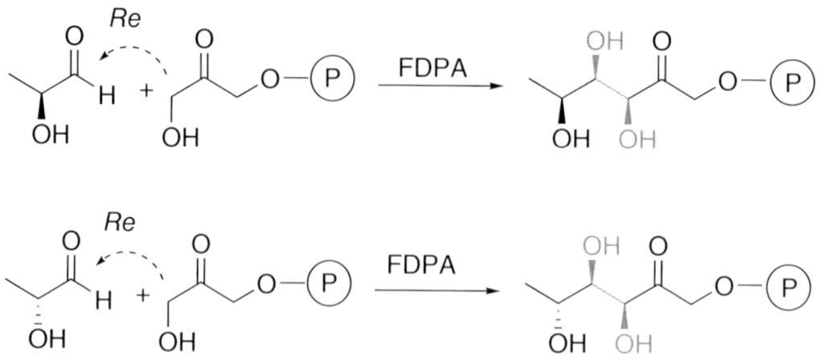

Figure 8.19 Diastereotope-selective addition using an aldolase enzyme.

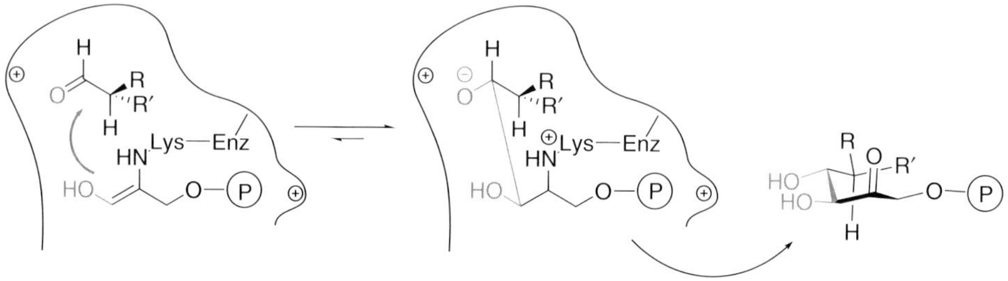

Figure 8.20 Effects controlling selectivity in an aldolase-catalyzed addition.

diastereotope selectivity is not controlled by the chiral center present in lactaldehyde, but it is the enzyme that shows a strictly predetermined preference for the absolute and relative configuration (2*S* and 2,3-*syn*) of the product (Figure 8.20).

## 8.6 Enantiomer Selectivity

Stereoselectivities associated with diastereomers (diastereomer and diastereotope selectivity, cf. Sections 8.4 and 8.5) are relatively "*simple*" cases of stereoselectivity. Diastereomers differ namely in all of their properties and achiral interactions and can be therefore relatively easily separated. The situation with stereoselectivity involving enantiomers is basically different. Since their scalar properties are identical, their separation is a complex task and selectivities affecting enantiomers require chiral interactions.

A possible method for the separation of enantiomers without chemical or biotransformations is based on physicochemical interactions with a chiral system (most often with s.c. resolving agents forming diastereomeric salts or complexes). Such methods fall under the general term **classic resolution.** Since classic resolutions have been exhaustively reviewed,[7] and our main topic is stereoselective reactions, in this work, such methods will not be discussed in detail.

Selectivities based on the difference in the rate of transformation of the individual enantiomers can be named, as proposed in Section 3.2, **enantiomer selectivity**. In the literature to this type of selectivity and processes based on it, several other names were given. In this work, we strive to call this type of selectivity consistently as *enantiomer selectivity*, while processes exploiting enantiomer selectivity will be termed *kinetic resolutions.*

### 8.6.1
### Degree of Enantiomer Selectivity in Irreversible Processes

**Kinetic resolution** is a process based on an enantiomer-selective reaction (Figure 8.21(a)). An advantage of this approach is that starting from a racemic mixture and using a reagent or catalyst with one sense of chirality, both enantiomers can be obtained. After completion of the transformation, one enantiomer of the substrate remains unchanged, while the other is transformed to a compound of different constitution, which can be then readily separated.

A shortcoming of simple kinetic resolution based on an enantiomer-selective reaction is that without additional procedures, in principle, no more than 50% can be obtained of the desired enantiomer (the useful enantiomer), while the undesired one is useless and can be regarded as a burdensome by-product.

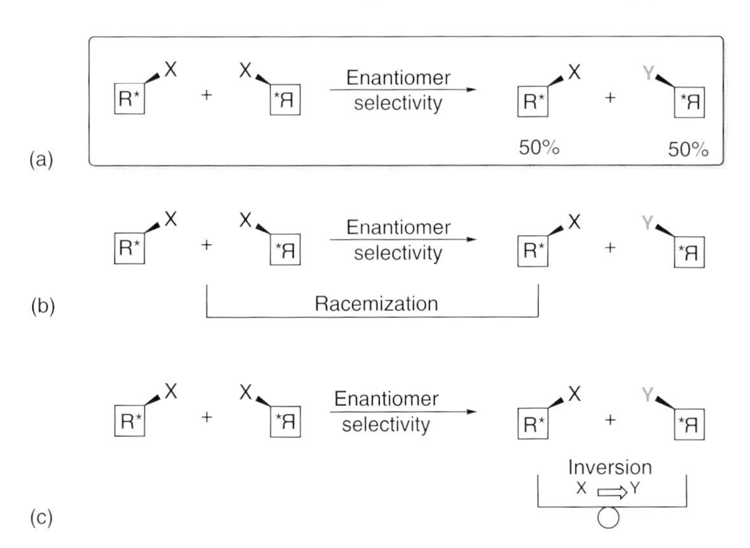


**Figure 8.21** An irreversible enantiomer-selective transformation (a) and enhancing the amount of the desired product by racemization of the unreacted enantiomer (b) or its transformation involving a configurational inversion (c).

7) About this topic, apart from the monograph by Eliel and Wilen [4], the handbook by Kozma [5] is recommended.

There are two possibilities available to circumvent this problem. One is **racemization** of the unwanted enantiomer and recycling the racemic product into the enantiomer-selective procedure (Figure 8.21(b)). In this case, the separation of the products and racemization of the recovered unwanted enantiomer has to be carried out in separate operations. A more efficient solution of the problem is the s.c. **DKR** when the racemization of the unreacted substrate takes place *in situ* parallel to the enantiomer-selective process (it is important that this racemization should not affect the product formed).

The other option is that the remaining unwanted enantiomer is transformed to the desired enantiomer in a process involving a configurational inversion (Figure 8.21(c)). Such processes involving also an inversion process have to be accomplished in separate steps.[8] If the product of the kinetic resolution process remains unaffected by the inversion reaction of the remaining enantiomer, separation of the mixture obtained in the kinetic resolution need not be separated.

The degree of the **enantiomer selectivity** ($E$) of the reaction shown in Figure 8.21 can be best characterized by the ratio of the two reaction rates:

$$E = k_R/k_S \tag{8.2}$$

For the characterization of the degree of selectivity, the enantiomeric excess of the product and the unreacted substrate ($ee_S$ and $ee_P$, respectively) is insufficient because in a reaction of a given enantiomer selectivity $ee_S$ and $ee_P$ are also dependent on the conversion ($c$) (Figure 8.23). The number characteristic for enantiomer selectivity ($E$) can be calculated as a function of any two of the threesome of enantiomeric excess of product ($ee_P$), remaining substrate ($ee_S$), and conversion ($c$) [6, 7] (Eqs. (9.3)–(9.5)).[9]

$$E_{c,ee_P} = \ln[1 - c(1 + ee_P)]/\ln[1 - c(1 - ee_P)] \tag{8.3}$$

$$E_{c,ee_S} = \ln[(1 - c)(1 - ee_S)]/\ln[(1 - c)(1 + ee_S)] \tag{8.4}$$

$$E_{ee_S,ee_P} = \ln[(1 - ee_S)/(1 + ee_S/ee_P]/\ln[(1 + ee_S)/(1 + ee_S/ee_P)] \tag{8.5}$$

The value of enantiomer selectivity ($E$) in genuinely irreversible-selective processes is constant and, contrary to the enantiomeric excess of the substrate and product fractions ($ee_S$ and $ee_P$, respectively), it is independent of conversion ($c$).

Although it is true with any type of selectivity that in kinetic processes selectivity cannot be exploited for an efficient differentiation between heterotopic groups/faces if the process is reversible, this phenomenon will be discussed in connection with enantiomer-selective processes in this section. In Figure 8.22, reversible and irreversible enantiomer-selective processes are shown.

8) Apparently, inversion running parallel to the enantiomer-selective transformation must be selective in relation to the unwanted enantiomer.

9) In the course of kinetic resolution by biocatalysis, occasionally, a concurrent nonselective chemical reaction or racemization during workup may decrease the enantiomeric excess of the product as compared to the inherent selectivity of the biotransformation.

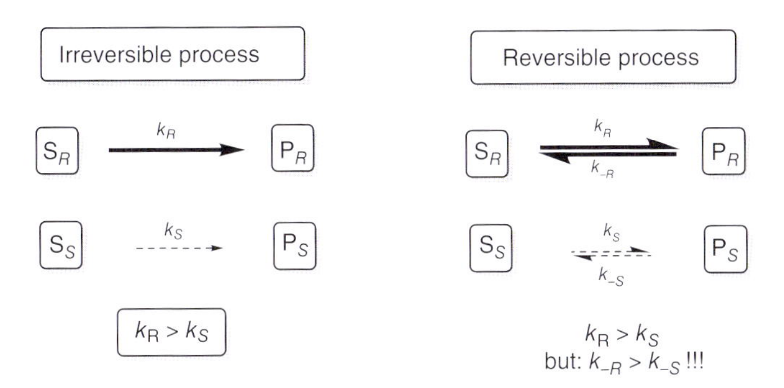


**Figure 8.22** Reversible and irreversible enantiomer-selective processes.

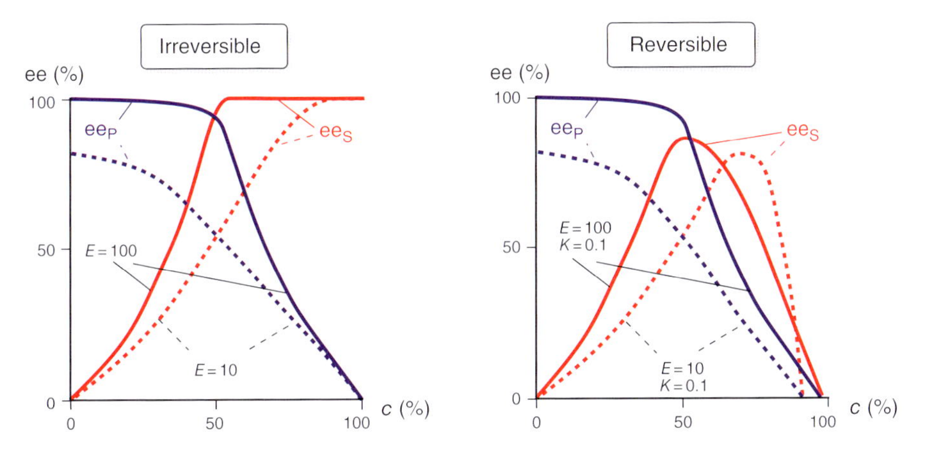


**Figure 8.23** Enantiomeric excess values of substrate (*S*) and product (*P*) in kinetic resolution under irreversible and reversible conditions.

On inspection of Figure 8.22, it is apparent why it is essential that enantiomer-selective processes should be irreversible. Substrates and products are pairwise both enantiomers. Let us assume that in a reversible process it is the *"forward"* reaction of the *R* enantiomer of the substrate that reacts faster (the pertinent transition state being of lower energy), then it is also the *R* enantiomer of the product that is reverting faster. Consequently, in due course, the minor *S* enantiomer will be enriched in the product. This fact is demonstrated in Figure 8.23 in which enantiomeric excess values of substrate and product ($ee_S$ and $ee_P$) are plotted as a function of conversion (*c*).

Figure 8.23 clearly demonstrates that in a reversible process (in our example, the equilibrium constant is $K = 0.1$), complete enantiopurity of the substrate fraction cannot be achieved even with high enantiomer selectivities. Increasing the conversion does not improve the situation. The diagrams also show that selectivity of enantiomer-selective processes cannot be characterized solely by ee values of one of the fractions. If our objective is to obtain by kinetic resolution in an irreversible reaction either a single enantiomer or both enantiomers of the product in adequate purity (ee > 95%) and yield (>45%, calculated on the racemic substrate), the value of $E$ should exceed 100. Highly enantiomer-selective reactions ($E \gg 100$) are suitable to transform one enantiomer of a difficult-to-separate racemate to a constitutionally different but (under optimal conditions) enantiopure product that can be then readily separated from the remaining pure enantiomer of the unreacted substrate.

In other words, enantiomer selectivity can be exploited in kinetics-controlled process. Reversibility allows thermodynamic control in which the product ratio depends on the relative *Gibbs* free energies of the products. Since *Gibbs* free energies of enantiomers are identical, any degree of reversibility in long term ($c = 100\%$) results in racemic mixtures.

## 8.6.2
## Kinetic Resolution Using Chemical Systems

Since with chemical catalysts and reagents usually only poor enantiomer selectivities can be achieved, the following consideration may be important. If our objective is restricted to the preparation of only one of the enantiomers in high purity, we can exploit the fact that in irreversible enantiomer-selective reactions, the enantiomeric excess of the remaining material can be (on the expense of yields) relatively high even when selectivity is poor (Figure 8.24). The figure demonstrates that with a modest selectivity of $E = 25$, a substrate fraction of ee = 98% can be isolated in about 40% yield. Under conditions of rather poor selectivity ($E = 4$), it is possible to prepare a substrate fraction of ee = 98%, although the yield drops to 7%. This demonstrates that, if it is possible, the procedure should be conducted in a way that the required enantiomer should be the unreacted substrate.

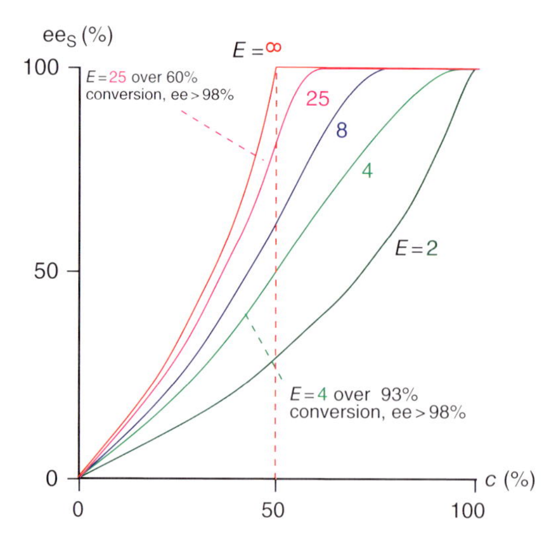


**Figure 8.24** Enantiomeric excess (ee$_S$) of the unreacted substrate as a function of enantiomer selectivity ($E$) in kinetic resolution.

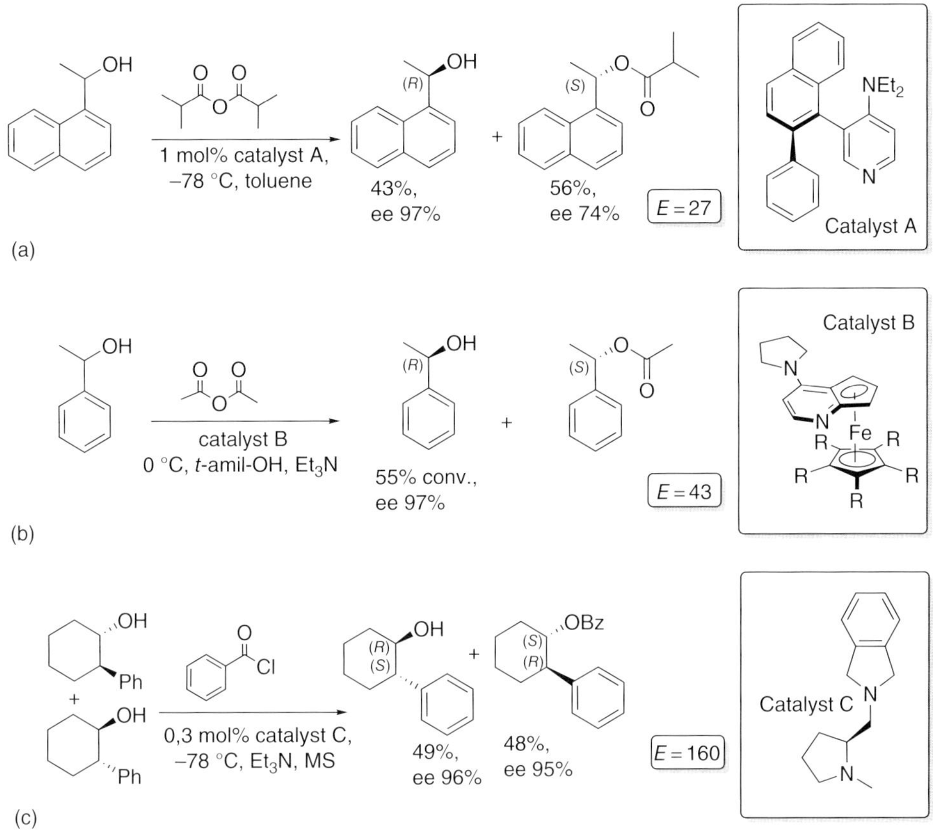

Figure 8.25 Kinetic resolution of racemic alcohols in chemical systems by acylation.

As an example for kinetic resolution employing a chemical system, base catalyzed acylation of racemic secondary alcohols is presented (Figure 8.25). Acylation of racemic 1-(naphth-1-yl)ethanol with isobutyric anhydride was catalyzed with an axially chiral pyridine derivative. When carried out at −78 °C, the reaction proceeded with acceptable selectivity ($E = 27$); the unreacted enantiomer could be recovered in good yield (43%) and high enantiomeric excess (ee = 97%) (Figure 8.25(a)).

Acylation of 1-(naphth-1-yl)ethanol with acetic anhydride proceeded – when catalyzed with the planar chiral iron complex – with a selectivity of $E = 43$ and gave at moderately low temperature (0 °C) and 55% conversion the unchanged substrate, in high enantiomeric excess (ee = 97%) (Figure 8.25(b)).

High selectivity has been achieved even with a relatively simple catalyst (Figure 8.25(c)). Catalyzing the acylation of *trans*-2-phenylcyclohexanol with a small amount of a chiral proline derivative at low temperature, very high enantiomer selectivity ($E = 160$) could be realized; both the ester and the unchanged alcohol were isolated with >95% ee.

The above examples demonstrate that with chemical systems, high enantiomer selectivities ($E$) can be realized also in kinetic resolutions. A disadvantage of chemical systems is, however, that even with fully selective systems, enantiomeric excess is limited by that of the catalyst (reagent). Since the usual chiral catalysts (reagents) are often synthetic products, they are, in contrast to biocatalysts, often not totally enantiopure.

#### 8.6.2.1 Parallel Kinetic Resolution with Chemical Systems

If a pair of reagents is available in which the same chiral element, but of opposite configuration, is present, with the use of such quasi-enantiomeric reagents, an s.c. **parallel kinetic resolution** is feasible. An interesting feature of this approach is that if both enantiomers of the substrate are reacted in separate reactions with the quasi-enantiomeric reagents, enantiomeric excess of the product can be higher than what can be reached with one of the two "normal" kinetic resolutions separately. With chemical systems of lower enantiomer selectivity, parallel resolution may be especially useful (Figure 8.26).

The advantage of the procedure is that, although enantiomer selectivities with either of the quasi-enantiomeric $N$-alkoxycarbonyl-pyridinium salts are poor ($E = 41$ and 42, respectively), the corresponding carbonates are formed with 49% conversion and high enantiomer excesses (ee = 88% and de = 95% respectively). Note for comparison that in traditional kinetic resolution, in order to be able to realize ee = 95% with 49% conversion, a selectivity parameter of $E = 125$ would have been necessary.

### 8.6.3
### Kinetic Resolution with Biocatalysts

Since biocatalysts are by their nature enantiopure biopolymers (most often partially or fully purified enzymes or whole cell systems), enantiomeric excess of

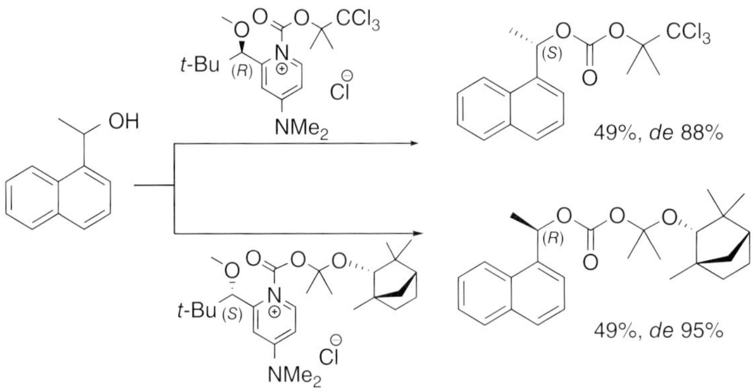

**Figure 8.26** Parallel resolution of racemic 1-(naphth-1-yl)ethanol by acylation with quasi-enantiomeric chiral reagents.

**Figure 8.27** Kinetic resolution of *N*-acyl-amino acids by hydrolysis with acylase-I.

the catalyst does not pose any problems concerning the enantiomeric excess of the products. Therefore, enantiomeric excess of the products is only dependent on the selectivity of the biocatalytic process itself.[10)]

It is thus not surprising that kinetic resolution is a field where biocatalysts are widely applied. A classic example is kinetic resolution of *N*-acylamino acids with acylase-I. This enzyme acts preferentially on L-amino acid derivatives, the configuration of which is analogous to that of natural amino acids. In some cases, it was possible to enhance the yield of the process beyond 50% or even near to 100% by racemizing the nonreacted enantiomer of the substrate (Figure 8.27).

The above example also demonstrates that application of kinetic resolution by means of hydrolases is a very versatile method because hydrolases are capable to transform a very wide range of substrates most often with high enantiomer selectivity. This can be understood when we consider the mechanism of action of serine hydrolases (Figure 8.28).

The family of **serine hydrolases**[11)] comprises among others proteases performing the hydrolytic cleavage of proteins, carboxylesterases catalyzing the cleavage of esters in the liver, lipases doing the hydrolysis of oils as well as thioesterases participating in the transformation of thioesters. A common feature of them is that, in the catalytic process, a serine side chain is added to the carbonyl group of the bond of the substrate to be cleaved. In this way, a covalent intermediate, an acyl-enzyme, is formed (Figure 8.28**C**). It can be seen how the addition of serine is aided by the histidine and aspartic acid/glutamic acid side chains, forming a **catalytic triade**.

The substrate ($R^1$–CO–Z–$R^2$) first forms a complex with the enzyme (Figure 8.28**A**), followed by the formation of a tetrahedral intermediate ($THI_1$) by addition of the catalytic serine side chain of the enzyme forming an enzyme–substrate complex (**B**). After the exit of the leaving group ($HZR^2$), an acyl–enzyme complex is formed (**C**). Under natural conditions, that is, in aqueous media, the complex reacts with water (HNu = HOH) and leads via a second tetrahedral intermediate ($THI_2$) to the product–enzyme complex (**D**), from which finally the product dissociates (**E**). It can also be gleaned from the

10) In the course of biocatalytic kinetic resolutions during workup, occasionally, a concurrent nonselective chemical reaction or racemization may decrease the enantiopurity of the product.

11) Serine hydrolases are very common enzymes, for example, more than 1% of the human genome codes for this family of enzymes. In most of them, a Ser–His–Asp/Glu assembly constitutes the catalytic triade.

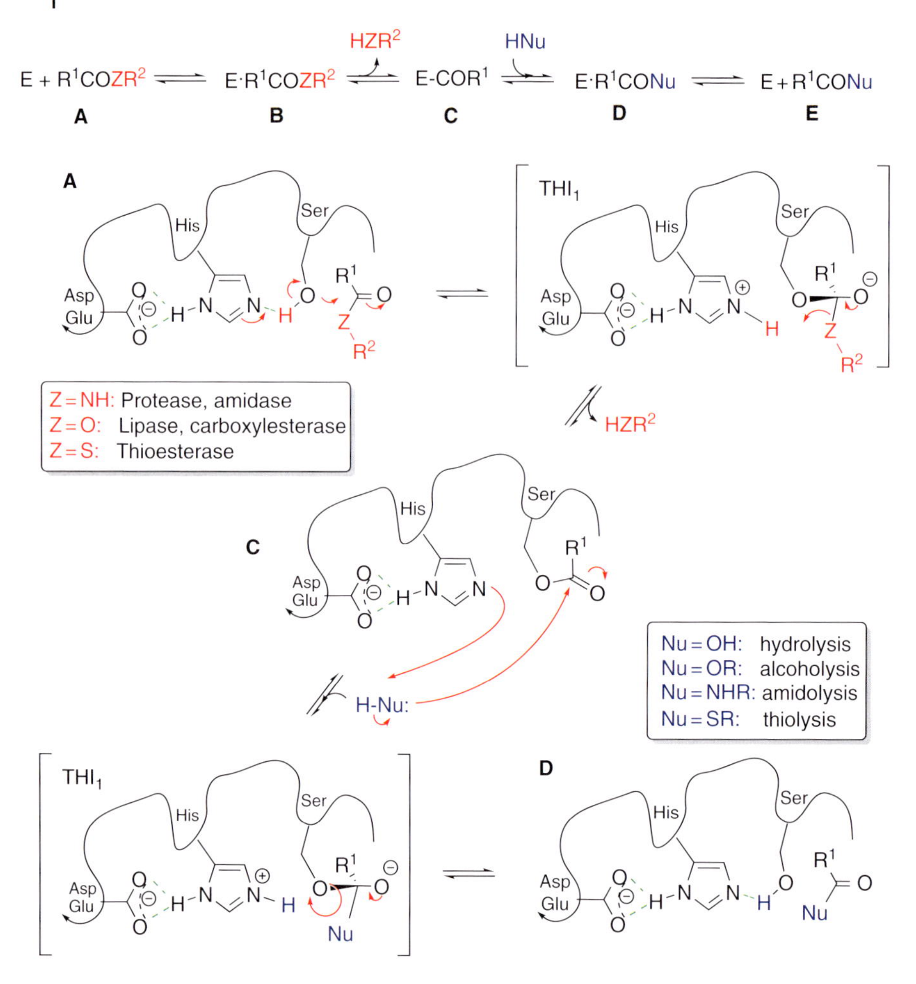


**Figure 8.28** Mechanism of action of serine hydrolases.

figure that every step of the catalytic process is reversible. This is completely understandable if we consider the catalytic character of enzymes, since a catalyst exerts its accelerating activity by lowering the energy of the transition state but does not influence the state of the equilibrium of the process.

The fact that in aqueous media hydrolytic reactions are practically irreversible can be traced back to other reasons (Figure 8.29). One of them is that water, serving also as the solvent, is present in very high concentration [~55 M]. In the near-neutral medium of aqueous solutions, the other important factor is that organic acids are mainly present in a deprotonated form (as acids their p$K_a$ value is around 5, i.e., much lower than that of water, being in this medium p$K_a$ ~ 7) Thus, in the deprotonation equilibrium, the product of hydrolysis, that is, the acid becomes a carboxylate ion and in this state it is unable to participate anymore in the enzymatic reaction.

$$E + R^1COZR^2 \;+\; HOH\;[\sim 55\,M] \;\rightleftharpoons\; E + R^1COOH \;+\; HZR^2$$

$$E + R^1COOH \xrightleftharpoons{HOH} E + R^1COO^- \quad H_3O^+$$

**Figure 8.29** Reasons for the "irreversibility" of enzyme-catalyzed processes in aqueous media.

Hydrolases retain their activity also in nonaqueous media[12] and it is understandable from the mechanism shown in Figure 8.28 why hydrolases can be applied in such a versatile way. Exploiting the reversibility of hydrolysis in organic media of low water content, it is possible to form esters with alcohols using lipases and carboxylesterases. The reaction can be made even irreversible if the eliminated water is removed by some technique.

Figure 8.28 also explains why serine hydrolases are often capable to catalyze unnatural transformations as well. For example, proteasess are often used for ester hydrolysis in media of low water content or even for esterification and transesterification. For synthetic purposes, lipases proved to be so far the most versatile enzymes. Apart from hydrolysis in aqueous media (HNu = HOH), they are suitable to perform alcoholysis in media of low water content, for thiolysis (HNu = HSR), or amidolysis (HNu = $H_2NR$).

In order that a nucleophile other than water should be able to participate in the transformations, a practically anhydrous medium is required. Moreover, if one wishes to utilize any type of selectivity of a "reverse" process carried out in an anhydrous medium, total irreversibility of the transformation (e.g., kinetic resolution) must be provided for.

With hydrolases, alcoholysis (acylation from the aspect of the alcohol) can be carried out with esters in which the leaving alcohol is a weak nucleophile (e.g., trifluoroethyl, trichloroethyl, or phenyl esters) (Figure 8.30(a)) or not nucleophilic at all (e.g., vinyl or isopropenyl esters (b)). In the latter case, it is of advantage that the liberated vinyl and isopropenyl alcohols are practically nonexistent, since the enol–oxo equilibrium is completely shifted in favor of acetaldehyde and acetone, respectively.

Note that while the strategy for shifting equilibrium by reducing nucleophilicity of the leaving alcohol can be used in hydrolase-catalyzed reactions with any nucleophile (Figure 8.30(a)), this measure fails with vinyl or isopropenyl esters (b) and amines as nucleophiles, since the oxo compounds arising from the former condense with amines to Schiff's bases.

It is important (and in principle, this is true not only for biocatalytic processes) that if a kinetic resolution is carried out with the same catalyst (reagent) but into

12) Apart from hydrolases, many other enzymes remain active also in two-phase aqueous–organic systems and in systems of very low water content.

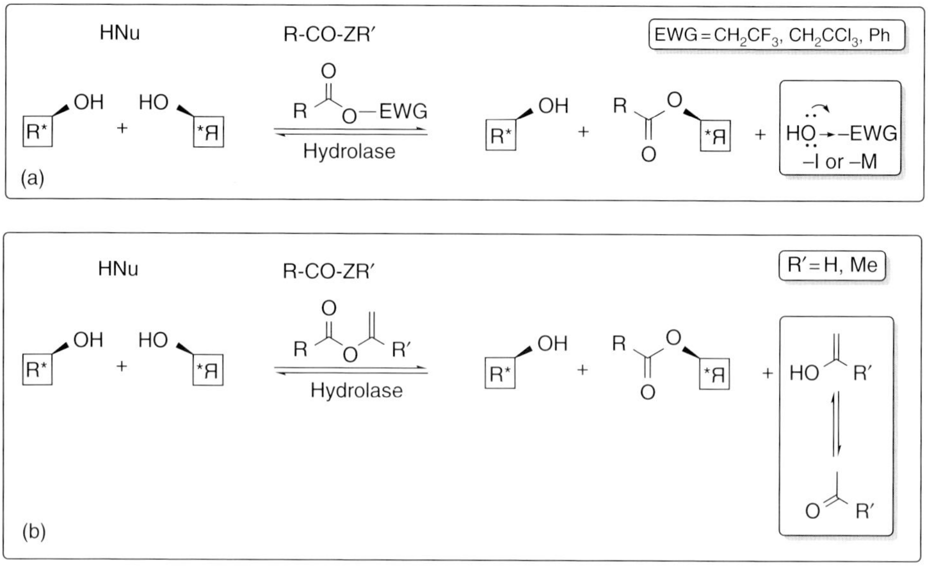

Figure 8.30 Possibilities to realize the "irreversibility" of hydrolase-catalyzed processes in nonaqueous media.

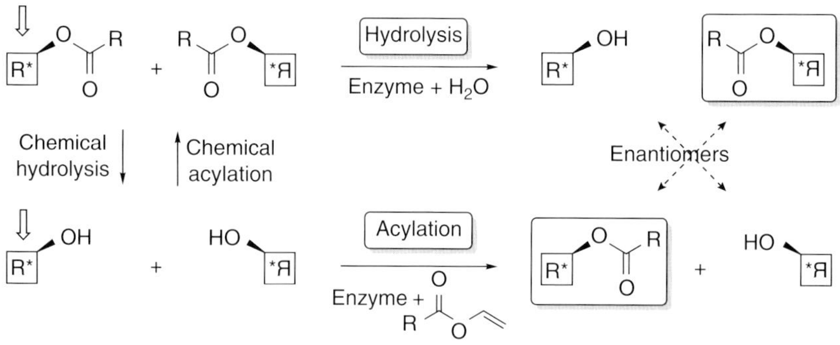

Figure 8.31 Kinetic resolutions by "opposite" reactions using the same catalyst.

the "opposite" direction, then it is possible to select which enantiomer should be present in the product mixture in which form (Figure 8.31).

The outcome of the process in Figure 8.31 is understandable since if a given catalyst (e.g., a hydrolase) prefers in a given direction one particular enantiomer [e.g., during hydrolysis, the enantiomer marked with an arrow (⇓)], then it is not surprising that in the "*opposite*" reaction it is also the same enantiomer that is preferred [e.g., in case of acylation with a hydrolase, the one marked with an arrow (⇓)]. Since racemic compounds can be interconverted via hydrolysis or acylation using simple chemical reagents, this property can be well exploited for controlling which enantiomer goes to which fraction. The procedure can be of importance

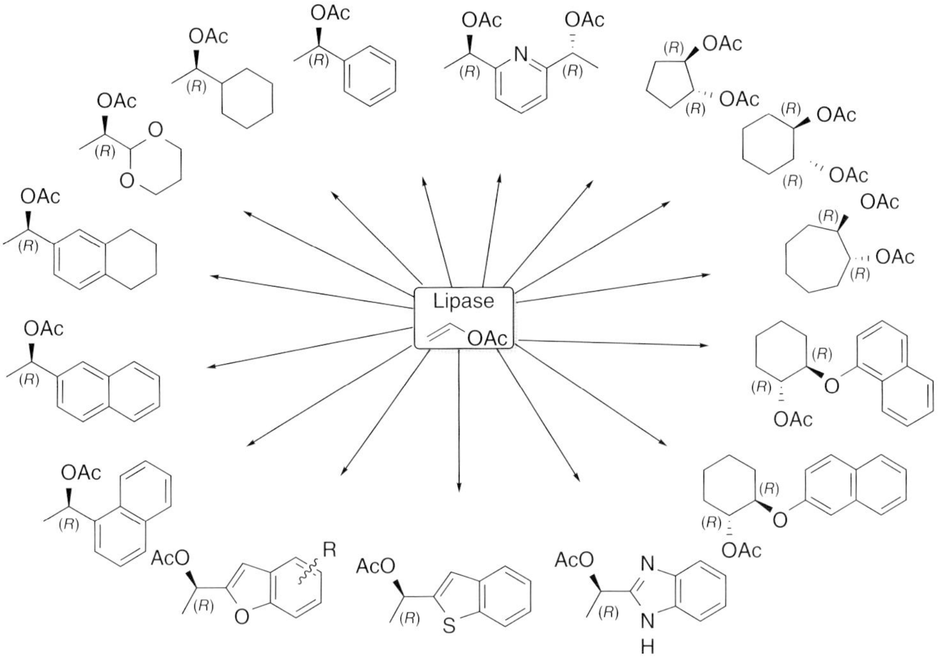

**Figure 8.32** Kinetic resolution of several racemic alcohols by lipase-catalyzed acylation (only the product of the fast reacting enantiomer is depicted).

with transformations of poor enantiomer selectivity, since as we have seen it is the unreacted fraction in which higher enantiomer purity is developing.

The versatile application of lipases for synthetic purposes is demonstrated by their use in kinetic resolutions (Figure 8.32). In this figure, results of kinetic resolutions using vinyl acetate as acylating agent and catalyzed by a lipase isolated from a thermophilic filamentous fungus are shown along with the enantiopure (ee > 98%) acetates obtained. It follows from the principle of kinetic resolution that using this enzyme, the unreacted fraction of these alcohols is of opposite configuration.

It follows from the principles shown in Figure 8.31 that the same enzyme is suitable also for the preparation of acetates of opposite configuration (and of course for that of the corresponding alcohols of opposite configuration) if one starts from the racemic form of the esters shown in Figure 8.32 and kinetic resolution is carried out by hydrolysis or alcoholysis with some simple alcohol (e.g., methanol, ethanol, and propanol).

Naturally, kinetic resolution with enzymes other than hydrolases can also be accomplished. For example, by oxidation of a girochiral alcohol, an almost enantiopure ketone can be prepared using horse liver alcohol dehydrogenase (HLADH), while the remaining alcohol can also be recovered in high enantiomeric excess (Figure 8.33).

HLADH
NAD+
OH HO OH O
ee > 98% ee > 98%

**Figure 8.33** Kinetic resolution of a racemic girochiral alcohol by HLADH-catalyzed oxidation.

*Acinetobacter* sp.
*Whole cells*
40%, ee > 97% 37%, ee > 97%

**Figure 8.34** Enantiomer and regioselective microbial *Baeyer–Villiger* oxidation of a bicyclic ketone.

#### 8.6.3.1 Parallel Kinetic Resolution with Biocatalysis

Microbial *Baeyer–Villiger* oxidation of a racemic bicyclic ketone intermediate of prostaglandin syntheses proceeds in an enantiomer and partially regioselective manner (Figure 8.34). Both products can be isolated in good yield and enantiopure. The process is essentially a parallel kinetic resolution, since transformation of both enantiomers is taking place but with different regioselectivity. In other words, enantiomer and 'selectivity take place concurrently.

### 8.6.4 Dynamic Kinetic Resolution (DKR)

The drawback of pure kinetic resolution is that in its pure form, the desired enantiomer can only be recovered in 50% yield maximum, while the unwanted enantiomer arises also in 50% yield and is most often a useless by-product. According to what has been discussed earlier, this disadvantage can be eliminated if the non-desired enantiomer is racemized *in situ* in a way that this process should not affect the product already formed. This procedure is the s.c. dynamic kinetic resolution (DKR) (Figure 8.35).

Note that this strategy cannot be applied in a simple way when the molecule contains more than one chiral element (Figure 8.36).

In case of compounds containing a single element of chirality, racemization is often feasible (Figure 8.36(a)). Since usually transformation (typically epimerization) of a single chiral element is possible at a time, the transformation would not lead to the enantiomer but to the diastereomer of the original (Figure 8.36(b)). It is crucial from the point of view of DKR how fast racemization takes place compared to the rate of kinetic resolution. If racemization is too slow and enantiomer selectivity incomplete, the less-reactive enantiomer will also be transformed. In DKR, ideally, the starting material should remain racemic throughout the process, which means that the rate of racemization must significantly exceed that of kinetic resolution.

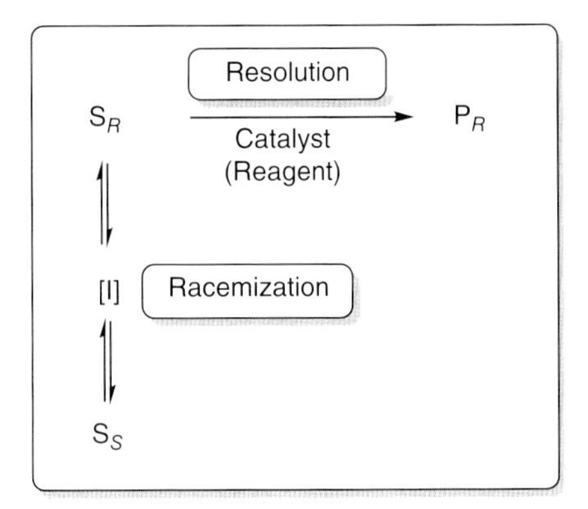


**Figure 8.35** The principle of dynamic kinetic resolution.

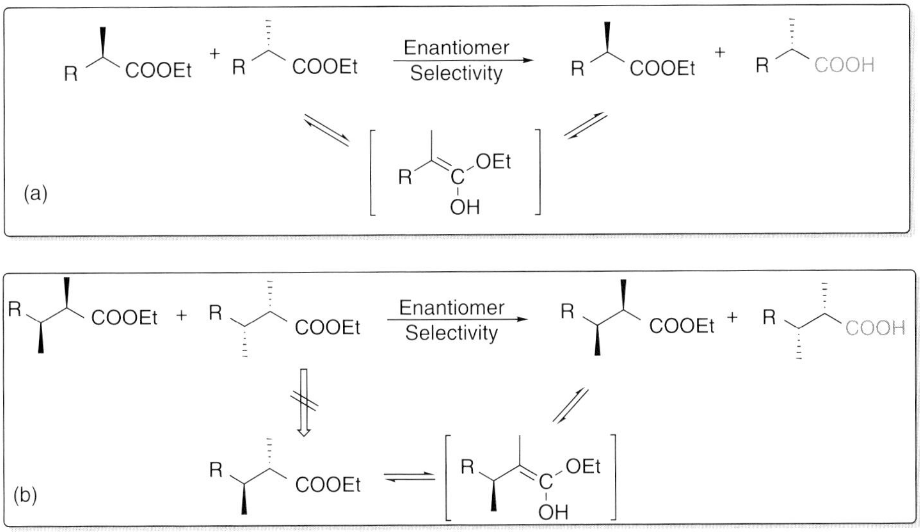

**Figure 8.36** An enantiomer-selective hydrolysis with concurrent racemization (DKR) involving a single center of chirality (a) and epimerization of the remaining enantiomer in case of two centers of chirality (b).

#### 8.6.4.1 Dynamic Kinetic Resolution Using Chemical Systems

DKRs can be realized also with chemical systems. An example is that of a β-oxoester carried out with a chiral catalyst as shown in Figure 8.37. Enantiomers of β-oxoesters undergo racemization by a fast enolization process. Owing to the high selectivity of the reaction, ee of the product is 99%. The most important contaminant is the *anti* diastereomer, although a *syn:anti* ratio of 96.4 : 3.6 is quite acceptable. The ligand of the ruthenium catalyst is a slightly modified version of BINAP, in which instead of the phenyl groups, 3,5-xylyl groups are found.

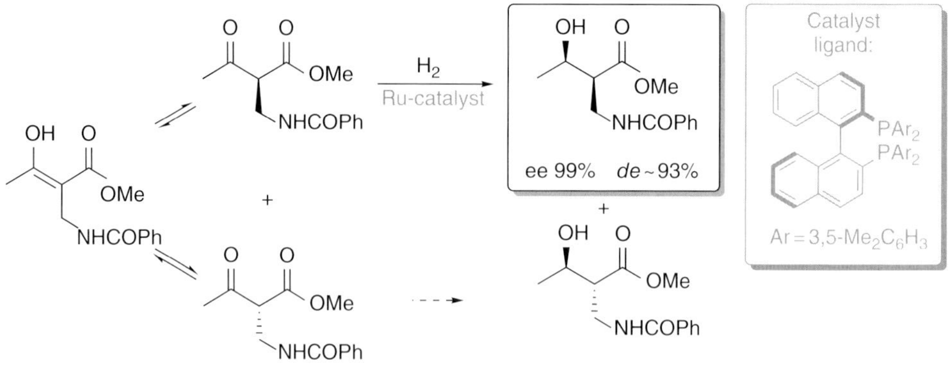

**Figure 8.37** Chemocatalytic dynamic kinetic resolution (DKR) by enantiomer-selective hydrogenation of an α,β-oxoester.

#### 8.6.4.2 Dynamic Kinetic Resolutions with Biocatalysis

A typical case of DKR is when biocatalytic kinetic resolution is combined with a chemical or biocatalytic process effecting the racemization of the substrate. In Section 8.6.4, we have shown that with biocatalysts, highly enantioselective kinetic resolutions can be accomplished. It is thus evident that there have been many procedures elaborated for the combined application of biocatalysis and chemocatalysis. The crucial point is the rate and efficiency of chemical racemization.

Several methods of racemization can be combined with kinetic resolutions involving biocatalysis. The most typical methods are listed as follows:

Base-catalyzed enolization
Acid-catalyzed $S_N1$-type reactions
Ring-opening reactions (ring–chain tautomerism)
Metal-catalyzed redox reactions
Other chemical methods
Biocatalytic methods.

#### 8.6.4.3 Dynamic Kinetic Resolution (DKR) Involving Base-Catalyzed Racemization

The principle of DKR was utilized in the preparation of the anti-inflammatory and analgesic drug (*S*)-ketorolac (Figure 8.38).

First, it was found that lipase-catalyzed hydrolysis of the racemic ester proceeded with preferential transformation of the *R* enantiomer, while proteases (e.g., from *Streptomyces griseus*) catalyzed *S*-selective hydrolysis. At near-neutral pH (8.0), kinetic resolution was highly selective and the product was isolated at 50% conversion with >96% ee. On increasing the pH, the substrate racemized and DKR became possible. This example demonstrated that the purity of the product did not solely depend on the selectivity of the enzyme-catalyzed process, owing to significant deterioration of enantiomeric excess by competing nonenantioselective chemical hydrolysis at the higher pH value.

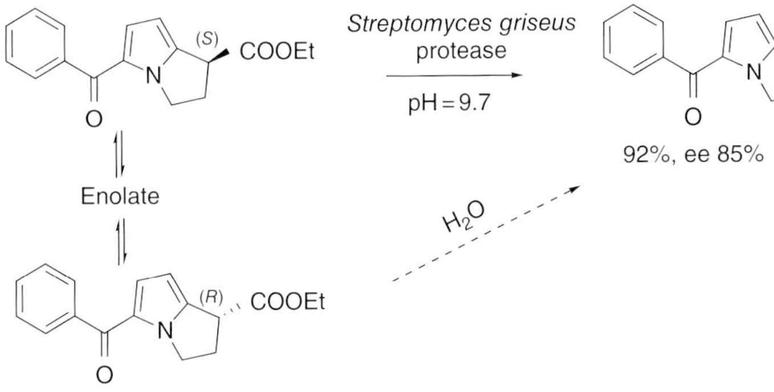

Figure 8.38 Utilization of dynamic kinetic resolution in the preparation of (S)-ketorolac.

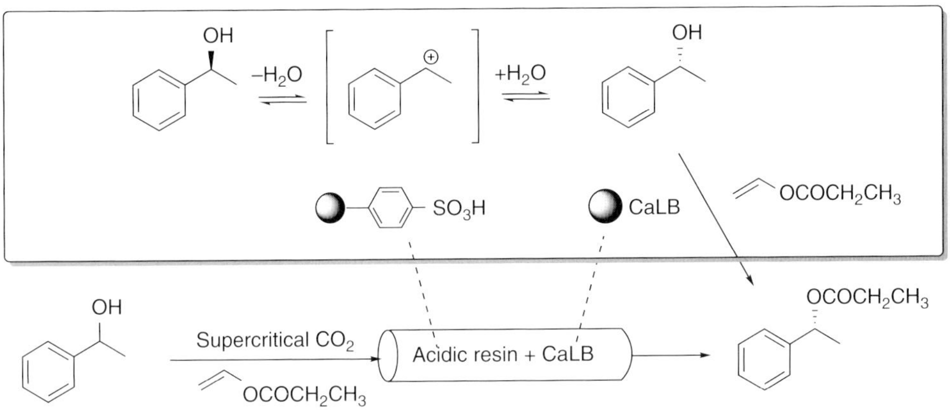

Figure 8.39 Dynamic kinetic resolution by acid-catalyzed racemization.

#### 8.6.4.4 DKR Involving Acid-Catalyzed Racemization

DKR of racemic 1-phenylethanol was accomplished in continuous operation by the parallel application of enzymatic acylation and acid-catalyzed racemization in a supercritical solvent medium (Figure 8.39).

DKR was performed in a continuous flow reactor. For resolution, an immobilized lipase (*Candida antarctica* lipase B, CaLB) was employed, while racemization was carried out by an acidic resin in the same reaction space. In the case of acid catalysis, elimination and ether formation associated with the cation formed in an $S_N1$ reaction and chemical acylation must also be accounted for. This is apparent when we compare the results of kinetic resolution (conversion 49% and ee > 98%) with that of the dynamic process (conversion 75% and ee 89%).

#### 8.6.4.5 DKR Combined with Racemization by Ring Opening

In ring–chain tautomerism and similar processes in the course of ring opening, the central atom of a chiral center takes up transiently an $sp^2$ hybrid state and this can be exploited for DKR (Figure 8.40).

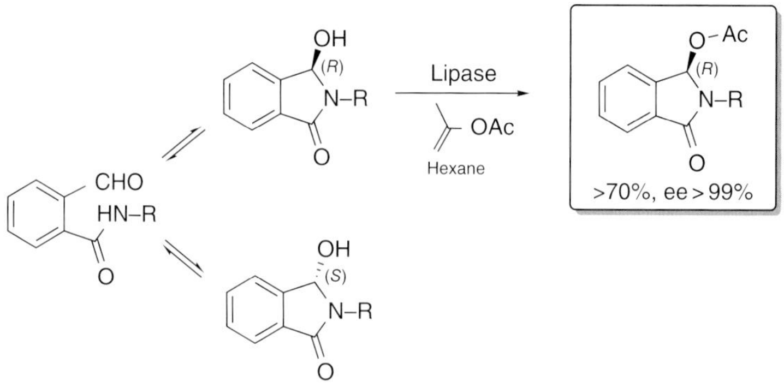

**Figure 8.40** Dynamic kinetic resolution with acid-catalyzed racemization.

Figure 8.40 demonstrates that a series of *N*-acyl hemiacetals of high enantiomeric excess could be prepared by kinetic resolution in hexane at 60–70 °C using various lipasess (lipase AK, lipase PS, and lipase QL) and isopropenyl acetate as acylating agents. The DKR character of the process was secured by racemization via ring opening.

#### 8.6.4.6 DKR Involving Metal-Catalyzed Racemization

Enzyme-catalyzed kinetic resolution can be carried out parallel to multiple metal- and heavy-metal-catalyzed reactions, and therefore such processes can be utilized to carry out racemization as well. Metal-/heavy-metal-catalyzed racemizations are of two main types: racemization is effected by metal-catalyzed redox reactions (Figure 8.41(a)) or complexes of allyl compounds with heavy metals (Figure 8.41(b)). In Figure 8.41, M stands for the metal/heavy metal. On formation of the allyl complex from the original ligand, a nucleophile ($Nu^-$) is liberated.

An example for DKR carried out via heavy-metal-catalyzed racemization is the preparation of the acetate of the aggregation pheromone (*R*)-sulcatol from the racemic alcohol by means of racemization catalyzed by a lipase enzyme (CaLB) and a ruthenium complex (Figure 8.42). The active catalyst is formed from the

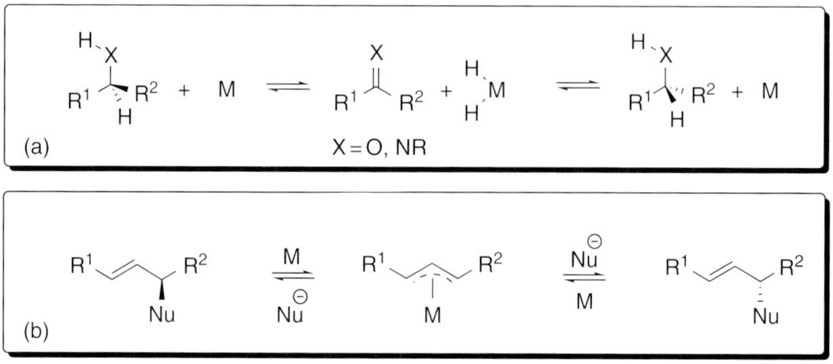

**Figure 8.41** Metal-catalyzed racemizations useful in dynamic kinetic resolutions.

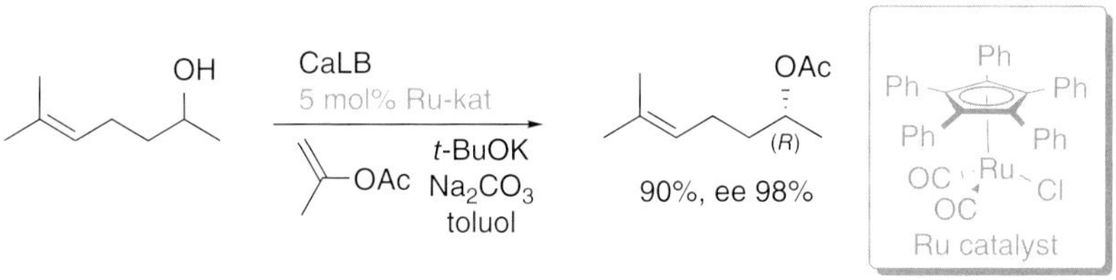

**Figure 8.42** Preparation of (*R*)-sulcatol acetate by DKR via ruthenium-catalyzed racemization.

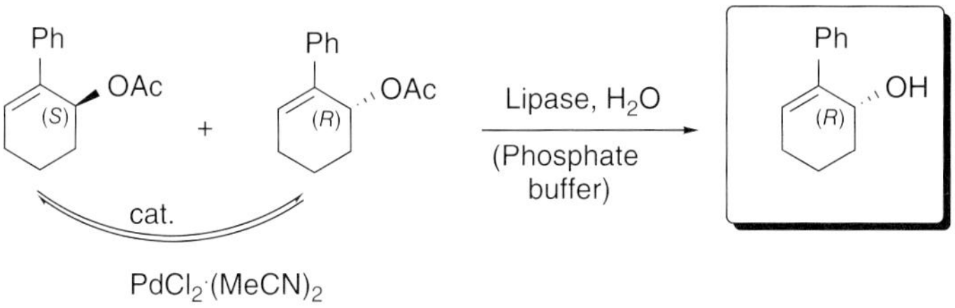

**Figure 8.43** Chemoenzymatic DKR by the Pd(II)-catalyzed racemization of allyl acetates.

chlorine containing ruthenium complex by replacing the chloride ion by a *t*-Bu group.

With allyl complexes of heavy metals (Figure 8.41(b)), the problem emerges that both terminal atoms of the allyl complex are equally ready to accept the attack of the nucleophile, and therefore racemization may be accompanied by rearrangement. A possible way to avoid this problem is to realize the reaction with a symmetrical allyl derivative (Figure 8.43).

This approach was applied to the chemoenzymatic DKR of allyl acetates by a combination of lipase-catalyzed hydrolysis and Pd(II)-catalyzed racemization (Figure 8.43). In the procedure carried out in a phosphate buffer at 40 °C, the remaining (*S*)-allyl acetate racemized under the action of $PdCl_2(MeCN)_2$ complex in a way that the product, that is, the (*R*)-alcohol, could be recovered in good yield and high enantiomeric excess.

#### 8.6.4.7 Dynamic Kinetic Resolutions Utilizing Other Chemical Racemization Processes

Apart from those already described, there are various other racemization processes that can be used in chemoenzymatic DKRs. An example is the lipase-catalyzed hydrolysis of methyl α-bromo-phenylacetate, which becomes, in the presence of bromide ions, a DKR process. Bromide ions react namely with the ester in an $S_N2$-type reaction and are therefore capable to transform the unreacted ester by inversion of configuration, while under the same conditions, the product does not undergo substitution (Figure 8.44).

#### 8.6.4.8 Dynamic Kinetic Resolution Comprising Biocatalytic Racemization

Industrial manufacturing of L-lysine by DKR on a multiton scale was developed using inexpensive α-amino-ε-caprolactam and employing only microbial biotransformations (Figure 8.45).

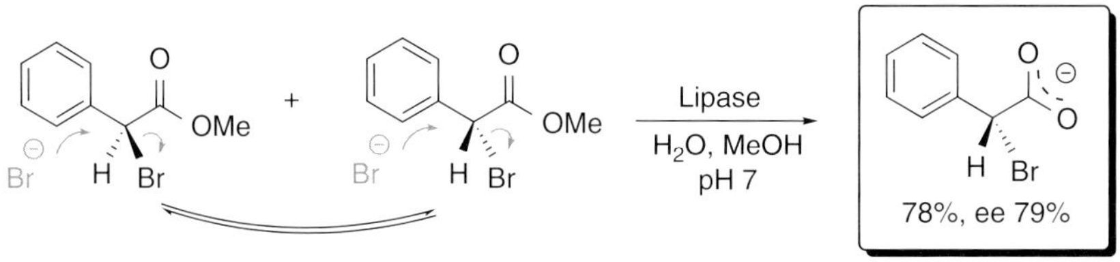

Figure 8.44 Chemoenzymatic DKR using bromide ion-catalyzed racemization.

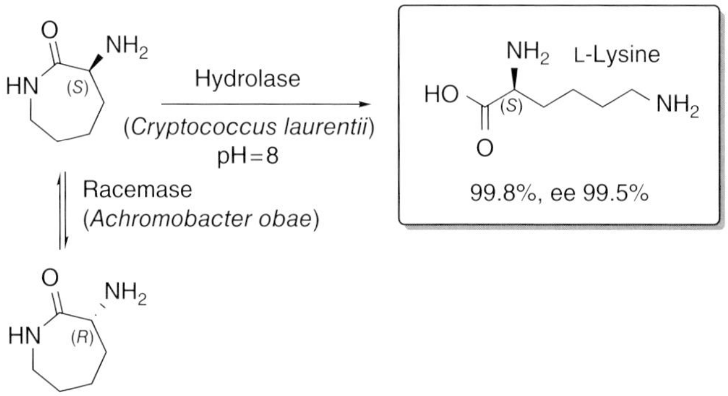

Figure 8.45 Manufacturing of L-lysine using purely biocatalytic dynamic kinetic resolution.

Enantiomer-selective hydrolysis of α-amino-ε-caprolactam is performed with very high selectivity by *Cryptococcus laurentii* hydrolase, while racemization of the unreacted (*R*)-lactam can be accomplished with a racemase from *Achromobacter obae*. This procedure is very efficient and permits the manufacturing of very pure L-lysine (ee 99.5%) in practically quantitative yield (99.8%).

### 8.6.5 Crystallization-Induced Dynamic Resolution (CIDR)

In the previous chapter, we demonstrated that in a kinetically controlled enantiomer-selective transformation (kinetic resolution), when coupled to racemization of the unreacted enantiomer, it is possible to transform the entire quantity of a racemic mixture to a given enantiomer. Note that in classic resolutions based on thermodynamic equilibrium and solubility, the yield of a given enantiomer is also limited to 50% of the racemic starting material. However, when the crystallization process is coupled to racemization, this limitation can also be avoided.

In Figure 8.46, enantiomer-selective variants of crystallization-induced dynamic resolution (CIDR) are presented. In its pure form, the process can be realized also without the use of a chiral auxiliary (Figure 8.46(a)), when seemingly without any external intervention one or the other enantiomer is crystallizing preferentially from the mixture. Such a process was observed by Barton and Kirby during the preparation of (−)-narwedine (Figure 8.47).

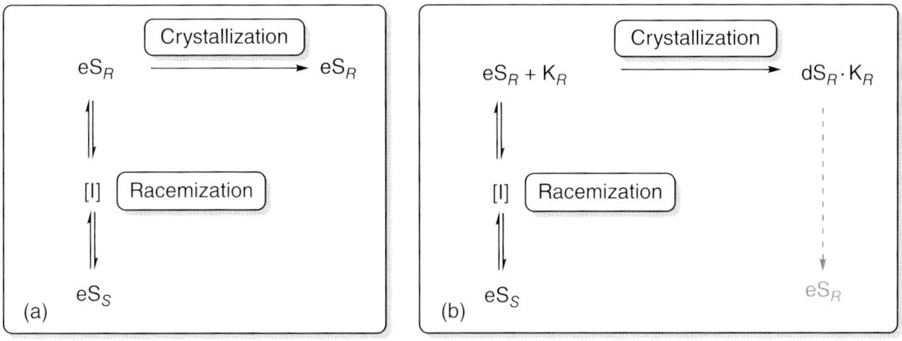

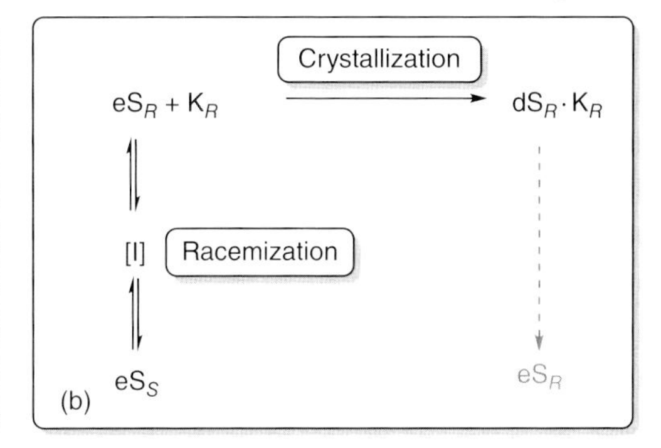


**Figure 8.46** Principle of crystallization-induced dynamic resolution (CIDR).

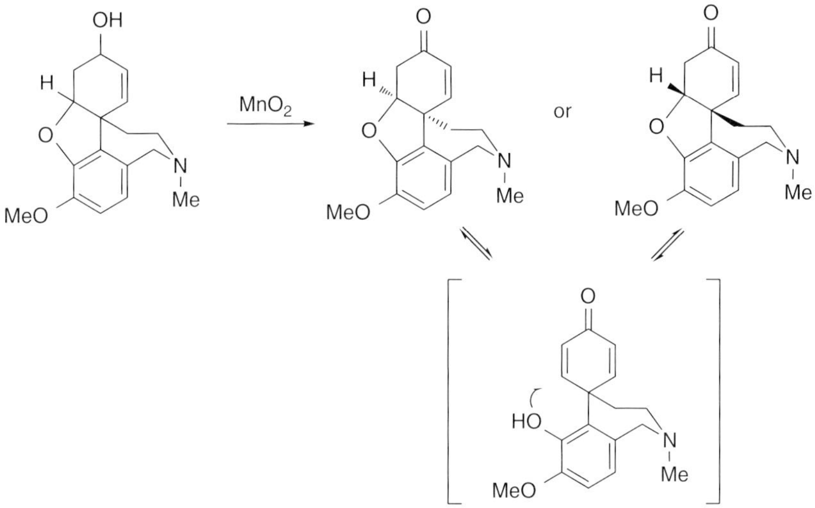

**Figure 8.47** A "pure" example of crystallization-induced solution dynamic kinetic resolution observed in the spontaneous resolution of narwedine.

On oxidation of natural (−)-galanthamine, (−)-narwedine was obtained. Crystallization of the product from acetone gave a product with negative rotation, while that from ethanol gave one with a positive one. Traces of (−)-galanthamine also induced the crystallization of the levorotatory product. Both (−)- and (+)-narwedine racemized in ethanol or in the presence of silica gel. A tricyclic phenol formed by ring opening was supposed to be the intermediate of racemization.

A more common case of crystallization-induced dynamic resolution (asymmetric transformation of the second kind) is salt formation with a chiral additive. In a process developed for the preparation of (*R*)-proline (Figure 8.48), racemization is induced by heating in (±)-butyric acid at 80 °C in the presence of a catalytic amount butyraldehyde. Thereby, an imine intermediate is formed that mediates racemization by the cleavage of the α-proton. On salt formation with unnatural

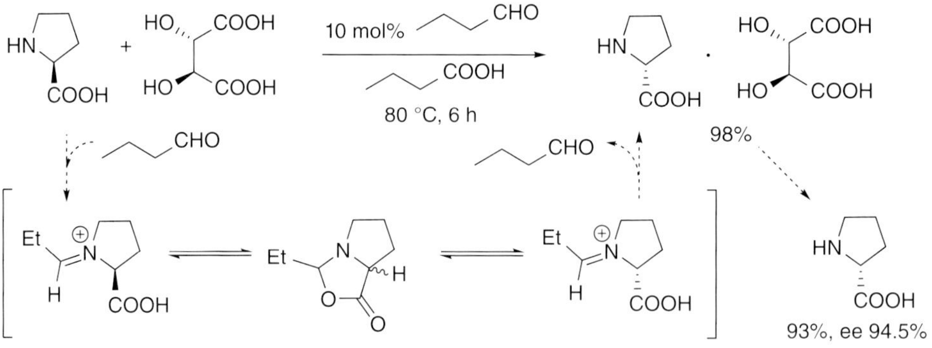

Figure 8.48 Preparation of (*R*)-proline by crystallization-induced dynamic resolution.

(*R,R*)-tartaric acid, the salt of (*R*)-proline crystallizes from the solution of the racemic substrate. From the salt, (*R*)-proline can be liberated in good yield and ee 94.5%.

### 8.6.6 Kinetic Resolution Followed by Configurational Inversion

A method other than DKR to transform the unwanted enantiomer to the useful one is **configurational inversion** (Figure 8.21, (c)). This multistep process, containing an inversion step, is exemplified in Figure 8.49.

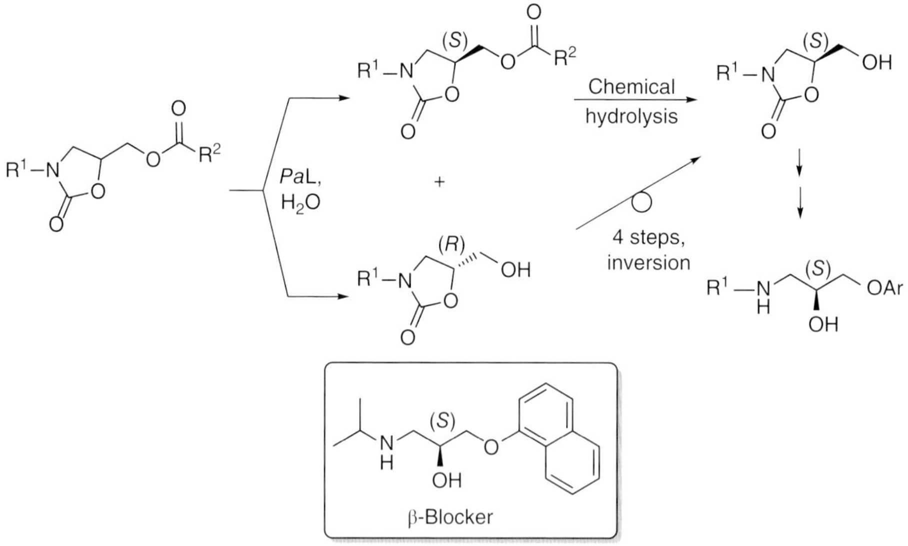

Figure 8.49 Complete conversion of the unchanged substrate combining a biocatalytic kinetic resolution and a multistep inversion.

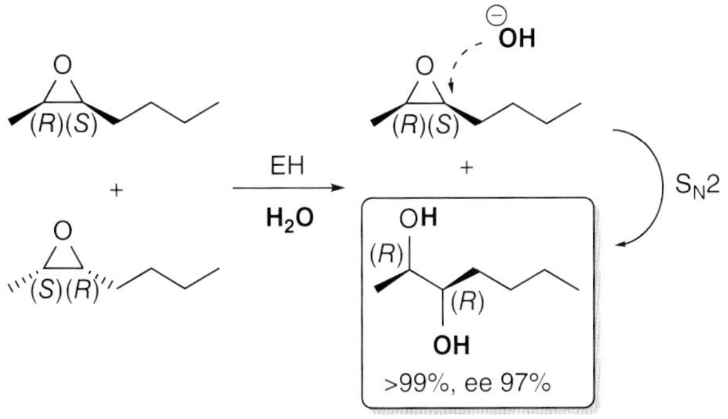

Figure 8.50 Complete conversion of a racemate by kinetic resolution using an epoxy-hydrolase (EH) followed by transformation of the substrate fraction in one step.

Multistep inversion combined with kinetic resolution was applied for the preparation of an (*S*)-aminodiol, an intermediate of β-blockers. The method was based on the fact that racemic oxazolidine esters are suitable substrates for multiple enzymes. Thus, *Pseudomonas aeruginosa* lipoproteine lipase (PaL) proved to be an efficient and very selective enzyme for a wide range of substituted oxazolidines. Therefore, hydrolytic kinetic resolution was carried out with this enzyme and the product, an (*R*)-alcohol, and the remaining (*S*)-ester were separated by adsorption methods followed by chemical hydrolysis of the latter to the (*S*)-alcohol. The undesired (*R*)-alcohol was converted in a four-step process, comprising an inversion step, to the useful (*S*)-alcohol.

Figure 8.50 presents a rare example when following kinetic resolution the unreacted enantiomer was transformed to the useful one in one step.

Kinetic resolution of racemic *cis*-2,3-epoxyheptane was carried out with an epoxy-hydrolase (EH) isolated from a *Nocardia* bacterium strain leading to (2*R*,3*R*)-heptane-2,3-diol. In a basic medium, the remaining (2*R*,3*S*)-epoxide underwent regioselective chemical hydrolysis giving also (2*R*,3*R*)-heptane-2,3-diol. Combined enzymatic and chemical hydrolysis thus provided the (*R*,*R*)-diol with enantiomeric excess of 97% and in quantitative yield. This method that is often called **deracemization** was successfully applied to the transformation of other epoxides too.

## 8.7 Enantiotope Selectivity

Enantiotope selectivity involves the selective transformation of either the enantiotopic faces (Figure 8.51(a)) or the enantiotopic groups (Figure 8.51(b)). In such reactions, a prochiral moiety of an achiral molecule is transformed to a chiral structural element (most often to a center of chirality) and in case of complete

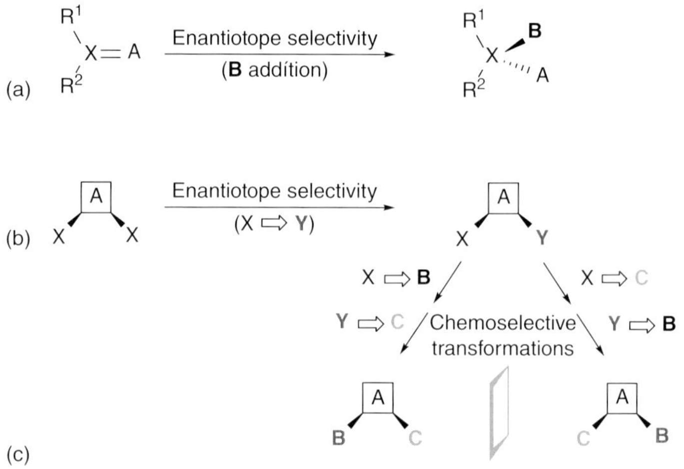

**Figure 8.51** Selective transformation of enantiotopic faces (a) and groups (b) and enantiodivergent transformations of the products (c).

selectivity, an achiral substrate is transformed to a pure enantiomer. A problem may arise when only one enantiomer of the chiral catalyst or reagent is available and this very enantiomer promotes the formation of the unwanted enantiomer of the product. The solution to this problem may often be that the primary product is transformed in additional steps in a *stereodivergent manner* to an optional enantiomer of the end product (Figure 8.51(c)).

Different types of enantiotope selectivity and the strategy of further enantiodivergent transformations will be exemplified later.

Similarly to considerations discussed with enantiomer selectivity (Section 8.6.1), the *irreversibility* of the processes is here of prime importance too (Figure 8.52).[13)]

In an irreversible process, the degree of enantiotope selectivity ($E$) is also equal to the ratio of the pseudo first-order rate constants of the competing processes (Figure 8.52). From the value of $E$, enantiomeric excess can be determined in a simple way (Figure 8.52). The equations show clearly that in irreversible enantiotope-selective reactions, enantiomeric excess of the product ($ee_p$) is constant and independent of conversion. This is in contrast to enantiomer-selective processes, in which enantiomeric excess of the product fraction ($ee_p$) was dependent on the conversion (Section 8.6.1)

13) The starting material is achiral and the products are enantiomers and thus of equal energy. Therefore, if the reactions are reversible and in the forward reaction it is the transformation leading to the *S* enantiomer that is accelerated by a chiral catalyst (the pertinent transition state being of lower energy), then in the reverse reaction also it is the *S* enantiomer that reverts faster. Consequently, in the product, the minor *R* enantiomer becomes enriched and high enantiomer purity cannot be achieved.

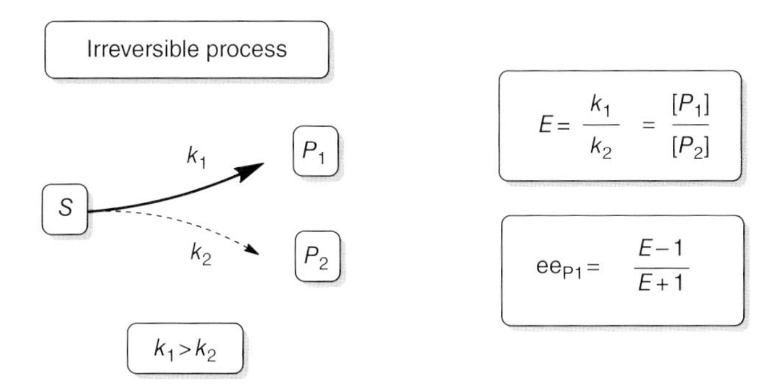


**Figure 8.52** Selectivity (*E*) and enantiomeric excess of the product ($ee_{P1}$) in an irreversible enantiotope-selective process.

### 8.7.1
### Enantiotope Selectivity in Chemical Systems

In chemical systems, enantiotope selectivity can manifest itself when the transformation of the achiral (and prochiral) substrate is carried out in either in a chiral environment (e.g., in a nonracemic chiral solvent with any type of reagent or catalyst) or in an achiral solvent using a nonracemic chiral reagent or catalyst. On discussing enantiomer selectivity in chemical systems, we pointed out (Section 8.6.2) that while in biocatalytic processes, enantiopurity of the biocatalyst need not be cared about, in chemical systems, enantiomeric excess of the product is limited not only by the selectivity of the process but also by the enantiomeric excess of the catalyst.

Note, however, that there is necessarily no linear correlation between ee values of catalyst and product (Figure 8.53(b)). Noyori and coworkers found that in the addition of diethyl zinc to benzaldehyde catalyzed by (−)-3-exo-(dimethylamino)isoborneol [(−)-DAIB], (*S*)-1-phenylpropanol was obtained in 92% yield and ee = 95% even when the enantiomeric purity of the catalyst was very poor (ee 15%) (Figure 8.53(a)). This reaction showed a significant positive deviation (positive nonlinearity effect, NLE) from the usually linear correlation between the enantiomeric excess of catalyst and product ($ee_{cat}$ and $ee_{prod}$ respectively) (Figure 8.53(b)).[14] This positive NLE can be explained by the fact that the catalyst is present not only as the catalytically active DAIB.Zn complex, but also as the catalytically inactive dimer. Among the dimers, the heterochiral (−)-(+)-$DAIB_2$ complex is the most stable, and thus the racemic part of the catalyst is fixed in this catalytically inert state, while the proportion of the active (−)-DAIB.Zn complex is significantly enhanced (Figure 8.53(c)).

14) In such cases, when a significant positive NLE is experienced, for obtaining a product of high enantiomer purity, it is not necessary to apply a catalyst of high enantiomer purity. This is true also the other way round: in case of a significant negative NLE, a slight contamination of the catalyst decreases significantly the enantiomer purity of the product.

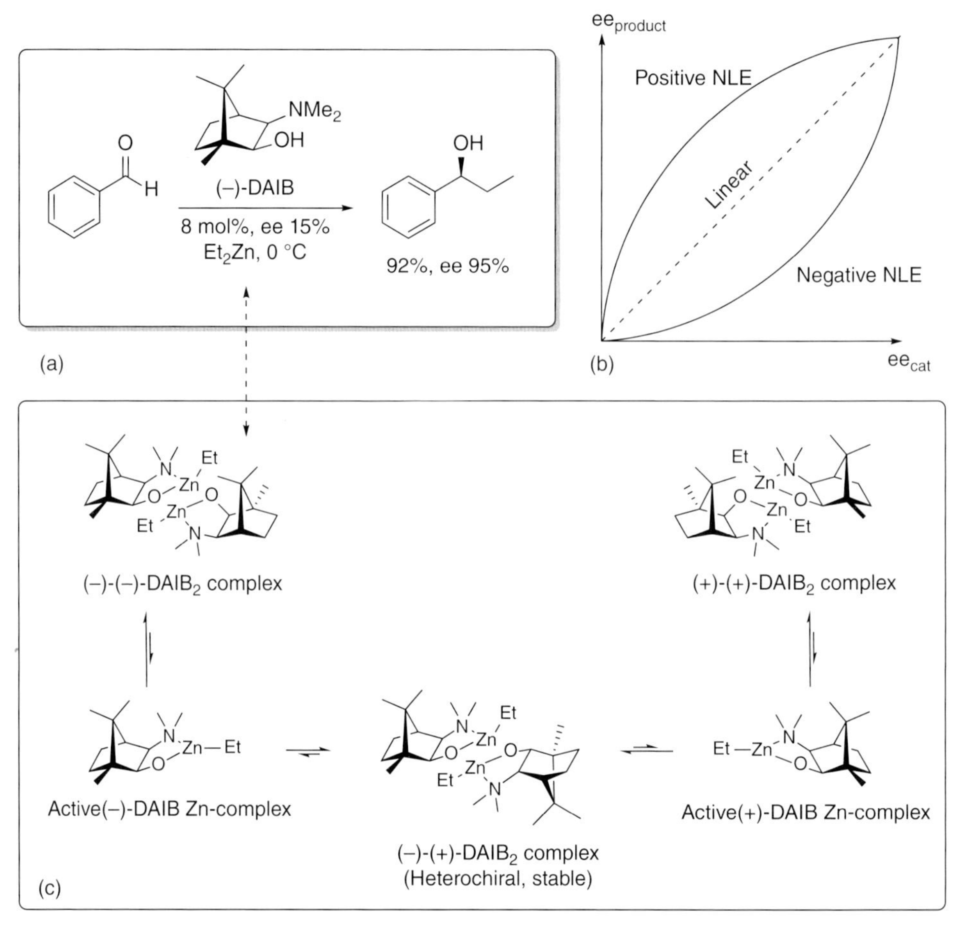


**Figure 8.53** Alkylation of benzaldehyde catalyzed by (−)-DAIB (ee 15%) and interpretation of the nonlinear correlation of enantiomeric excess of catalyst and product.

There is an extensive catalog of enantioselective chemical transformations carried out with chiral reagents and catalysts (several examples are quoted in Part IV). In this context, quality of reagent/catalyst is of prime importance. In Figure 8.54, without claiming to be comprehensive, some molecules are presented which, as ligands of metals/heavy metals, induce significant chiral induction. These and similar molecules provided with proper functional groups, called **organocatalysts,** may be suitable in themselves, that is, without complexation to metals to function as catalysts. As can be gleaned from the figure, relatively rigid chiral systems of $C_2$ symmetry are especially hopeful candidates to accomplish chiral induction. $C_2$ symmetry decreases the number of possible catalyst–substrate arrangements reducing the number of possible competing diastereomeric transition states counteracting selectivity.

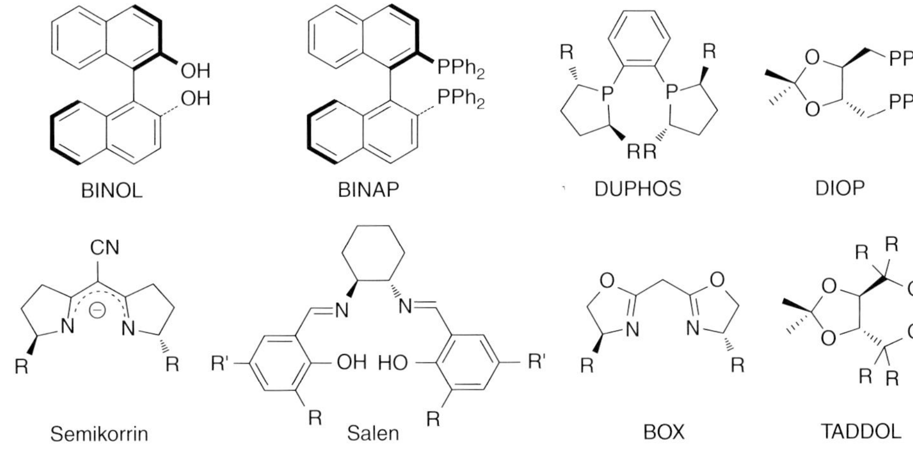

**Figure 8.54** Chiral molecules with $C_2$ symmetry used as catalysts or ligands in catalysts/reagents in enantiotope-selective reactions (most often both pure enantiomers are available).

### 8.7.2 Enantiotope Selectivity in Biocatalytic Systems

As discussed earlier (Section 8.6.2), in contrast to stereoselective reactions implemented with chemical systems, in which the enantiomeric excess of the product is controlled not only by the selectivity of the transformation but also of the enantiomeric excess of the catalyst/reagent, with biotransformations only selectivity is relevant, because complete enantiomeric excess of the catalyst can be taken as granted.

#### 8.7.2.1 Transformations of Enantiotopic Groups by Biocatalysis

A typical example for the biocatalytic transformation of prochiral compounds is the enantiotope-selective oxidation of glycerol to (*S*)-glyceraldehyde with HLADH (Figure 8.55). In this process, selective oxidation of the *pro-S* hydroxyl group takes place; the direction of the reaction and its irreversibility is secured by the addition of the oxidized form of $NAD^+$ cofactor.

As with prochiral compounds, enantiotopic groups of open-chain *meso*-compounds are also amenable to selective biocatalytic transformations, as exemplified by hydrolysis of a *meso*-dimethyl ester by means of a carboxylesterase

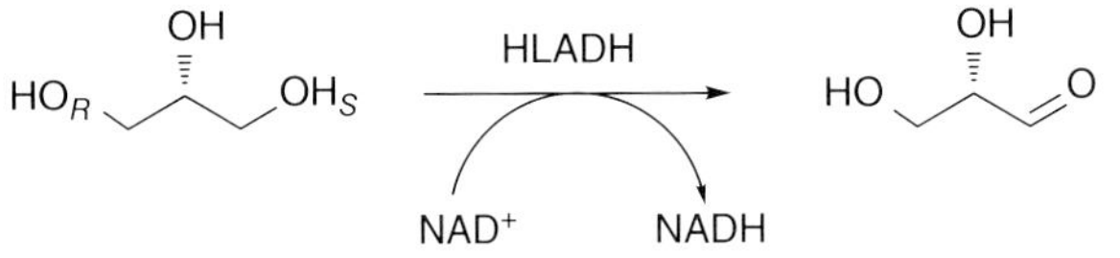

**Figure 8.55** Enantiotope-selective enzymatic oxidation of a prochiral diol.

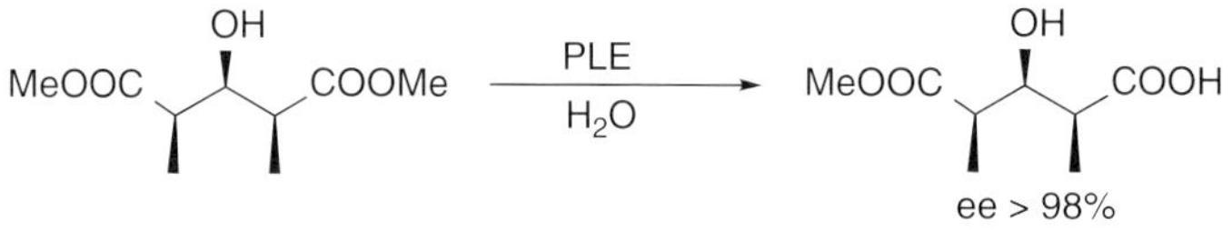

**Figure 8.56** Enantiotope-selective enzymatic hydrolysis of a *meso*-diester.

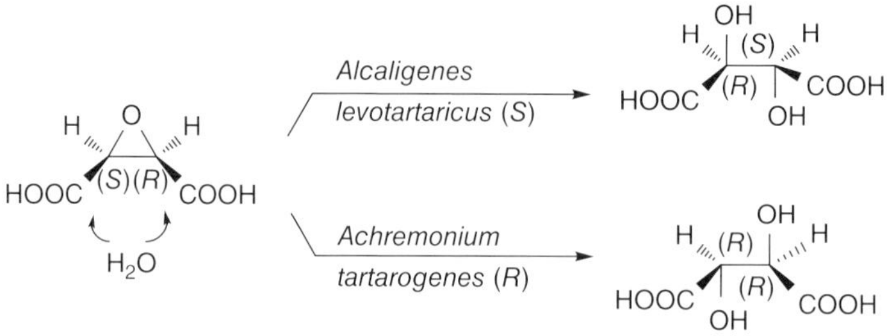

**Figure 8.57** Ring opening of a prochiral epoxide with epoxy hydrolases of opposite enantiotope selectivity.

isolated from pig liver (PLE) (Figure 8.56). Irreversibility of the reaction is provided here by deprotonation following enantiotope-selective hydrolysis.

Selective transformation of enantiotopic groups of cyclic *meso*-compounds by biotransformations is also feasible. Figure 8.57 illustrates that starting from the same substrate but using different biocatalysts, both enantiomers of the product can be obtained. The two microorganisms used produce epoxide hydrolase enzymes of opposite enantioselectivity, leading to antipodal products (Figure 8.57).

Nonsymmetrically substituted sulfides having two pairs of nonbonding pairs of electrons can also be considered as prochiral compounds, since the latter are in an enantiotopic relationship. Therefore, it is obvious that the microbial oxidation of an achiral iron ferrocenyl sulfide gives, depending on the strain selected, selectively an (*R*)- or (*S*)-sulfoxide (Figure 8.58).

Selective biotransformations based on differentiation of enantiotopic faces have been known for long. A classic example is the reduction of achiral ketones

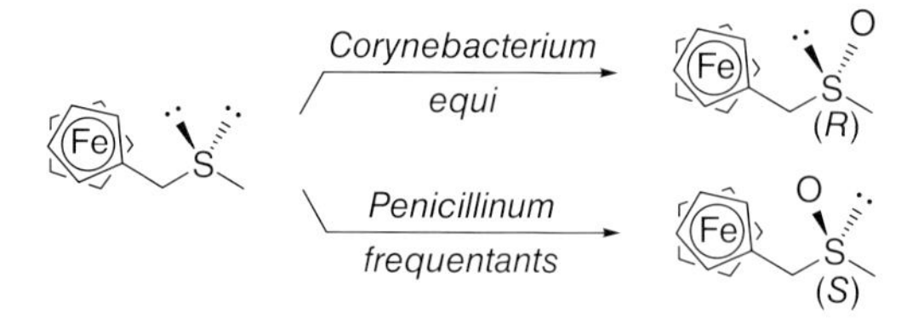


**Figure 8.58** Transformation of enantiotopic faces by biocatalysis. Enantiotope-selective microbial oxidation of a prochiral sulfide.

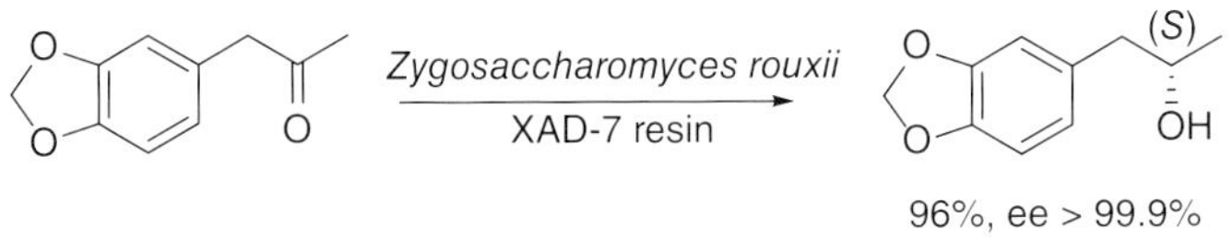

Zygosaccharomyces rouxii
XAD-7 resin
(S)
OH
96%, ee > 99.9%

**Figure 8.59** Enantiotope-selective microbial reduction of a prochiral ketone.

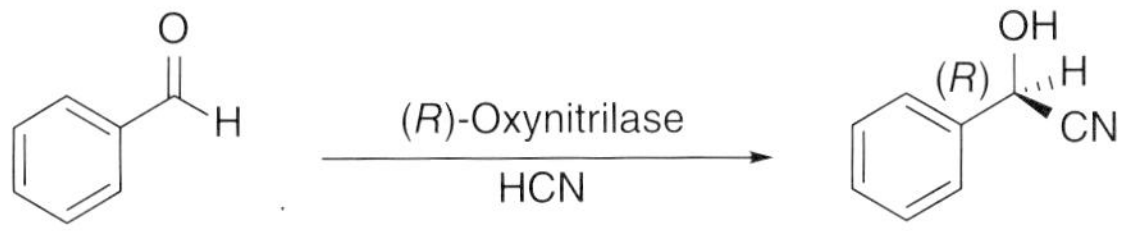

(R)-Oxynitrilase
HCN
(R)
OH
H
CN

**Figure 8.60** Enzyme-catalyzed enantioselective addition of hydrogen cyanide onto benzaldehyde.

catalyzed by yeast. Fermentative reduction of the ketone in Figure 8.59 with the yeast *Zygosaccharomyces rouxii* gives the (*S*)-alcohol in very good yield. The product is an intermediate in the synthesis of the drug Talampanel.

Selective biotransformations based on the differentiation of enantiotopic faces are not restricted to reductive transformations of carbonyl compounds. A lyase called (*R*)-oxynitrilase catalyzes the selective addition of hydrogen cyanide onto aldehydes (Figure 8.60). The enzyme permits the preparation of (*R*)-mandelonitrile of high enantiomeric excess from benzaldehyde.

Differentiation of enantiotopic faces in biotransformations is also possible in the reduction of carbon–carbon double bonds (Figure 8.61).

The above example draws our attention to the fact that in this case also the geometry of the double bond can exert significant influence on the outcome of the reaction. It is apparent that in reduction of (*E*)- and (*Z*)-α,β-unsaturated esters, double-bond geometry affects not only the degree of selectivity but also its direction. Reduction of substrates of differing geometry takes place from opposite faces and gives rise thereby to enantiomeric products (Figure 8.61). It

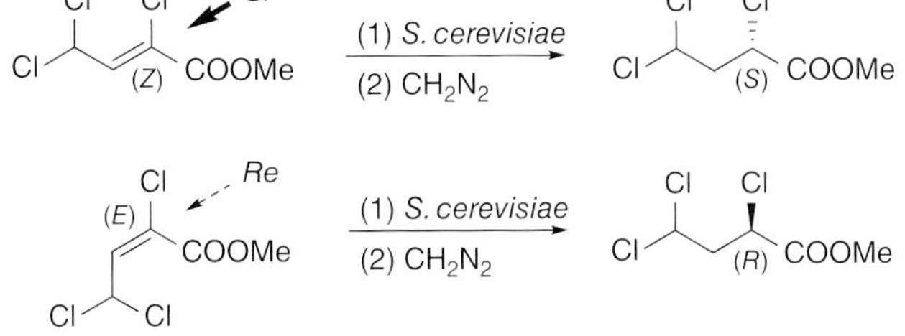

Cl Cl Si
Cl (Z) COOMe
(1) S. cerevisiae
(2) $CH_2N_2$
Cl Cl
Cl (S) COOMe

Cl Re
(E) COOMe
Cl Cl
(1) S. cerevisiae
(2) $CH_2N_2$
Cl Cl
Cl (R) COOMe

**Figure 8.61** Microbial reduction with alternative enantiotope selectivity of prochiral *E*/*Z* double bonds.

seems that enzymic hydrolysis to the free acid prior to reduction is required, since with nonhydrolysable esters, reduction does not take place at all.

### 8.7.3
### Consequences of Opposing Enantiotope-Selective Transformations

When discussing enantiomer-selective reactions in Section 8.6.1, we have demonstrated that it is generally true that if kinetic resolution is carried out with the same catalyst (reagent) but in "*opposite*" direction, then we can choose which enantiomer should be present and in which fraction of the product mixture (Figure 8.31). This principle not only applies to kinetic resolutions but is also true with enantiotope-selective reactions in general.

Realization of this principle is demonstrated by lipase-catalyzed transformation of prochiral diols (Figure 8.62). It is apparent that if in the course of the hydrolysis of the diacetate a given enzyme preferred the hydrolysis of the *pro-R* ester group (⇒), then the same enzyme would generally promote the esterification of the same prochiral hydroxyl group (⇒). As a consequence, by changing the direction of the process, one can select that using a given biocatalyst which enantiomer of a chiral product should be obtained.

The principle of "*inversion*" of enantiotope-selective processes was utilized very successfully in lipase-catalyzed transformation of *meso*-diols and the corresponding diacetates (Figure 8.63). Irreversibility of the acylation of *meso*-diol at its *pro-R* hydroxyl catalyzed by lipase-AK was implemented using vinyl acetate as acylating agent and led to enantiopure (*R*)-monoacetate in quantitative yield. Hydrolysis, catalyzed also by lipase-AK of the *meso*-diacetate prepared by chemical acylation, also involved the *pro-R* acetoxy group and led in very good yield to the enantiopure (*S*)-monoacetate.

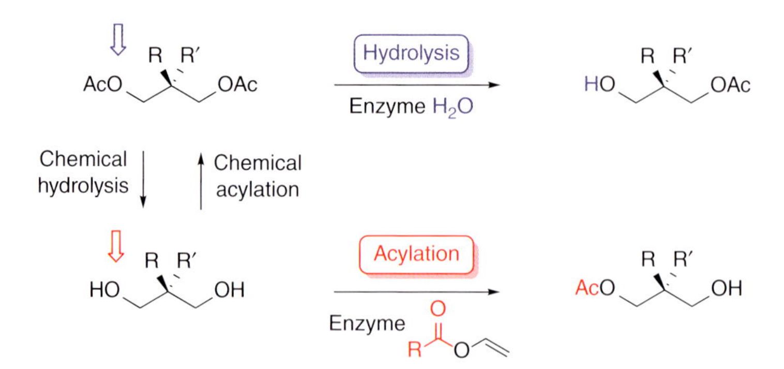


**Figure 8.62** Inversion of enantiotope selectivity using the same biocatalyst in a reaction of "opposite" direction.

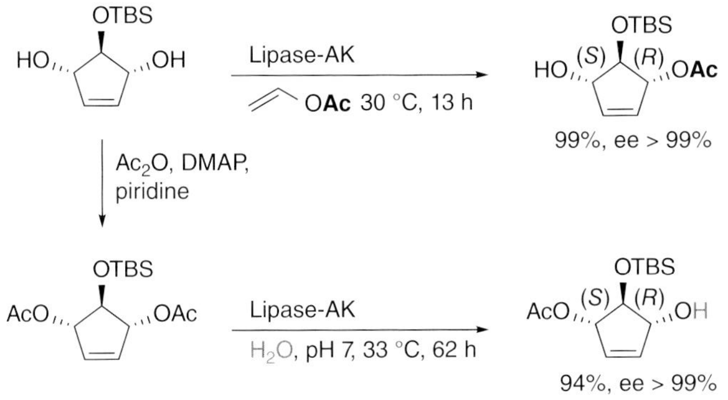

**Figure 8.63** "Inversion" of enantiotope selectivity in the preparation of cyclic monoacetate enantiomers.

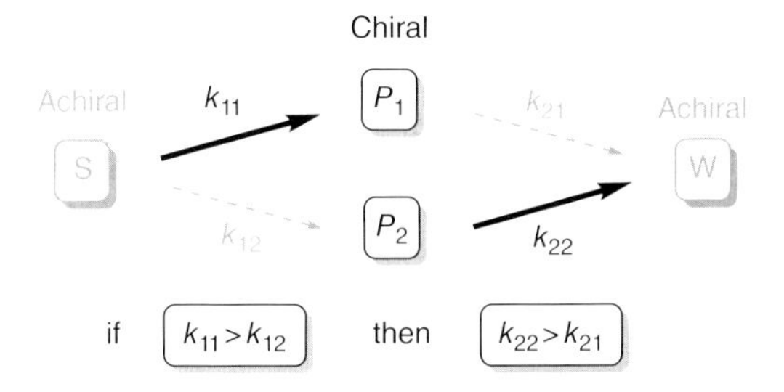


**Figure 8.64** Irreversible enantiotope-selective process involving further transformation of the product.

## 8.7.4 Kinetic Amplification

If in an enantiotope-selective process the product can undergo further transformations, the simple model becomes much more complex, even when the condition of irreversibility is completely valid (Figure 8.64).[15]

It can be easily understood that if the preference of the chiral catalyst/reagent is retained in the second step, in a consecutive process (i.e., an enantiomer-selective reaction) the enantiomer formed faster in the first step will now be transformed slower, since in the second step it is the disfavored group that remained intact (Figure 8.64). This phenomenon can be utilized in the s.c. **kinetic amplification**.[16]

15) In reactions based on enantiotope selectivity, selective transformations of one of the enantiotopic groups permit further reaction since the constitution of the two groups is identical and therefore from a chemical point of view possess the same properties. Since in the primary product, one group is still present unchanged, the possibility of its further transformation is understandable. See Ref. [8].

16) In the literature, this method is also mentioned as kinetic resolution coupled with desymmetrization.

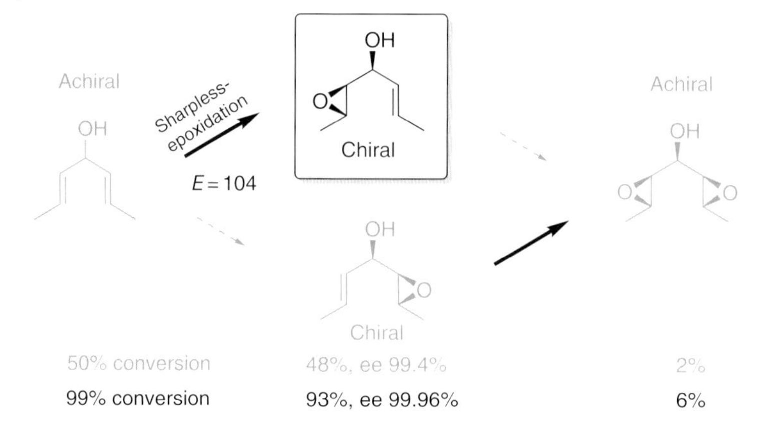


**Figure 8.65** Kinetic amplification in the Sharpless epoxidation of a bis(allyl alcohol).

Kinetic amplification was put to use in the *Sharpless* epoxidation of a prochiral bis(allyl-alcohol) (Figure 8.65).

It can be seen that since already in the first step selectivity is very high ($E = 104$), the product is formed even at low conversion with high enantiomer excess (ee = 99.4%).

Although in irreversible enantiotope-selective reactions, enantiomeric excess of the product is independent from conversion, in the second step, the two mono-functional products give an achiral bifunctional product. The second step, which is essentially a kinetic resolution, involves the preferred transformation of the minor enantiomer. In this way, it was possible that after 99% conversion, the chiral product could be isolated enantiopure (ee 99.96%) and in high yield (93%).

This is even more conspicuous in cases when the selectivity of the first step is moderate (Figure 8.66).

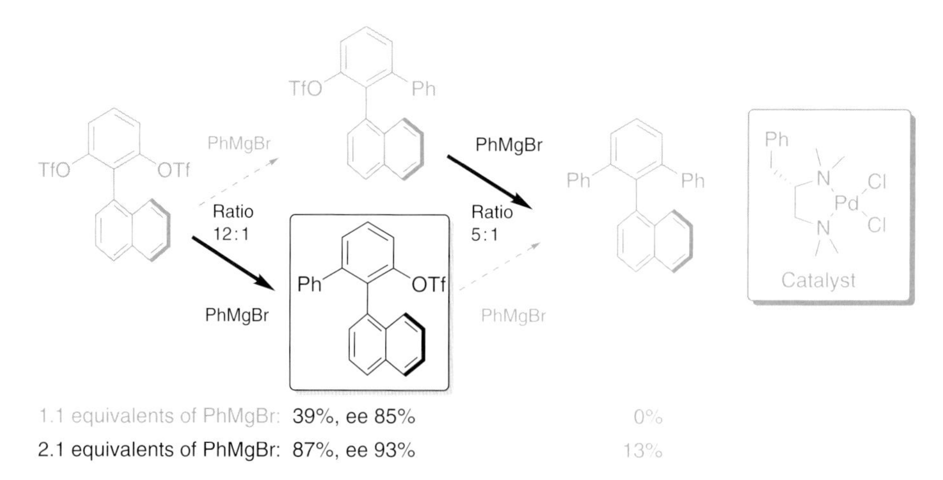


**Figure 8.66** Kinetic amplification in the enantiotope-selective synthesis of atropisomers.

The first step of enantiotope-selective *Kumada* reactions using a *Grignard* reagent proceeds with moderate selectivity ($E = 12$), and permits, when using 1.1 equivalents of the reagent and exploiting the selectivity of only the first step, the preparation of the product with a modest (ee = 85%) enantioselectivity. If by increasing the conversion, selectivity of the second step is let to dominate as well, enantiomeric excess improved impressively (at 87% conversion and 2.1 equivalents of reagent to ee = 93%). Enantiomeric excess can be further enhanced at higher conversion, but of course, at the expense of decreasing yield.

### 8.7.5 Enantiodivergent Reactions Following the Transformation of Enantiotopic Groups

In Section 8.5.3, we have shown how an enantiodivergent strategy can be employed when in a diastereotope-selective process aided by a chiral auxiliary both enantiomers of which are not available, or one of them prohibitively expensive, but nevertheless both enantiomers of the end product are needed. It has also been mentioned that a similar strategy can be applied when starting from the products of enantiotope-selective reactions (Figure 8.51(c)). This can be necessary when the catalyst aiding the transformation is not available as both enantiomers (characteristic, e.g., for biocatalysts).

Figure 8.67(a) and (b) demonstrates how it is possible starting from the product of an enantiotope-selective reaction to produce both enantiomers of the desired end product from this single enantiomer by simply changing the order of two chemoselective reactions.

Following the selective hydrolysis of the prochiral or *meso*-dicarboxylic ester (Figure 8.67(a)), the ester function of the chiral half ester was reduced to an alcohol leading by acidification to one enantiomer of a lactone. The other enantiomer of the same lactone can be prepared starting from the same half ester, if first the free acid is reduced, and after hydrolysis of the ester the product is cyclized by acidification.

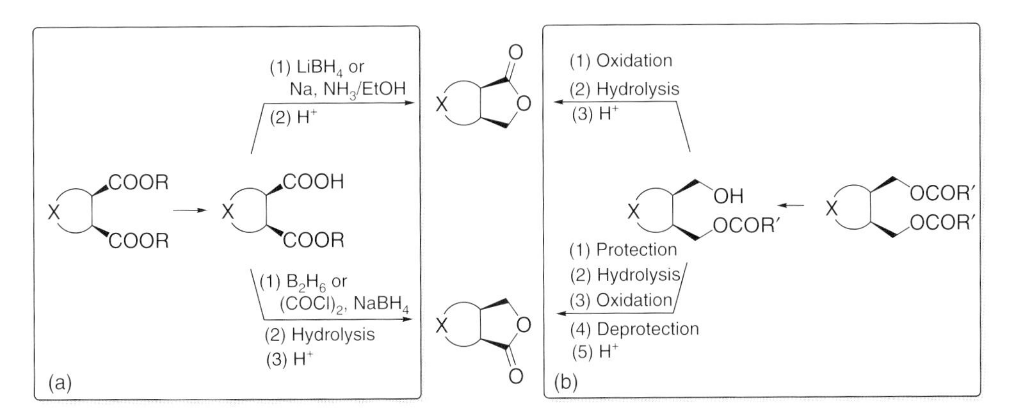


**Figure 8.67** Enantiodivergent lactone formation from half esters obtained in selective reactions.

In a similar way, after the selective hydrolysis of the prochiral or *meso*-diol diester (Figure 8.67(b)), the alcohol function of the chiral half ester is oxidized to acid; then after hydrolysis, ring closure by acidification provides one enantiomer of the lactone. For arriving to the other enantiomer first the free hydroxyl of the half ester is protected, the ester hydrolyzed to the alcohol, followed by oxidation to the acid. Finally, by removing the protective group and acidification, the other enantiomer of the lactone is obtained.

## 8.8 Combination of More Than One Type of Selectivity

In chemical reactions or biotransformations, more than one type of selectivity may concurrently occur.

Chymotrypsin (CTR)-catalyzed hydrolysis of the triester in Figure 8.68 is an example for the concurrent manifestation of chemo- and enantiotope selectivities. Selective hydrolysis of the *pro-S* ester function gave in good yield a nearly enantiopure product, while the benzoate ester function remained intact.

Baker's yeast reduced both enantiomers of the racemic bicyclic ketone in Figure 8.69, that is, enantiomer selectivity was absent, but diastereotope selectivity was high. The diastereomeric products were separable giving two alcohols of high enantiomeric excess. After reoxidation, the products were used to synthesize cyclobutane analogs of penicillins.

Parallel multiple selectivity was observed in the reduction of the cyclic triketone depicted in Figure 8.70. Its reduction with baker's yeast was regio-, enantiotope, and diastereotope selective. While reduction proceeded with almost complete stereoselectivity (the *pro-R* carbonyl group remained intact and the

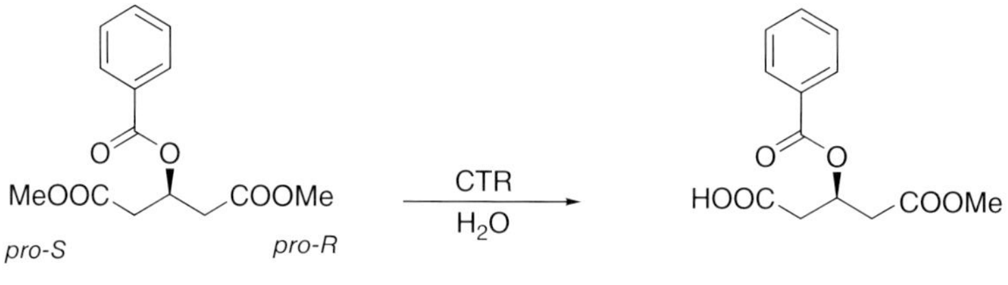

**Figure 8.68** Simultaneous chemo- and enantiotope selectivity in the hydrolysis of a triester.

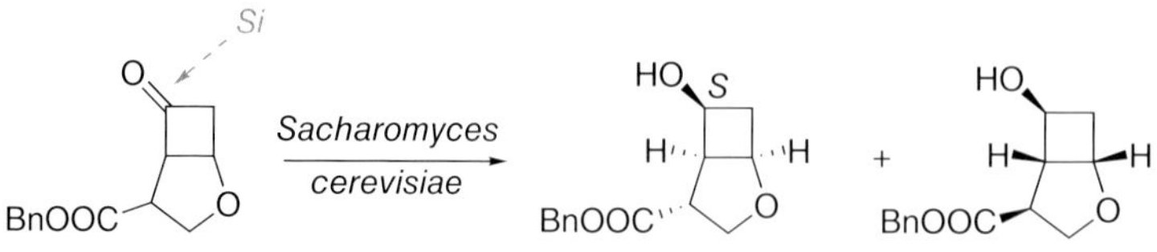

**Figure 8.69** Diastereotope selectivity in the reduction of both enantiomers of a racemic ketone.

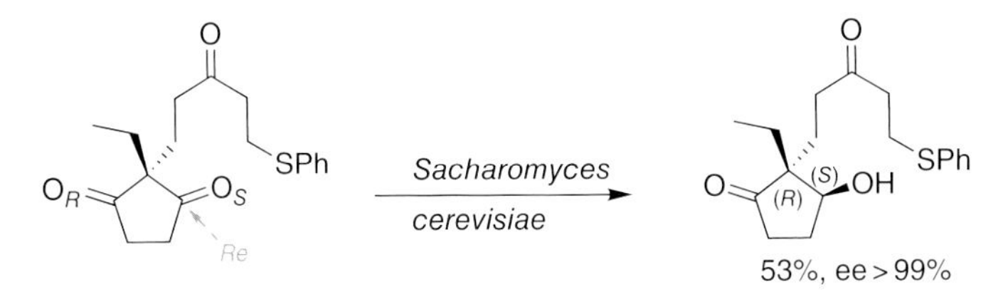


**Figure 8.70** Simultaneous regio-, diastereotope, and enantiotope selectivities manifested in the reduction of a racemic triketone.

*pro-S* carbonyl group was only attacked at its *Re* face), regioselectivity was lower, and as a by-product also 13% of a diol was isolated owing to reduction of the side chain.

## References


1. Vigneron, J.P., Dhaenens, M., and Horeau, A. (1973) Nouvelle methode pour porter au maximum la purete optique d'un produit partiellement dedouble sans l'aide d'aucune substance chirale. *Tetrahedron*, **29**, 1055–1059.
2. Anderson, N.G. (2005) Developing Processes for Crystallization-Induced Asymmetric Transformation. *Org. Proc. Res. Devel.* **9**, 800–813.
3. Weinges, K., Klotz, K.-L., and Droste, H. (1980) Asymmetrische Synthesen, V. Optisch aktive 5-Amino-4-phenyl-1,3-dioxane und deren Einfluß auf die Stereoselektivität der asymmetrischen Strecker-Synthese. *Chem. Ber.* **113**, 710–721.
4. Eliel, E.L., Wilen, S.H., and Mander, L.N. (1994) *Stereochemistry of Organic Compounds*, John Wiley & Sons, New York.
5. Kozma, D. (2002) *CRC Handbook of Optical Resolutions via Diastereomeric Salt Formation*, Boca Raton, CRC Press.
6. Chen, C.S., Fujimoto, Y., Girdaukas, G., and Sih, C.J. (1982) Quantitative analyses of biochemical kinetic resolutions of enantiomers. *J. Am. Chem. Soc.* **104**, 7294–7299.
7. Rakels, J.L.L., Straathof, A.J.J., and Heijnen, J.J. (1993) A simple method to determine the enantiomeric ratio in enantioselective biocatalysis. *Enzyme Microb. Technol.* **15**, 1051–1056.
8. Bergens, S. and Bosnich, B. (1987) Asymmetric Synthesis. Kinetic Amplification of Enantiomeric Excess. *Comm. Inorg. Chem.* **6** (2), 85–90.

# Problems to Part III

### Problem 7.1.1

The enzyme triose-phosphate isomerase is catalyzing the isomerization equilibrium between D-glyceraldehyde-3-phosphate and dihydroxyacetone-phosphate.

a) What type of selectivity is that the isomerase enzyme is producing exclusively D-glyceraldehye-3-phosphate (no L-isomer is produced)?
b) What type of selectivity is that the enzyme is transforming exclusively D-glyceraldehyde-3-phosphate to dihydroxyacetone phosphate (does not catalyze isomerization of the L-enantiomer)?

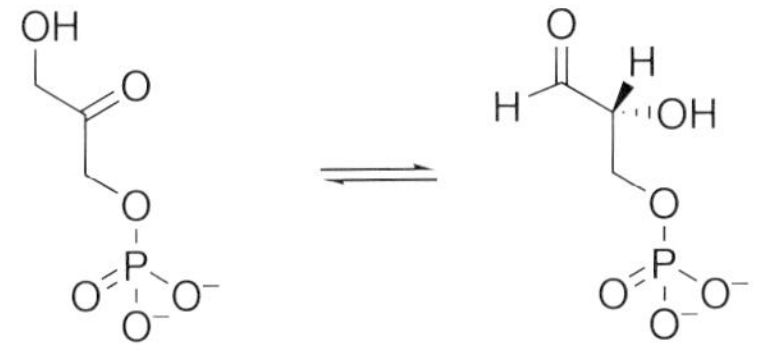

### Problem 7.2.1

The enzyme HMG-CoA reductase produces by reduction of HMG (*R*)-mevalonic acid, an intermediate of terpene syntheses. What type of selectivity can be ascribed to the fact that no (*S*)-mevalonic acid is formed in this reaction?

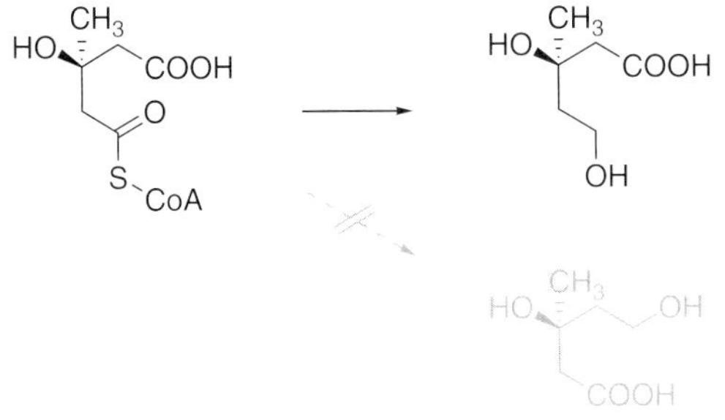

*Stereochemistry and Stereoselective Synthesis: An Introduction,* First Edition.
László Poppe and Mihály Nógrádi.

Companion Website: www.wiley.com/go/poppe/stereochemistry

**Problem 7.2.2**

With the aid of the lipoxygenase enzyme system, arachidonic acid is transformed to $LTA_4$, while the same substrate is transformed by the cyclooxygenase enzyme system to $PGG_2$.

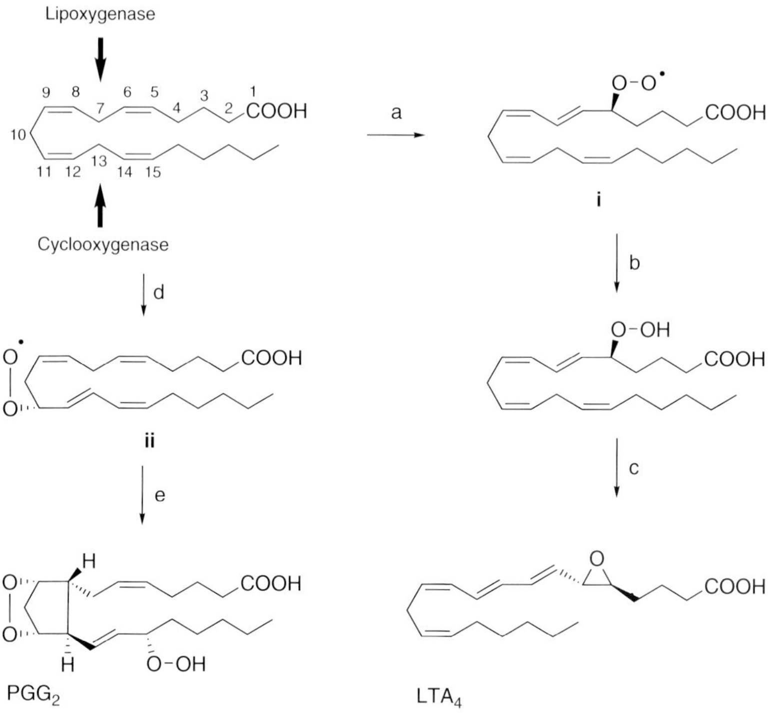

Lipoxygenase: (a) –H, $+O_2$, (b) $+e^-/+H^+$, (c) –H, –OH
Cyclooxygenase: (d) –H, $+O_2$, (e) $+O_2$, $+e^-/+H^+$

- What type of selectivity is manifested by the fact that the first enzyme system cleaves a hydrogen atom from C7, while the second one from C13 as the first step of this chain of reactions?
- What type of selectivity is manifested by the fact that both enzyme systems are cleaving the *pro-S* hydrogen atom?
- What type of selectivity is manifested by the fact that the delocalized radicals arising by the cleavage of hydrogen atoms (C5–C9 and C11–C15, respectively) react with oxygen exclusively at carbons C5 and C11?
- What type of selectivity is manifested by the fact that in the reactions on entering the oxygen atoms, only a single stereoisomer of the peroxy radical (**i** and **ii**, respectively) is formed?
- What type of selectivity is manifested by the fact that while the peroxy radical **i** is reduced with the intermediacy of the first enzyme system, peroxy radical **ii**

is entering a cyclization reaction with the aid of the second enzyme system and another molecule of oxygen?

### Problem 8.1.1

Is the solvolysis of (1*R*,2*R*)-2-(tosyloxy)cyclohexyl acetate in acetic acid stereospecific, in the course of which (1*R*,2*R*)-cyclohexane-1,2-diyl diacetate and (1*S*,2*S*)-cyclohexane-1,2-diyl diacetate are formed in a ratio of 50–50%? Why is (1*R*,2*S*)-cyclohexane-1,2-diyl diacetate not forming in this reaction?

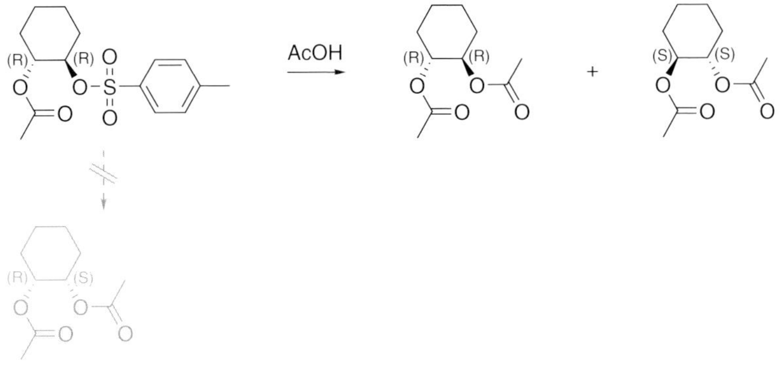

(R) (R) AcOH (R) (R) + (S) (S)

(R) (S)

### Problem 8.2.1

Are the following reactions of malic acid studied by Walden stereospecific or stereoselective reactions?

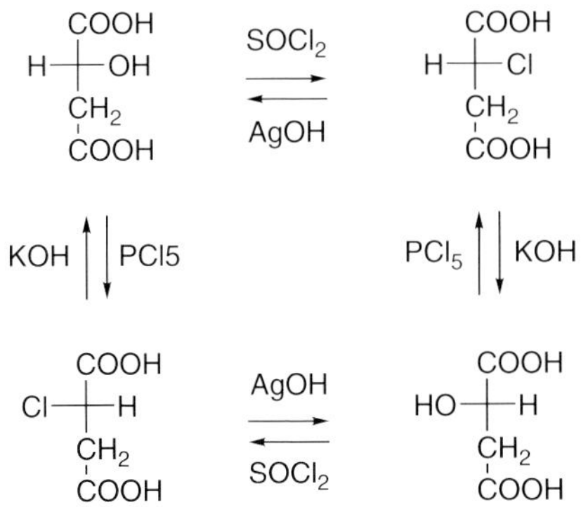

COOH / H–OH / $CH_2$ / COOH $\underset{AgOH}{\overset{SOCl_2}{\rightleftharpoons}}$ COOH / H–Cl / $CH_2$ / COOH

KOH / PCl5 ; $PCl_5$ / KOH

COOH / Cl–H / $CH_2$ / COOH $\underset{SOCl_2}{\overset{AgOH}{\rightleftharpoons}}$ COOH / HO–H / $CH_2$ / COOH

### Problem 8.3.1

Preparation of L-threonine from crotonic acid passes the steps shown below. Addition of hypobromic acid ($Br_2$, $H_2O$) to crotonic acid, followed by treatment with cc. ammonia solution gives DL-allothreonine. The amino group of the product is protected by benzoylation, the carboxyl group by esterification, and protected allothreonine is reacted with thionyl chloride to give a dihydrooxazolecarboxylic acid with a DL-threonine configuration. After alkaline hydrolysis of the ester group, the acid is resolved with brucine. The precipitating salt has an L-threonine configuration that is converted to L-threonine hydrochloride by acidic hydrolysis. Characterize the selectivity of each individual steps.

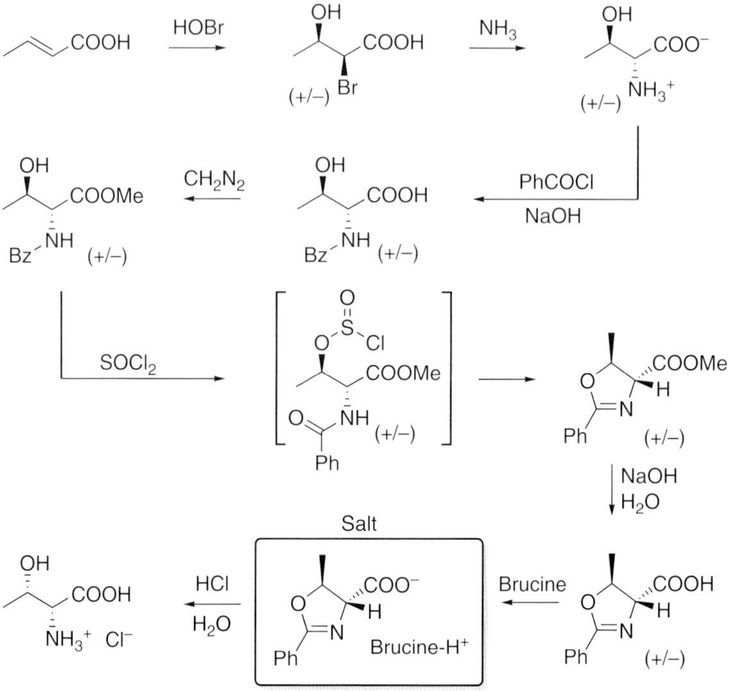

**Problem 8.4.1**

Explain the diastereomer selectivity observed in the following reaction. What is the explanation of $k_1 \ll k_2$?

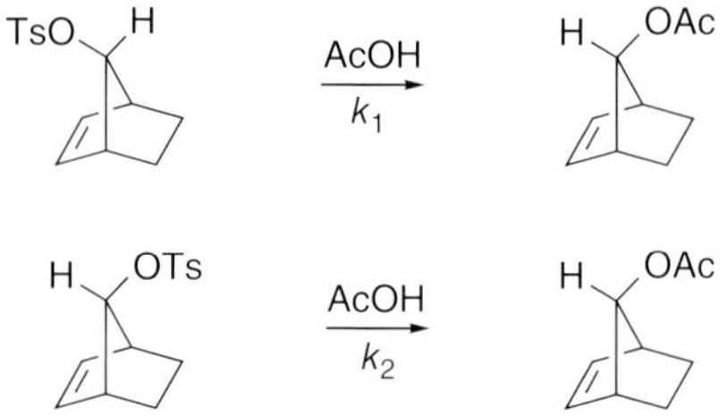

**Problem 8.5.1**

From the intermediate of the preceding reactions, there is the possibility for the formation of two diastereomeric products. Explain the observed diastereotope selectivity!

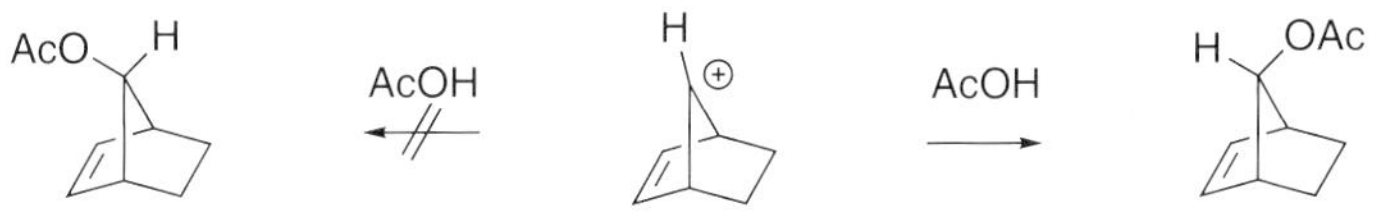

AcO H — AcOH (crossed-out arrow) ← H ⊕ — AcOH → H OAc

**Problem 8.5.2**

Are the two diastereomeric products of the following reaction formed in an equal or an unequal proportion?

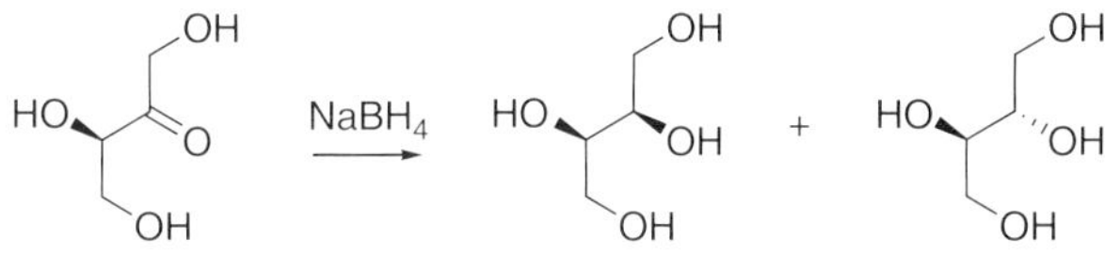

HO, OH, O, OH — $NaBH_4$ → HO, OH, OH, OH + HO, OH, OH, OH

**Problem 8.5.3**

What types of selectivity do you identify in the course of the following synthesis carried out with chirality transfer?

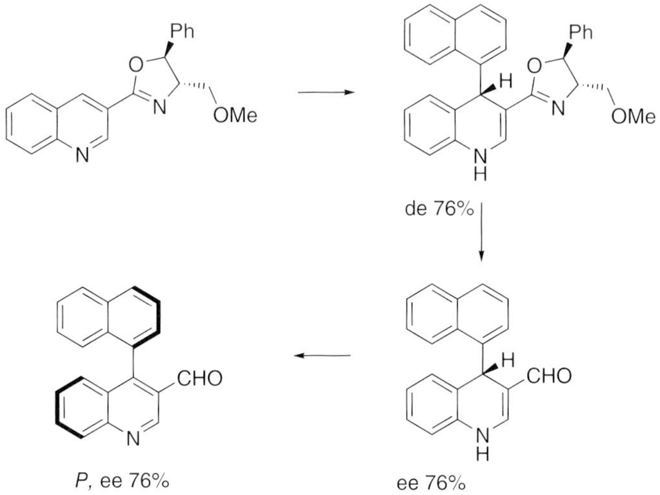

N, O, N, Ph, OMe → H, O, N, Ph, OMe, N H

de 76%

H, CHO, N H

ee 76%

CHO, N

*P*, ee 76%

**Problem 8.5.4**

Stereoselective synthesis of the sesquiterpene below was carried out from a naphthalene-1-one derivative by double-chirality transfer. In the first step, the ketone was reduced with a chiral reagent followed by forming a chromium complex and removal of the hydroxyl group serving as chiral auxiliary. The chiral-chromium complex was alkylated in several lithiation steps; finally, the chromium complex was decomposed.

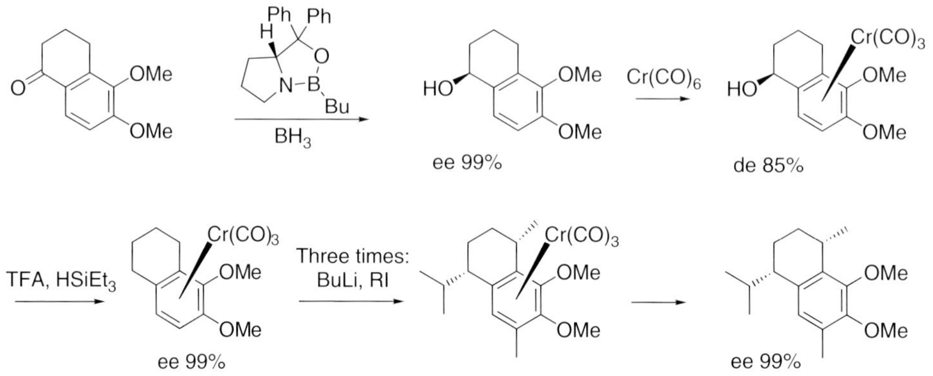

- What type of selectivity is presenting the reduction of the ketone?
- What type of selectivity is present in the complexation reaction with chromium?
- What type of selectivity is present in the two alkylation reactions affecting the ring?

**Problem 8.5.5**

In the previous synthesis, the hydroxy-chromium complex of de 85% yielded a reduced chromium complex of ee 99%. Were the diastereomeric chromium complexes separated prior or after the removal of the hydroxyl group?

**Problem 8.5.6**

Give an explanation for the facts that while the optical rotation of an aqueous solution of D-glucose is +52.7, that of its freshly prepared aqueous solution of from D-glucose crystallized from its aqueous solution with the addition of acetic acid is 112.2, and further that of a freshly prepared aqueous solution of D-glucose crystallized from an aqueous solution with the addition of ethanol is +18.7!

**Problem 8.5.7**

What type of selectivity do you assign to reactions in the first phase of the *Calvin* cycle? What is the role of $Mg^{2+}$ cation in manifestation of selectivity?

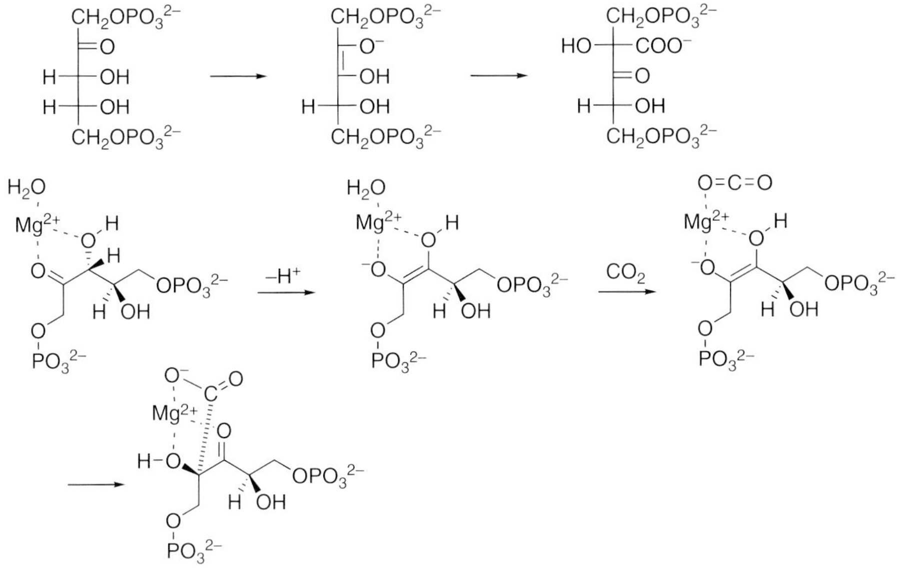

**Problem 8.6.1**

Based on the data given below, calculate enantiomer selectivity ($E$) of *Sharpless* epoxidation [Ti(O*i*Pr)$_4$, L-(+)-DIPT, *t*-BuOOH] for the individual reactions!

| Substrate | Unreacted substrate | |
|---|---|---|
| | Yield (%) | ee (%) |
| 1-Cyclohexylbut-2-en-1-ol | 48 | 94 |
| 1-(Cyclohex-1-enyl)ethanol | 47 | 98 |
| Non-1-en-3-ol | 46 | 86 |

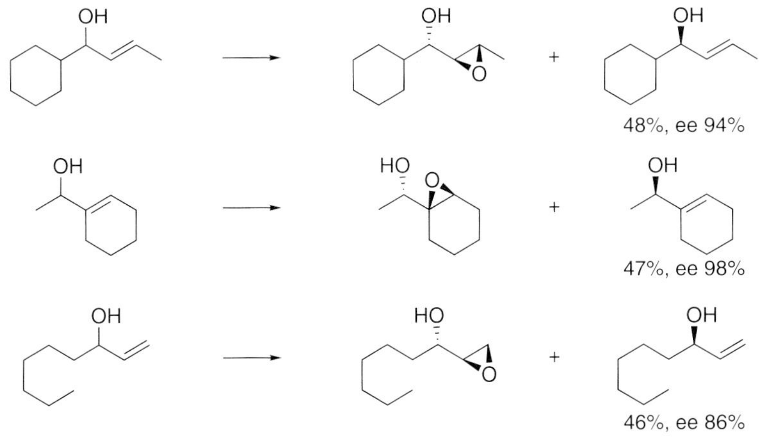

48%, ee 94%

47%, ee 98%

46%, ee 86%

**Problem 8.6.2**

Calculate the enantiomer selectivity ($E$) for the reaction catalyzed by Rh-catalyst (A) based on ee data for the product and the unreacted substrate!

Rh-catalyst (A)

OMe

52%, ee 72%

41%, ee 93%

Rh-catalyst (B)

47%, ee 84%

45%, ee 88%

Using another Rh-catalyst (B), both enantiomers are transformed resulting in different products, each. Calculate the quantity and enantiomeric composition of the unreacted substrate!

**Problem 8.6.3**

Methyl acetate or vinyl acetate should be preferred for performing lipase-catalyzed enzymic acylation of 6-methyl-5-hepten-2-ol? Explain your answer!

Lipase

### Problem 8.6.4

Phenylalanine ammonia-lyase (PAL) is transforming L-3-arylalanines to derivatives of cinnamic acid. The phenylalanine 2,3-aminomutase (PAM) from *Taxus canadensis* (*Tc*PAM) can transform L-3-arylalanines to (*R*)-3-aryl-$\beta$-alanines. Both the enzymes also catalyze the reverse reactions. (Phenyl)alanine racemase (PR) can transform both L- and D-3-arylalanines to nearly racemic arylalanines.

$COO^-$, $NH_3^+$ — PAL ⇌ — $COO^-$ + $NH_3$

$COO^-$, $NH_3^+$ — PAM ⇌ — $NH_3^+$, $COO^-$

$COO^-$, $NH_3^+$ — PR ⇌ — $COO^-$, $NH_3^+$

- What is the product of the reaction of cinnamic acid in 5 M ammonia solution in the presence of PAL?
- What is the product of the reaction of cinnamic acid in 5 M ammonia solution in the presence of a system consisting of PAL and *Tc*PAM?
- What are the products of the reaction of a racemic phenylalanine with PAL?
- What are the products of the reaction of racemic phenylalanine with *Tc*PAM?
- Compose a multienzyme system capable to transform racemic phenylalanine to (*R*)-phenylalanine close to 100% conversion (dynamic kinetic resolution)!

### Problem 8.6.5

Vitamin $B_6$ (pyridoxal phosphate) catalyzing among others isomerization of amino acid. *Candida antarctica* lipase B (CaLB) is transforming only L-amino acid esters to amides insoluble in water by ammonolysis to amides. Design processes for the transformation of racemic amino acids to L-amino acid amides by batch and continuous mode dynamic kinetic resolution.

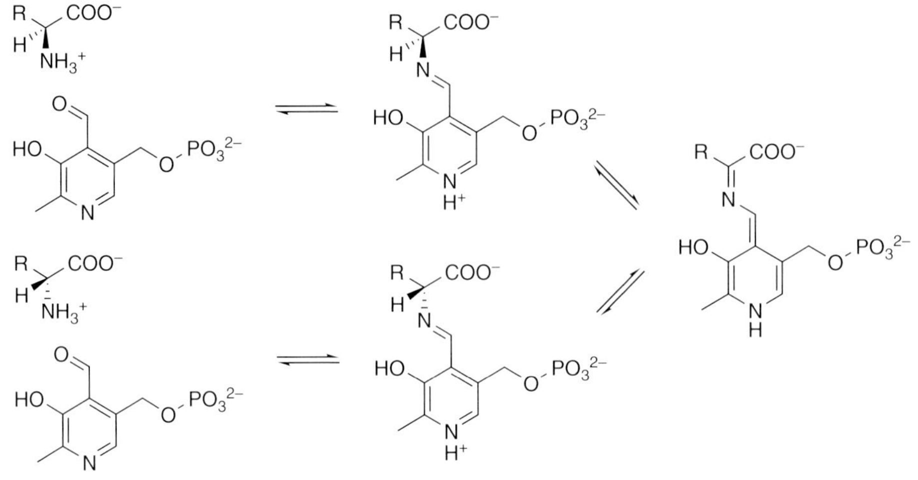

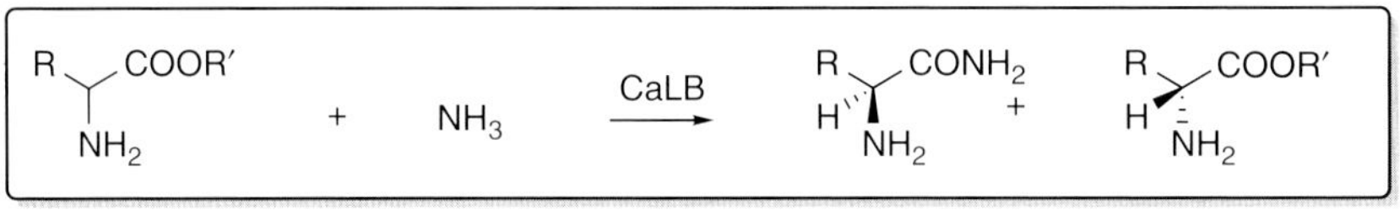

### Problem 8.7.1

(*R*)-Glycerol-3-phosphate necessary for the biosynthesis of triglycerides is formed in the living organisms in two ways: either by phosphorylation of glycerol with the aid of *glycerol kinase* or from dihydroxyacetone phosphate by reduction with NADH with the aid of *glycerol phosphate dehydrogenase*. What type of selectivity is characterizing the two enzymes?

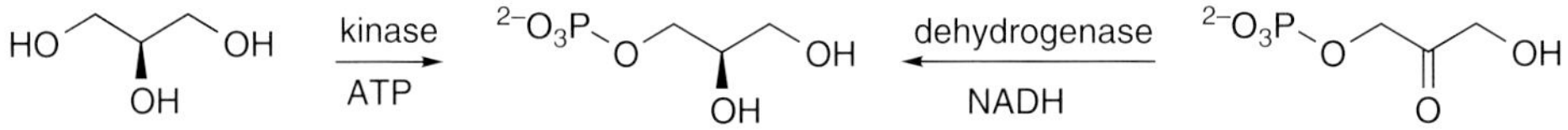

### Problem 8.7.2

Why is it advantageous to use catalysts of $C_2$ symmetry, such as derivatives of tartaric acid (*R,R* or *S,S*) as chiral catalysts?

### Problem 8.7.3

Triose-phosphate isomerase enzyme is transforming dihydroxyacetone phosphate to D-glyceraldehyde-3-phosphate via the intermediate enediolate. What type of selectivity characterizes the formation of the enediolate and its further transformation?

# Part IV
# Applications of Enantioselective Methods

In this part, chemical transformations are reviewed suitable for the generation of enantiomeric products in a nonracemic form. In this respect, one can distinguish (i) stoichiometric methods in which the source of chirality is present in the starting materials in an equivalent amount and (ii) kinetic methods in which the source of chirality is only added in a catalytic amount to the reaction mixture.

*Stereochemistry and Stereoselective Synthesis: An Introduction*, First Edition.
László Poppe and Mihály Nógrádi.

Companion Website: www.wiley.com/go/poppe/stereochemistry

# 9
# Stoichiometric Methods of Enantioselective Synthesis

One of the stoichiometric approaches is when chiral starting materials are employed for the preparation of a chiral end product. If the chiral source or part of it is not incorporated into the end product, the procedure is called *chirality transfer*.

Nowadays, a large number of chiral compounds at various prices are at the disposal of synthetic chemists. This assembly of compounds is called the *chiral pool*. Members of the chiral pool may originate from the following:

- ➢ Biological sources: most often obtained from plant materials by extraction and are typically the representatives of the following classes of compounds: terpenoids, alkaloids, amino acids, and carbohydrates.
- ➢ Members of the following classes of compounds can be produced *de novo* by fermentations using natural microorganisms or their genetically modified versions: hydroxy acids, amino acids, steroids, and antibiotics.
- ➢ Enantioselective synthesis: economic synthetic preparation of various chiral compounds of biological origin.

The simplest type of stoichiometric methods is when the chiral starting material is incorporated into the product with either the original or inverted configuration (Figure 9.1).

Often, the end product does not contain all of the chiral elements of the starting material. For example, the antibiotic negamycin (**5**) containing two centers of chirality is prepared from glucose (**4**) having four centers of chirality (Figure 9.2).

## 9.1
## Diastereotope Selective Methods

### 9.1.1
### Substrate-Controlled Selectivity

The first type of stereoselective synthesis is when into a molecule containing already one or more stereogenic elements, one or more additional stereogenic elements are introduced. Molecules containing opposite configuration of the

*Stereochemistry and Stereoselective Synthesis: An Introduction*, First Edition.
László Poppe and Mihály Nógrádi.

Companion Website: www.wiley.com/go/poppe/stereochemistry

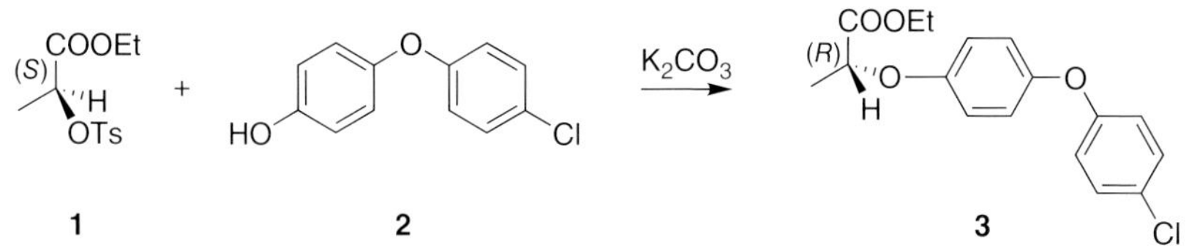

**Figure 9.1** A stereospecific $S_N2$ reaction.

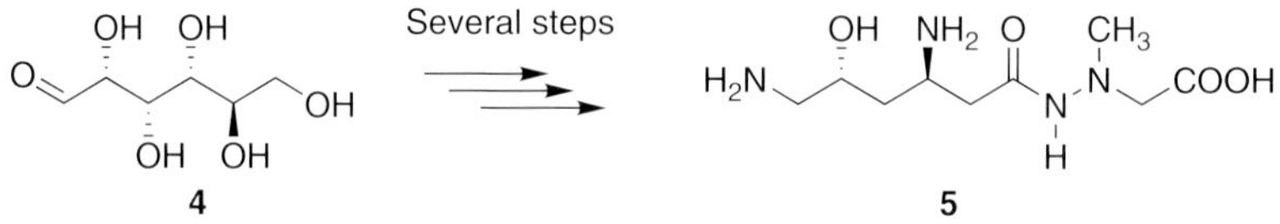

**Figure 9.2** Stoichiometric stereoselective synthesis of negamycin.

new stereogenic element(s) are related as diastereomers. Consequently, their free enthalpy of formation is different and also the free enthalpies of activation of reactions leading to them are also different. Therefore, both under kinetic and thermodynamic control, the probability of formation of the individual diastereomers will be unequal. In general, the larger this difference is, the larger is the steric or electronic interaction between the newly formed and the original stereogenic elements.

As a typical example, the initial steps of the total synthesis of the antibiotic rifamycin are quoted. Starting from (*S*)-3-benzyloxy-2-methylpropanal (**6**), the unsaturated ester **7** was prepared by means of a series of reactions not involving the already existing center of chirality. These reactions were dominated by the first manifestation of diastereotope selectivity: the newly formed double bond assumed a *Z* configuration. Reduction of the ester gave an allyl alcohol (**8**), which was oxidized with *m*-chloroperbenzoic acid to an epoxide (**9**). Here, a second diastereotope selective effect prevailed: the oxidant attacked the double bond exclusively from the less-hindered face. After desilylation, a regio- and diastereoselective alkylation gave the protected triol (**10**), which contained already three centers of chirality (Figure 9.3).

In substrate-controlled processes, the use of symmetrical reagents (such as *m*-chloroperbenzoic acid, $Me_2CuLi$) can secure high selectivity, since the source of asymmetry is inherent in the substrate.

### 9.1.2
### Diastereotope Selectivity Controlled by an Auxiliary Group – Enantioselective Synthesis

A more advanced variation of substrate-controlled syntheses is when the group imparting selectivity is only attached transiently to the substrate and after completion of the diastereoselective step, it is cleaved off from the product. When a new center of chirality comes into being, the overall process can be regarded as

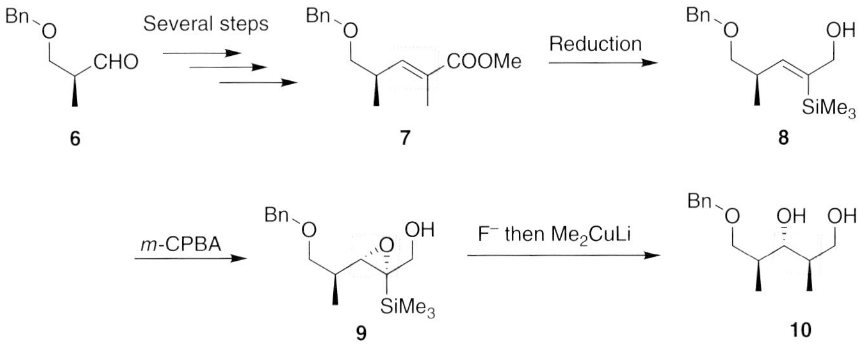

Figure 9.3 Initial steps of the total synthesis of rifamycin. (New stereogenic elements formed in a diastereoselective way are highlighted.)

enantioselective. Since the source of chirality is the chiral auxiliary, the method is often called **chirality transfer**. The chiral auxiliary group is generally derived from an inexpensive natural chiral molecule, for example, an amino acid derivative.

Enantioselective alkylation of carboxylic acids was realized by diastereotope selective reaction of amides formed with (*S*)-prolinol. Direction of diastereotope selectivity is much dependent on the structure of the auxiliary group. For example, with *N*-propanoyl-(*S*)-prolinol (**11**), it is the *Si*-face of the enolate (**12**) that was the preferred face of alkylation, while with *O*-ethyl-*N*-propanoyl-(*S*)-prolinol (**15**), alkylation from the *Re*-face was dominant. In this way, it was possible to prepare either enantiomers of 2-methylbutyric acid (**14** or **18**) in about ee 85% (Figure 9.4).

It is usually difficult to find a pair of highly efficient chiral auxiliary groups of opposite stereoselectivity. Thus, for the preparation of enantiomers of 2-methylhexanoic acid instead of prolinol, a more convenient auxiliary group with

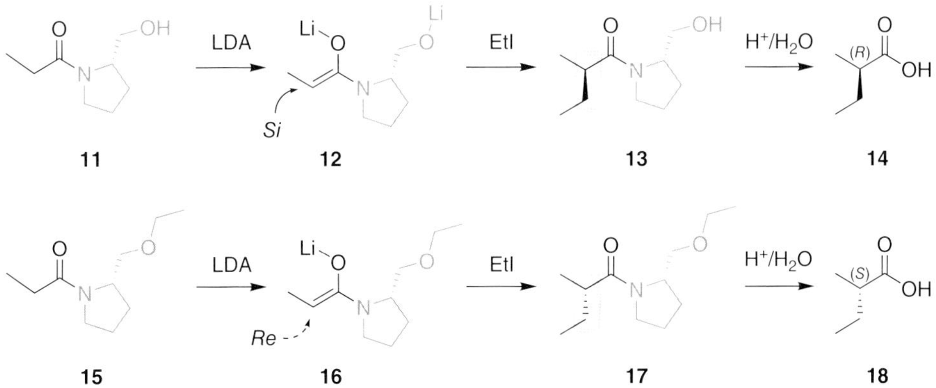

Figure 9.4 Stereoselective alkylations aided by (*S*)-prolinol as chiral auxiliary group. (New stereogenic elements formed in a diastereoselective manner are highlighted.)

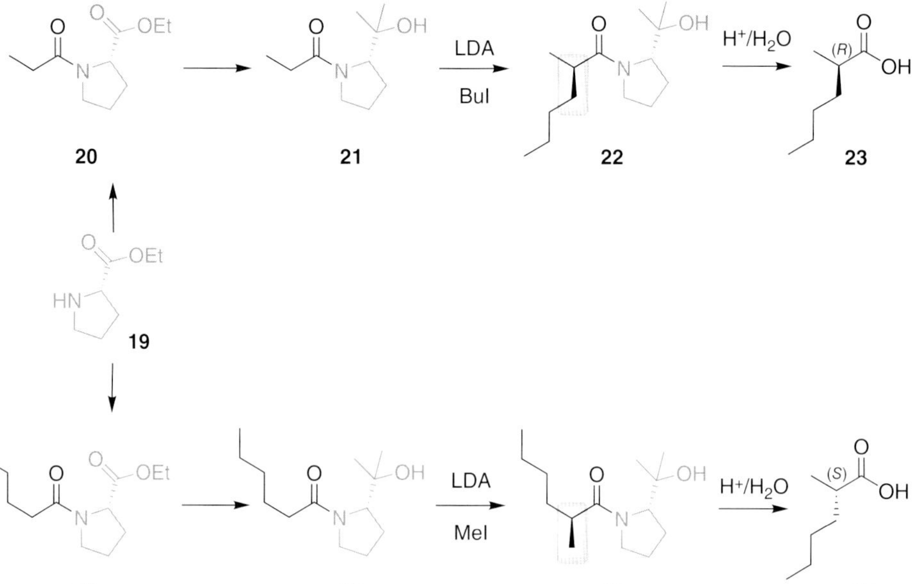

**Figure 9.5** Stereoselective alkylations using an auxiliary group derived from (*S*)-prolinol. (New stereogenic elements formed in a diastereoselective manner are highlighted.)

a *tert*-alcohol function, was used which could be finally removed from the end product by extraction with an acid. Proline ethyl ester (**19**) was first acylated with propanoyl chloride. The resulting amide (**20**) was transformed to a *tert*-alcohol function by *Grignard* reaction. The product (**21**) was alkylated with *n*-butyl iodide, followed by hydrolysis to give (*R*)-2-methylhexanoic acid (**23**) in 77% yield and ee 87% (Figure 9.5). When acylation was carried out with hexanoyl chloride, and after an analogous transformation of the amide **24** was alkylated with methyl iodide, after hydrolysis, (*S*)-2-methylhexanoic acid (**27**) was obtained in 96% yield and ee 75%.[1] An improved variant of the above synthesis employed an **enantiodivergent strategy** in which the same auxiliary group and the same building blocks were used to prepare both enantiomers (Figure 9.6).

Starting from the chiral oxazoline (**29**) readily prepared from (1*S*,2*S*)-2-amino-1-phenyl-3-methoxypropanol (**28**) and acetic acid by changing the order of the two alkylation steps, either (*R*)- or (*S*)-2-methylhexanoic acid (**23** or **27**) could be prepared with better than ee 70%. The source of stereoselectivity is that in the lithium azaenolates formed from compounds **30** and **32**, respectively, by treatment with lithium diisopropylamide (LDA) and in which the lithium cation is complexed by the oxazoline nitrogen and the oxygen atom of the methoxymethyl

1) Since the intermediates of the two series of reactions (**20–21** and **24–25**) are not enantiomers of each other, it is obvious that yields and enantiomeric excess values are different.

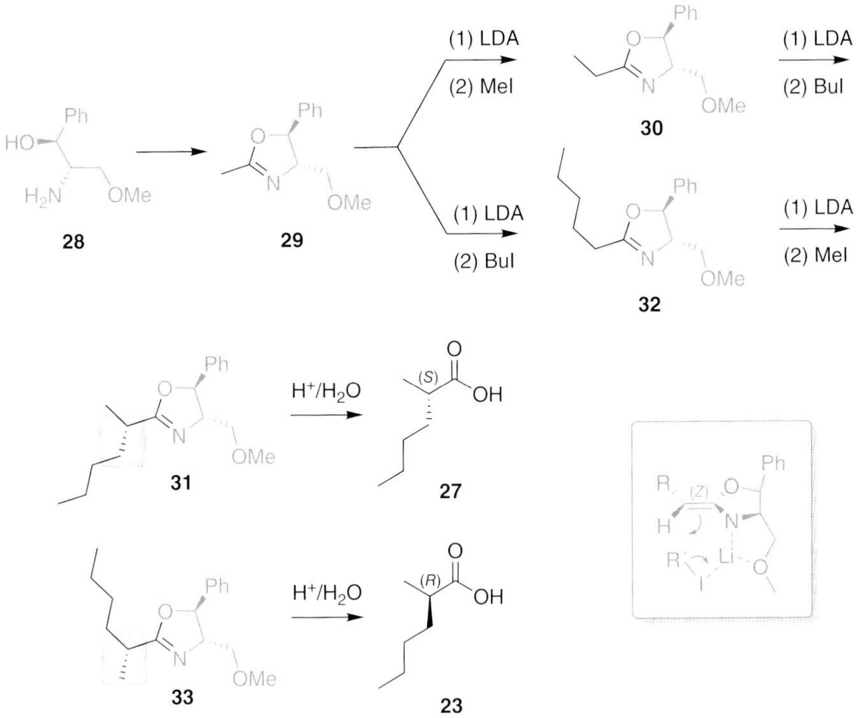

**Figure 9.6** Stereoselective alkylation employing an enantiodivergent strategy. (The new stereogenic elements formed in a diastereoselective manner are highlighted.) In the insert, the favored transition state of alkylation is shown.

side chain, binds the alkylating agent on the side of the methoxymethyl group of the ring.[2)]

In processes using chiral auxiliary groups, it is possible to achieve a high degree of selectivity using simple achiral substrates (e.g., propanoyl chloride, hexanoyl chloride, and acetic acid) and simple achiral reagents (e.g., LDA, methyl iodide, and butyl iodide), since the source of stereoselectivity is the auxiliary group. A disadvantage of this approach is that a stoichiometric amount of the auxiliary group is expended and if it cannot be recycled, the carbon efficiency of the method is poor.[3)]

When an auxiliary group contains a function capable of complex formation, higher selectivity can be achieved if it contains two such stereogenic elements.

2) Depending on the structure of group R, the *E/Z* ratio of the lithium azaenolate varies, and this ratio determines the diastereoselectivity of alkylation and thus the isomer ratio of the end product.
3) Carbon efficiency is defined as the percentage of all carbon atoms that are incorporated into the end product. For example, in the synthesis of 2-methylhexanoic acid shown in Figure 9.5, only 7 of the 14 carbon atoms are built into the end product, that is, carbon efficiency is but 50%. In the synthesis in Figure 9.6, after the final hydrolysis, aminoalcohol **28** can be recovered, improving thereby the cost efficiency of the process.

**Figure 9.7** Stereoselective alkylation using a chiral auxiliary group of $C_2$ symmetry. (New stereogenic elements formed in a diastereoselective manner are highlighted.)

Especially efficient are chiral auxiliary groups of $C_2$ symmetry containing two identical centers of chirality, such as (2*R*,5*R*)-2,5-bis(methoxymethyl)pyrrolidine (**34**). Using this auxiliary group in the synthesis of (*R*)- or (*S*)-2-methylhexanoic acid (**23** or **27**), a yield of 81% and ee 95% could be realized (Figure 9.7).[4)]

### 9.1.3 Double Asymmetric Induction. Concerted Diastereotope Selectivity of Chiral Molecules

A special case of substrate-controlled diastereotope selectivity is when two organic molecules are joined, both of which already contain stereogenic elements. Directing effect of the two substrates producing a new stereogenic element is either cooperating positively (*matched pair*) or in an antagonistic way (*mismatched pair*). As examples, five diastereotope selective aldol reactions are quoted. In the first three, only one of the partners is chiral, while in the last two, both are chiral.

Reaction of the chiral aldehyde **39** with the achiral boranyl enolate **40** provided in a 3 : 2 ratio the two diastereomers **41** and **42**. In the course of the diastereotope selective reaction, two centers of asymmetry are generated and therefore, in principle, four diastereomers could have been formed. However, owing to the fixed structure of the complexes formed in the transition states (**A** and **B**)[5)], only the products of *syn* configuration (**41** and **42**) could be derived. The asymmetric

4) High stereoselectivity can be attributed to the fact that in the lithium enolates generated from the acyl derivatives **35** and **37** *E*/*Z* isomerism is absent.

5) In the six-membered cyclic transition sate, stabilized by the complex bond between the boron atom and the oxygen of the carbonyl group, the bulky phenylsulfanyl group and the 4-(methoxycarbonyl)pentan-2-yl group with a branched chain take up equatorial positions and thus the methyl group of the enolate is forced into an axial position.

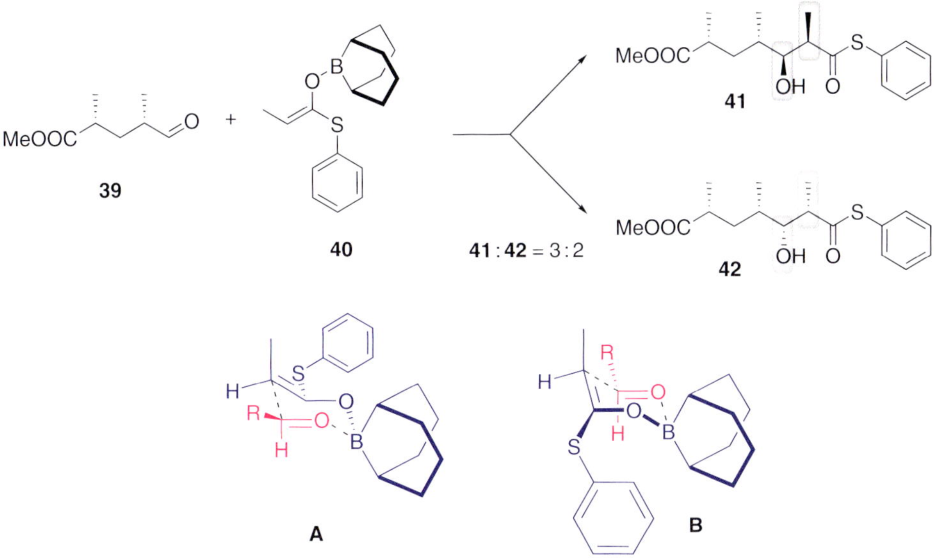

**Figure 9.8** Stereoselective aldol reaction of a chiral aldehyde and an achiral enolate. (The stereogenic elements newly formed in a diastereoselective manner were highlighted.) In the insert, the favored transition states are depicted; the two components are drawn in different colors.

centers of aldehyde **39** moderately promote the formation of diastereomer **41** with an *S*-configured hydroxyl group and an *R*-configured methyl group (Figure 9.8).

Reaction of the chiral boranyl enolate (*S*)-**43** with the achiral isobutyraldehyde **44** (Figure 9.9) is highly diastereotope selective and produces the diastereomers **44** and **45** in a ratio of 99 : 1. The stereogenic element of the boranyl enolate promotes the formation of a product with an *S*-configured hydroxyl and an *R*-configured methyl group.

Obviously, when as chiral component the antipodal boranyl enolate [(*R*)-**43**] was employed, the predominant diastereomer (**47**) contained an (*R*)-hydroxyl and an (*S*)-methyl group (Figure 9.9).

When the chiral aldehyde **39** preferring the (*S*)-hydroxy-(*R*)-methyl configuration in the product was reacted with the chiral boranyl enolate (*S*)-**43** with the same preference (*matched pair*), an efficient cooperative directing effect of both components produced exclusively the alcohol **48**, a compound containing not less than *five centers* of chirality (Figure 9.10).

If in turn the same aldehyde **39** was paired with the boranyl enolate (*R*)-**43** showing opposite preference (*mismatched pair*), the directing effect of the components counteracted each other and the diastereomers **49** and **50** formed in a ratio of 1 : 30 (Figure 9.10).

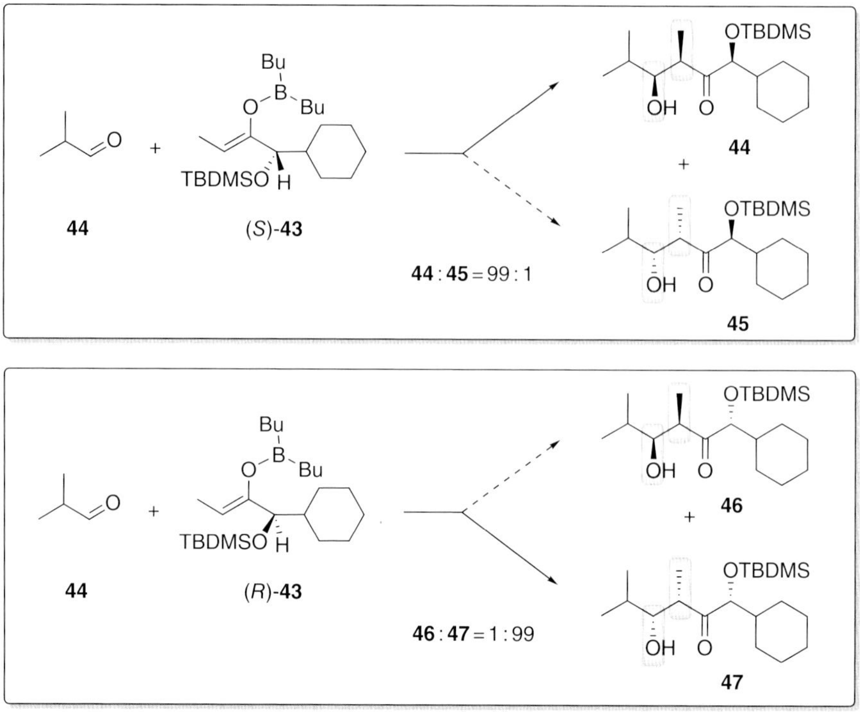

Figure 9.9 Stereoselective aldol reaction of an achiral aldehyde and a chiral enolate. (New stereogenic elements formed in a diastereoselective manner are highlighted.)

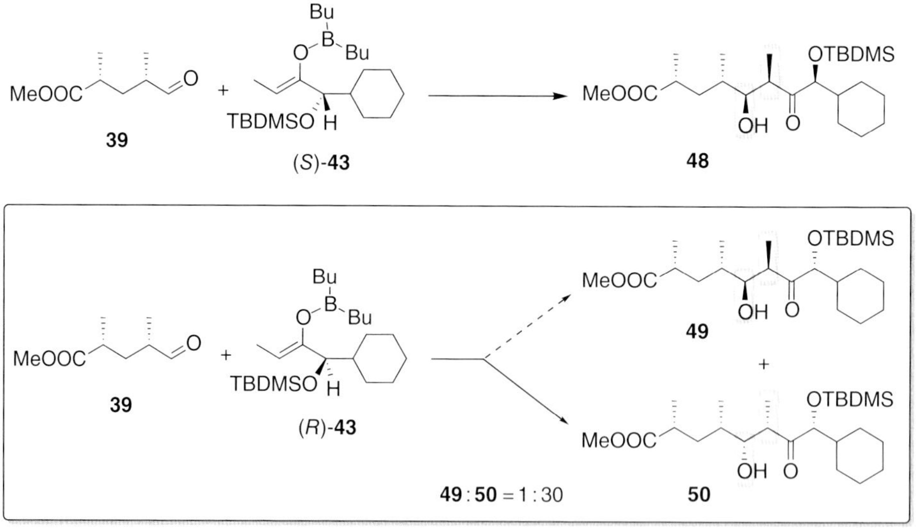

Figure 9.10 Stereoselective aldol reactions of chiral enolates with chiral aldehydes. The upper row shows a "matched pair" and the lower one a "mismatched pair." New stereogenic elements formed in a diastereotope selective manner are highlighted.

## 9.2 Enantiotope Selective Methods

Enantiotope selective methods are such methods that generate from a prochiral substrate a pair of enantiomeric products in a ratio different from 1 : 1. Diastereomeric relationship between the possible transition states of a reaction producing a new stereogenic element can be generated either when there is a stereogenic element already present in the substrate (asymmetric induction according to earlier usage) or using chiral reagents. When a new center of chirality is formed in an achiral substrate, a pair of enantiomers is produced and therefore we call this procedure enantioselective chirality transfer.

### 9.2.1 Reagent-Controlled Selectivity

Enantiotope selective reduction of aliphatic–aromatic ketones can be accomplished, for example, with chiral hydrides of aluminum and boron. The chiral

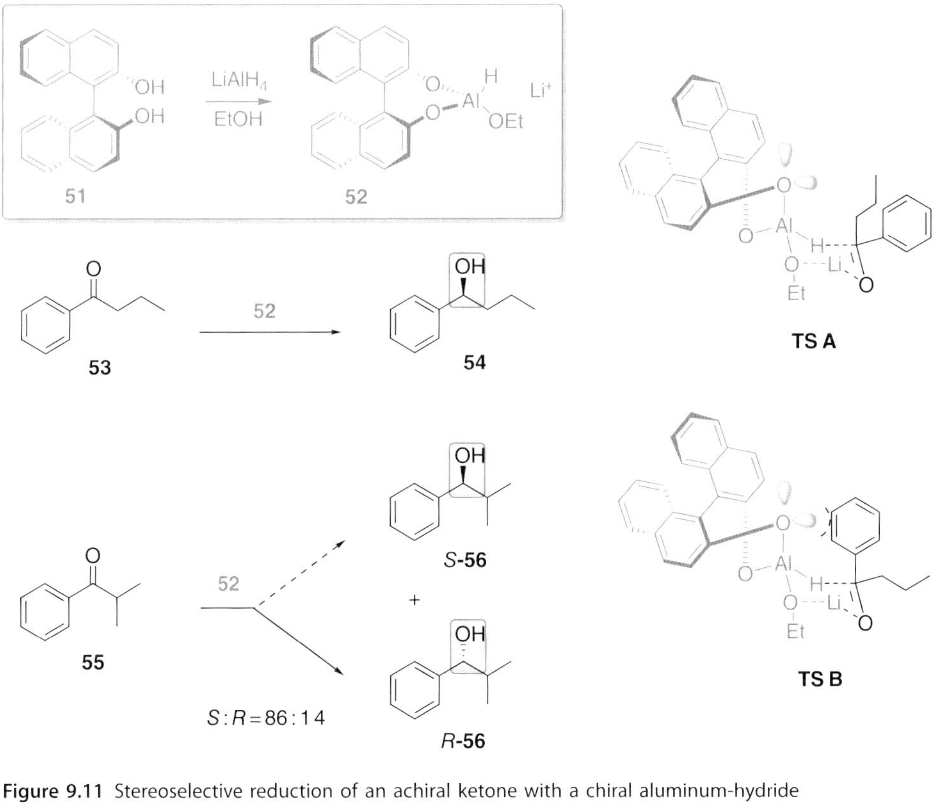

51 → ($LiAlH_4$, EtOH) → 52

53 → (52) → 54

55 → (52) → S-56 + R-56

S : R = 86 : 14

TS A

TS B

**Figure 9.11** Stereoselective reduction of an achiral ketone with a chiral aluminum-hydride reagent. The new stereogenic element formed in an enantiotope selective way is highlighted. (**TS A**) Favored transition complex. (**TS B**) Disfavored transition complex.

reagent attacks the opposite enantiotopic faces of the carbonyl group with unequal probability.

Reduction of 1-phenylbutan-1-one (**53**) with the complex reagent prepared *in situ* from the axially chiral (*S*)-1,1′-binaphthalin-2,2′-diol, an equivalent amount of ethanol, and lithium tetrahydridoaluminate (**52**, BINAL-H) gave in 78% yield and in high enantiomeric excess (*S*)-1-phenylbutan-1-ol (**54**). An explanation for the very high selectivity is that in the favored six-membered cyclic transition state (**TS A**), the aromatic ring takes up an equatorial disposition to avoid the electrostatic repulsion between the aromatic electron cloud and the lone pair of electrons at the axially oriented oxygen of the binaphthyl ligand (Figure 9.11).

If the bulkiness of the alkyl moiety of the ketone is increasing, it adopts more likely an equatorial position in the transition state, thereby decreasing the stereoselectivity. It is exemplified by the reduction of 1-phenyl-2-methylpropan-1-on (**55**) giving (*S*)-1-pheny-2-methylpropan-1-ol (**56**) in 68% yield and only ee 72% (Figure 9.11).

In reagent-controlled processes, high enantiotope selectivity can be realized with simple symmetrical substrates (e.g., butyrophenone), since the source of asymmetry can be found in the auxiliary group of the reagent. A disadvantage of the method is that the auxiliary group is consumed in a stoichiometric amount and if the recycling of the auxiliary group has not been solved, the carbon efficiency of the process is poor.

# 10
# Catalytic Methods of Enantioselective Synthesis

Chiral (*asymmetric*) catalysis is a procedure when the source of chirality (asymmetry) creating a diastereomeric relationship between the transition states of the reaction leading to one or more new stereogenic elements is a chiral compound present in the reaction mixture in a catalytic amount. It participates in the transition complex but is not consumed by the reaction. When at an achiral substrate a new center of chirality is generated, the possible products are enantiomers of each other, and therefore the process is called enantioselective catalytic chirality transfer.

## 10.1
## Chemical Catalysis

Chiral chemical catalysis is most often accomplished with complexes of chiral ligands with heavy metals (lanthanides) and *Lewis* acids. As chiral ligands most often molecules containing n-donor (phosphines, amines) and/or π-donor (aromatic) moieties of $C_2$ symmetry with two identical stereogenic elements are used. In this section, the following chiral chemical catalytic processes are discussed:

- Reductions: catalytic reduction of C=C and C=O bonds, hydrogenation, hydride transfer
- Oxidations: oxidation of C=C bonds, epoxidations, diol formation
- Formation of C–C bonds (*Heck* reaction, *Suzuki–Miyaura* reaction)
- Additions: onto C=C and C=O bonds (*Michael* addition, aldol reaction)
- Cycloadditions: (*Diels–Alder* and *Simmons–Smith* reactions).

### 10.1.1
### Chiral (Asymmetric) Catalytic Hydrogenation

An important group of enantiotope-selective catalytic reductions are homogeneous phase catalytic hydrogenations. As chiral ligands most often phosphines, as central metal atom rhodium or ruthenium, were used.

Homogeneous catalytic reduction of methyl (*Z*)-*N*-acetyl-cinnamate has been studied in much detail. Excellent results were achieved with rhodium

*Stereochemistry and Stereoselective Synthesis: An Introduction*, First Edition.
László Poppe and Mihály Nógrádi.

Companion Website: www.wiley.com/go/poppe/stereochemistry

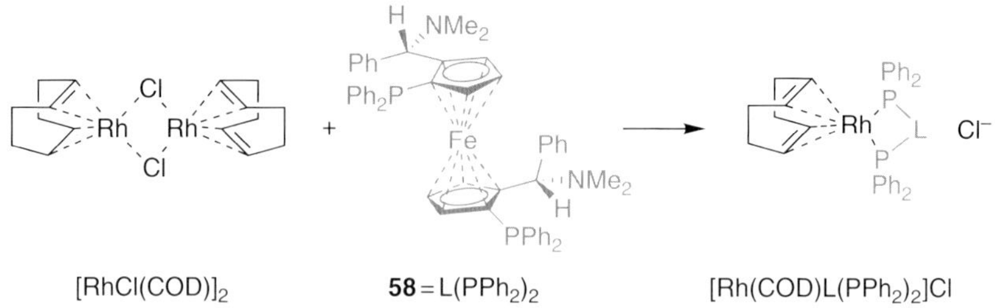

$[RhCl(COD)]_2$ + **58** = $L(PPh_2)_2$ → $[Rh(COD)L(PPh_2)_2]Cl$

**Figure 10.1** Preparation of a chiral rhodium complex catalyst.

complexed to ferrocenyl-phosphine ligands. The catalyst $[Rh(COD)L(PPh_2)_2]Cl$ is prepared *in situ* from commercially available rhodium complex $[RhCl(COD)]_2$ (COD = cycloocta-1,5-diene) and a chiral bidentate bisphosphine ligand, for example, **58** [=$L(PPh_2)_2$]. This ligand has two centers of chirality (both *S*) and a $C_2$ symmetry,[1] and is also axially chiral ($R_a$) (Figure 10.1).

About 1% solution of the catalyst is added to the solution of the substrate. In the hydrogen atmosphere, cyclooctadiene becomes saturated and driven out from the complex by the solvent. Thereafter, the substrate (MAC) substitutes the solvent and gives rise to a complex containing also the substrate $[Rh(MAC)L(PPh_2)_2]Cl$. Owing to the high asymmetry of the chiral ligand, the substrate binds in a stereoselective way to the complex. Activation of molecular hydrogen takes place at the central atom and leads to a dihydro derivative. Migration of the hydrogen atoms onto the substrate takes place within the complex $[RhH_2(MAC)L(PPh_2)_2]Cl$. In the course of the reductive elimination of the reduced end product, that is, methyl *N*-acetyl L-phenylalaninate (**59**), two molecules of the solvent (e.g., methanol, S) are bound as ligands to the complex: $[RhS_2L(PPh_2)_2]Cl$ (Figure 10.2).

The two solvent molecules are finally displaced by a new substrate molecule regenerating thus the reactive complex $[Rh(MAC)L(PPh_2)_2]Cl$. Thereafter, $[Rh(MAC)L(PPh_2)_2]Cl$, $[RhH_2(MAC)L(PPh_2)_2]Cl$, and $[RhS_2L(PPh_2)_2]Cl$ participate in the catalytic cycle (Figure 10.3).

Besides the catalytic saturation of C=C double bonds, another widely applied process is the stereoselective catalytic reduction of the carbonyl group of ketones. For this, most often the ruthenium complex of BINAP [(*S*)-1,1′-binaphthalene-2,2′-bis(diphenylphosphane)] (**60**), an axially chiral molecule of $C_2$ symmetry, is used as catalyst. When besides the carbonyl group to be reduced there is another n-donor group in the molecule [e.g., amino, hydroxyl, or another carbonyl group as, e.g., in ethyl acetoacetate (**61**)], no additional chiral ligand is necessary to achieve excellent (99%) stereoselectivity. With the reduction of simple aliphatic ketones [such as acetophenone (**63**)], a further chiral ligand (e.g., (*S,S*)-DPEN, (*S,S*)-1-diphenylethane-1,2-diamine, also of $C_2$ symmetry) is required to achieve

1) An advantage of $C_2$ symmetry is that binding to the central rhodium(I)-cation from any direction produces the same asymmetric complex.

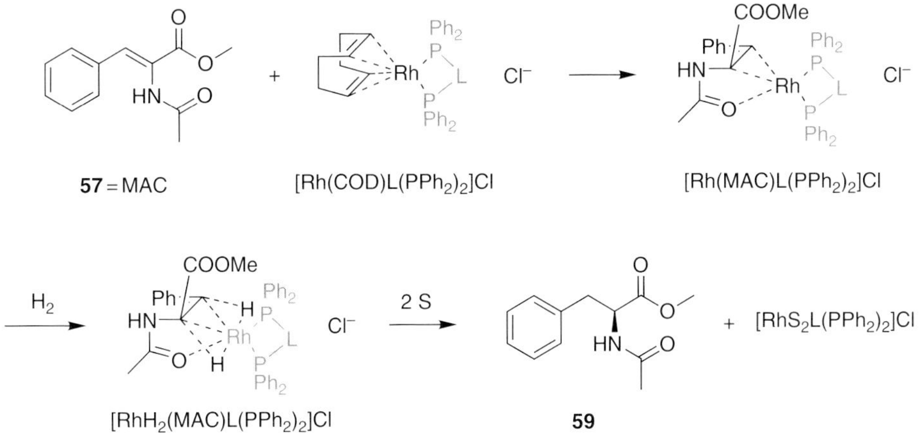

Figure 10.2 Hydrogenation with a chiral rhodium complex catalyst. ($L(PPh_2)_2$ = a chiral bis-phosphine ligand, S = solvent)).

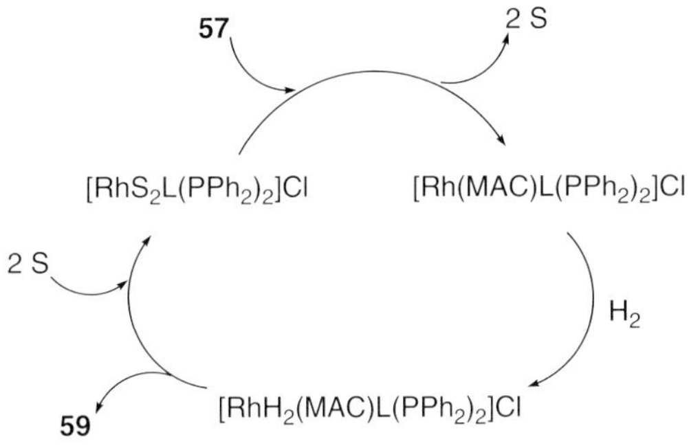

Figure 10.3 Catalytic cycle of the saturation of a double bond catalyzed by a chiral rhodium–phosphine complex.

a comparable degree of selectivity. With the reduction of simple dialkyl ketones, even this measure is not providing satisfactory stereoselectivity (Figure 10.4).

For the reduction of ketones containing further heteroatomic groups, one of the commercially available complexes, for example, (*S*)-[Ru(BINAP)$(OAc)_2$], is used to form the catalytic complex formed *in situ*, which contains two molecules of solvent. In the catalytic cycle, first a hydrogen molecule reacts with the complex furnishing the reactive hydrido complex [(*S*)-[RuHCl(BINAP)$S_2$] and hydrogen chloride. Solvent molecules are then replaced by the substrate (**61**) followed by transposition of the hydride ion from the central atom onto the substrate. At this point, the product is bound to the central metal atom as an alcoholate, while the hydride ion is replaced by a solvent molecule. Under the effect of hydrogen chloride present in the solution, the complex containing the product decomposes, and after protonation, the alcoholate is leaving the complex. By

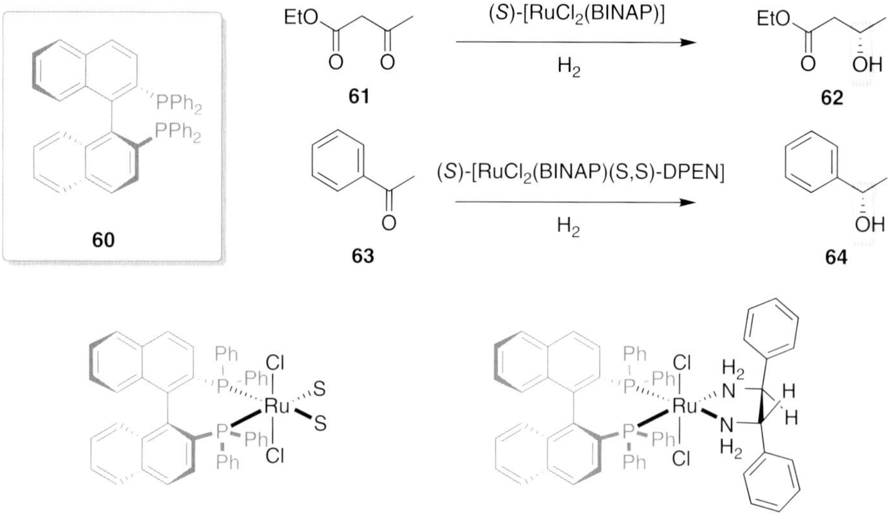

Figure 10.4 Enantiotope-selective reduction of ketones with a chiral ruthenium (S)-BINAP (60) complex catalyst. New stereogenic elements formed in a stereoselective way are highlighted. (S = solvent).

participation of the chloride ion and another solvent molecule, the starting complex [(S)-[$RuCl_2$(BINAP)$S_2$]] is recovered (Figure 10.5).

Let us consider the importance of the $C_2$ symmetry of the chiral ligand (S)-BINAP. The (S)-[Ru(BINAP)$(OAc)_2$] and (S)-[$RuCl_2$(BINAP)$S_2$] complexes show the same symmetry. Owing to this symmetry, the two acetate and two chlorine atoms, respectively, are homotopic. When the (S)-[$RuCl_2$(BINAP)$S_2$] complex reacts with a hydrogen molecule, the two homotopic chlorine atoms, shown in the figure on top or at the bottom, are replaced, and being homotopic, with an equal probability. The resulting two (S)-[RuHCl(BINAP)$S_2$] complexes are, however, identical and can be brought to superposition by rotation by 180° around the axis passing the ruthenium atom and the middle of the 1,1′-bond of the binaphthyl ligand. The same is true for the (S)-[$RuCl_2$(BINAP)$S_2$] complexes containing the substrate. It follows from the $C_2$ symmetry that whichever of the chlorine atoms is leaving in the course of hydrogen uptake, the result is the same, and therefore it is the same side from which the hydride ion attacks the substrate and the product is always of the same configuration (**62**) (Figure 10.6).

The other group of stereoselective catalytic reductions comprises homogeneous catalytic reactions involving hydride ion transfer. As reducing agents, boranes, silanes, or even organic hydrides can be applied (*transfer hydrogenation*).

For borane reductions, most often chiral boron containing heterocycles with a pyrrolidino[1,2-*c*][1,3,2]oxazaborolane skeleton (CBS) are used as chiral catalysts. The tertiary nitrogen atom of this skeleton establishes a dative bond with the

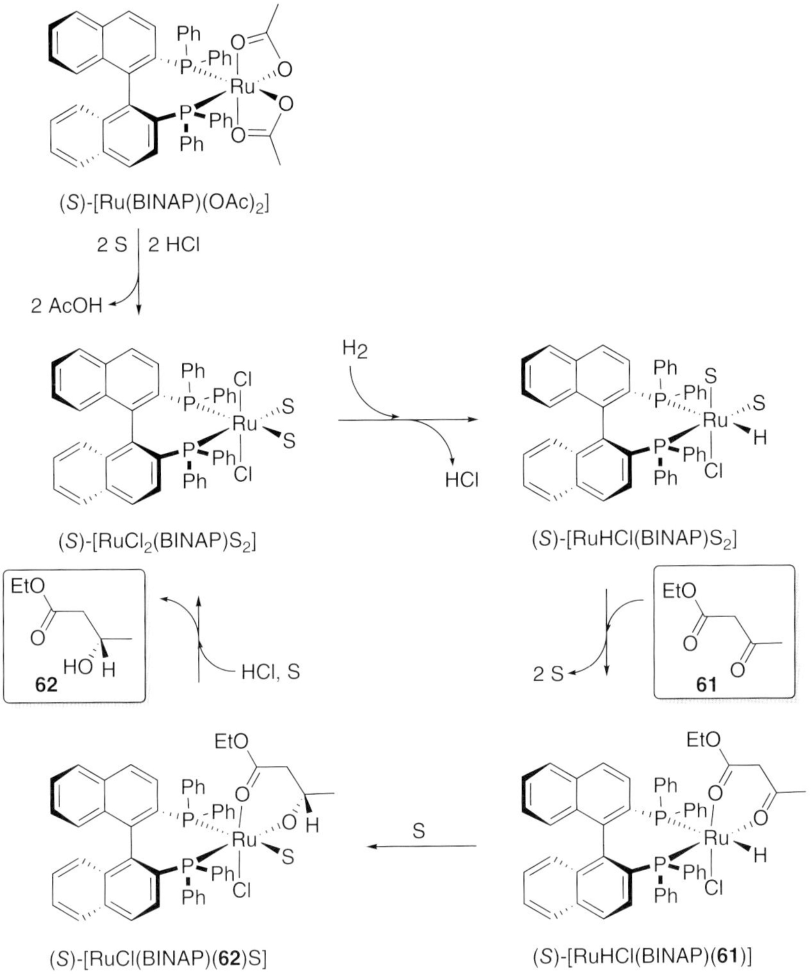

(S)-[Ru(BINAP)(OAc)$_2$]

2 S | 2 HCl

2 AcOH

$H_2$

HCl

(S)-[RuCl$_2$(BINAP)S$_2$]

(S)-[RuHCl(BINAP)S$_2$]

HCl, S

2 S

**62**

**61**

S

(S)-[RuCl(BINAP)(**62**)S]

(S)-[RuHCl(BINAP)(**61**)]

**Figure 10.5** Enantiotope-selective reduction of ethyl acetoacetate (**61**) with a chiral ruthenium complex catalyst. The catalytic cycle.

borane reagent (as, e.g., the commercially available $BH_3$·THF complex). The ketone substrate is bound to the CBS·$BH_3$ complex. In this, with the participation of one of the hydrogens of the borane (hydrogen bridge), a six-membered cyclic structure is formed, in which the substituent R attached to the boron atom of the oxazaborolane skeleton and the $R^1$ substituent of the ketone take up an axial orientation. The source of selectivity is the steric interaction between the two axial substituents. The more pronounced is the difference between the bulk of substituents $R^1$ and $R^2$; the more predominant is the complex in which the smaller substituent takes up the axial position (Figure 10.7).

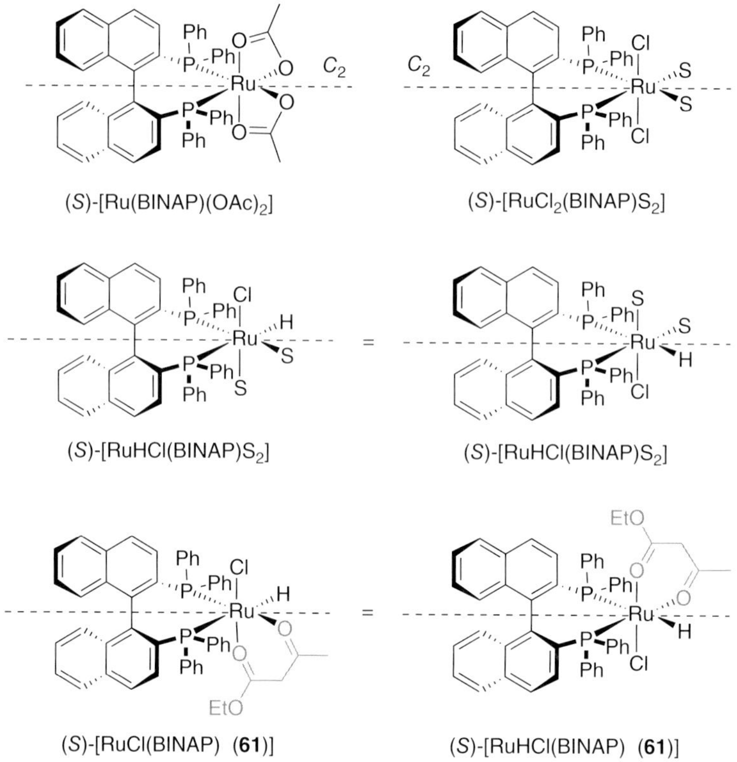

(*S*)-[Ru(BINAP)(OAc)$_2$] (*S*)-[RuCl$_2$(BINAP)S$_2$]

(*S*)-[RuHCl(BINAP)S$_2$] = (*S*)-[RuHCl(BINAP)S$_2$]

(*S*)-[RuCl(BINAP) (**61**)] = (*S*)-[RuHCl(BINAP) (**61**)]

**Figure 10.6** Enantiotope-selective reduction of ethyl acetoacetate (**61**) with a chiral ruthenium complex catalyst. Symmetry relationships of the complexes.

In order to increase the steric directing effect, one may increase the bulk of the R group or replace unsubstituted borane with a bulky borane derivative. This was the method by which enantiotope-selective reduction of 1-(triisopropylsilyl)oct-1-ine-3-one (**63**) to the alcohol **66** in 98% yield and 97% enantiomeric excess was accomplished. As catalyst, the oxazaborolane **64** prepared by reaction of **67** and **68** and as reagent benzo[1,3,2]dioxaborol (**65**) was used (Figure 10.8).

Enantiotope-selective reduction of the C=C bond of unsaturated esters (e.g., **69** and **70**) was realized with sodium tetrahydridoborate ($NaBH_4$) as reducing agent and a chiral cobalt complex prepared *in situ*. The products (**71** and **72**) were obtained with ee 94%. Their configuration depended on the stereostructure of the olefinic bond. The chiral ligand was the bis(pyrrol) derivative **73** (Figure 10.9).

As an example of enantiotope-selective hydrogen transfer, a variant of the *Meerwein–Ponndorf–Verley* reaction is quoted. As catalyst, instead of the usual aluminum triisopropoxide, a chiral samarium compound (**74**) was used. This method yielded on reduction of 2-choroacetophenone (**75**) (*R*)-1-(2-chlorophenyl)ethanol (**76**) (Figure 10.10).

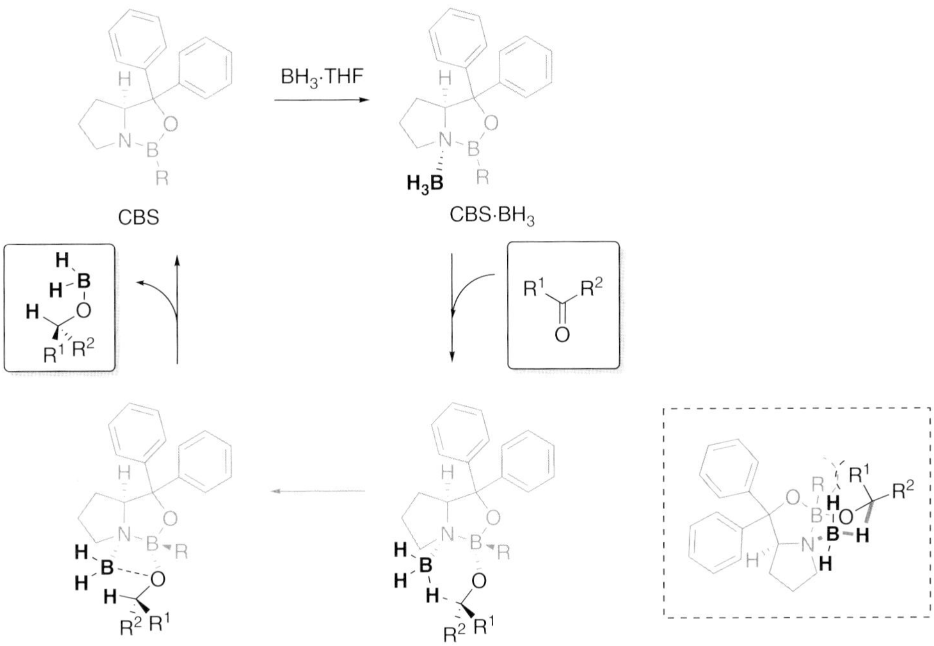

**Figure 10.7** Stereoselective reduction of ketones with the aid of a chiral catalyst having a pyrrolidino[1,2-*c*][1,3,2]oxazaborolane skeleton. The catalytic cycle. In the offset, steric interactions controlling stereoselectivity are shown.

### 10.1.2
### Enantiotope-Selective Catalytic Epoxidation

Among stereoselective oxidations, the best known is the *Sharpless* epoxidation of allylic alcohols. As oxidants peroxols, most often *tert*-butyl-hydroperoxide (TBHP) is used. The catalyst is prepared *in situ* from titanium tetraisopropoxide and diethyl (*R,R*)- or (*S,S*)-tartrate, compounds of $C_2$ symmetry. The advantage of the method is that the catalyst can be prepared *in situ* from inexpensive reagents of which both enantiomers are readily available (Figure 10.11), and therefore the configuration of the centers of chirality generated can be controlled at will.

The active dimeric titanium complexes, $[(R,R)\text{-DET}\cdot\text{Ti(OPr}^i)_2]_2$ and $[(S,S)\text{-DET}\cdot\text{Ti(OPr}^i)_2]_2$, respectively, which are formed from two moles of $\text{Ti(OPr}^i)_4$ and two moles of DET of the same configuration (accompanied by the loss of four moles of isopropyl alcohol) show also a $C_2$ symmetry.

$$2(R,R)\text{-DET} + 2\text{Ti(OPr}^i)_4 \rightarrow [(R,R)\text{-DET}\cdot\text{Ti(OPr}^i)_2]_2 + 4i\text{-PrOH}$$

or

$$2(S,S)\text{-DET} + 2\text{Ti(OPr}^i)_4 \rightarrow [(S,S)\text{-DET}\cdot\text{Ti(OPr}^i)_2]_2 + 4i\text{-PrOH}$$

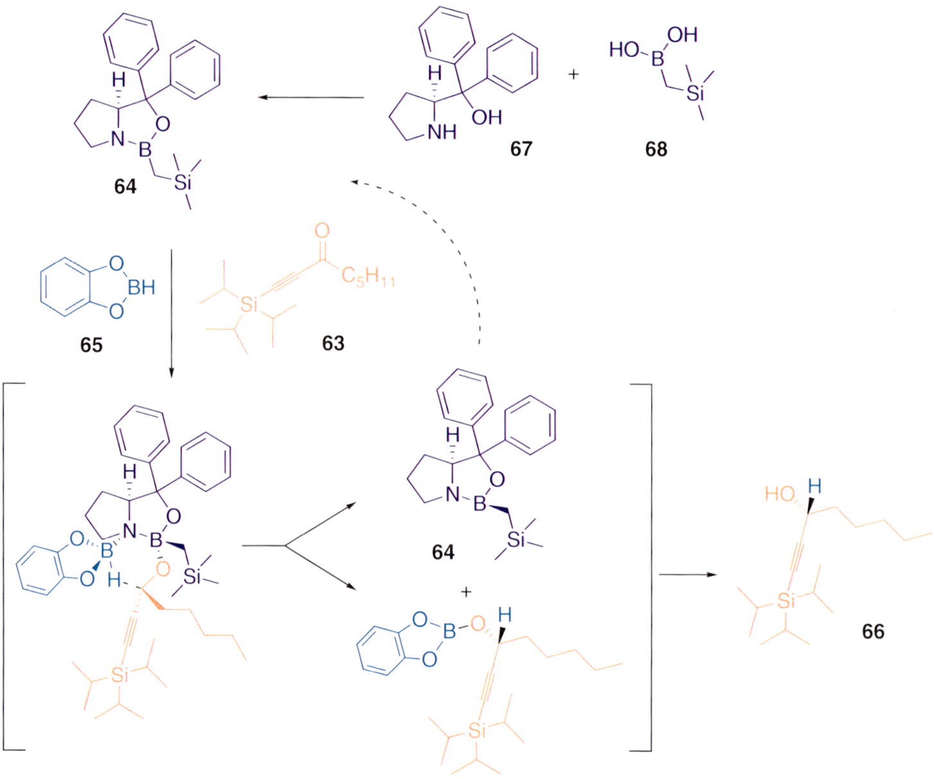

**Figure 10.8** Enantiotope-selective reduction of an aliphatic ketone (**63**) with a catalyst containing the chiral pyrrolidino[1,2-c][1,3,2]oxaborolane (CBS) skeleton (**64**). The individual components are drawn in different colors.

The catalytic process then takes place at one of the titanium atoms of the dimeric complex (Figure 10.11)[2] preceded by the elimination of two isopropyl alcohols by TBHP and the allyl alcohol, bound to the center as alcoholates. By choosing the appropriate enantiomer of DET, an achiral primary allyl alcohol (e.g., 77) can be converted to epoxides (e.g., **78** and **79**) with high enantiomeric excess (ee = 90–98%) (Figure 10.12).

When the primary allylic alcohol substrate is chiral (e.g., **80**), **double asymmetric induction** takes place (i.e., diastereotope-selective addition: cf. Section 9.1.3). The result depends on the configuration of DET. Without chiral catalysis, the preferred direction of epoxidation is *anti*, that is, the main product will be the epoxide **82** (70%). In the chiral catalytic process, (*R*,*R*)-DET promotes the formation of **81**, while (*S*,*S*)-DET that of **82** (Figure 10.13). If the directing effect of the catalyst and substrate are opposite, the diastereomer **81** is obtained in

2) Owing to $C_2$ symmetry, it is indifferent which of the titanium atoms will be the center of the reaction.

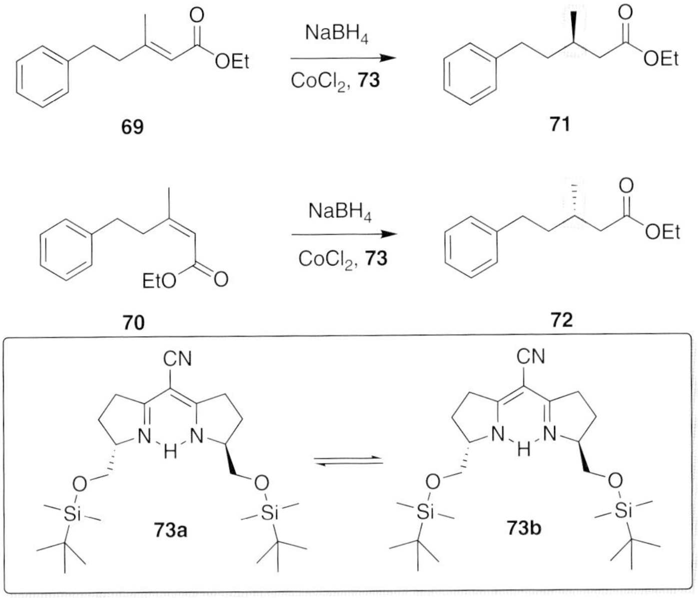

**Figure 10.9** Enantiotope-selective reduction of unsaturated esters **69** and **70** with the aid of a chiral cobalt complex catalyst. New stereogenic elements formed in a stereoselective way are highlighted. In the bottom row, two tautomers (**a** and **b**) of the chiral ligand (**73**) are shown that can be interconverted by a $C_2$ symmetry operation.

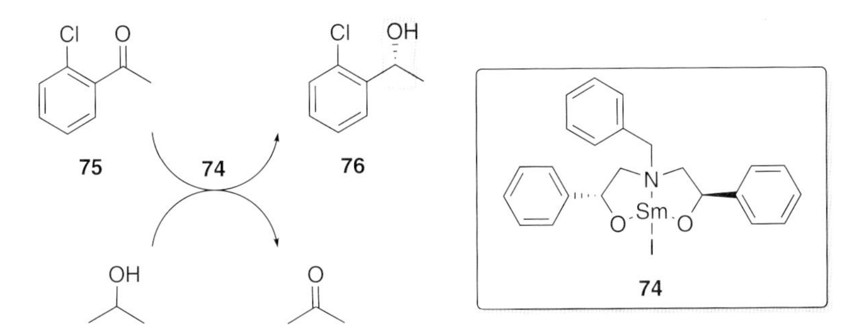


**Figure 10.10** Enantiotope-selective reduction of an aromatic ketone (**75**) with transfer hydrogenation effected by a chiral samarium catalyst (**74**). New stereogenic elements formed in a stereoselective way are highlighted.

about 90% purity. If, however, catalyst and substrate work in the same direction, the product (**82**) will be of 98% purity (Figure 10.13).

With racemic secondary allylic alcohols (e.g., **83**), a kinetic resolution process (see Section 8.6.2) – by a diastereotope-selective addition in an enantiomer-selective way – depending on the configuration of complexing diisopropyl tartrate (DIPT) epoxidation affects mostly only one of the enantiomers. With

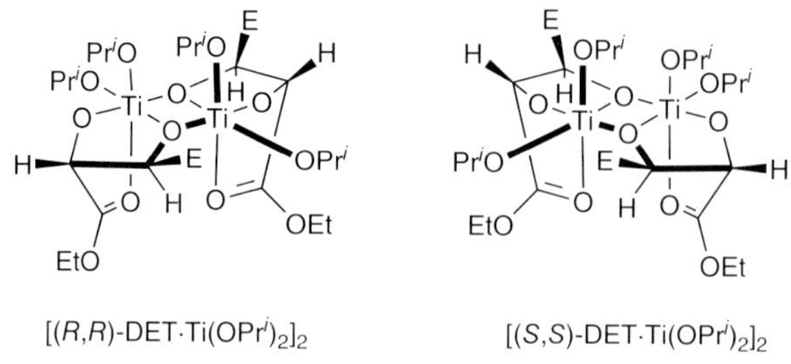
[(R,R)-DET·Ti(OPr^i)_2]_2
[(S,S)-DET·Ti(OPr^i)_2]_2

**Figure 10.11** Enantiomeric forms of the *in situ* prepared catalyst of *Sharpless* epoxidation. (Abbreviations: ethoxycarbonyl group, E = EtOOC, DET = diethyl tartrate.)

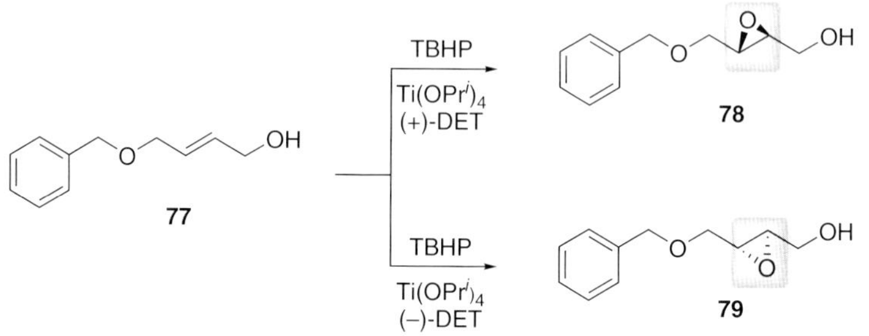
TBHP
Ti(OPr^i)4
(+)-DET
TBHP
Ti(OPr^i)4
(–)-DET
OH
77
78
79

**Figure 10.12** Enantiotope-selective epoxidation of a primary allylic alcohol (**77**) catalyzed by a chiral titanium complex. New stereogenic elements formed in a stereoselective way are highlighted.

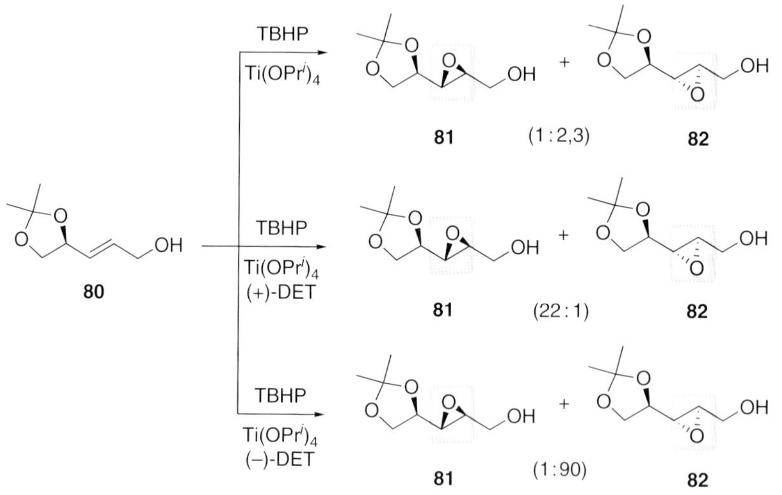
TBHP
Ti(OPr^i)4
(+)-DET
(–)-DET
80
81
82
(1:2,3)
(22:1)
(1:90)

**Figure 10.13** Stereoselective epoxidation of a chiral allylic alcohol (**80**) catalyzed by a chiral titanium complex. The new stereogenic elements formed in a diastereoselective way are highlighted. **Double asymmetric induction**. Upper row: Without a chiral catalyst (substrate selectivity). Middle row: Opposite selectivity of substrate and catalyst (mismatched pair). Lower row: Cooperating substrate and chiral catalyst selectivity (matched pair).

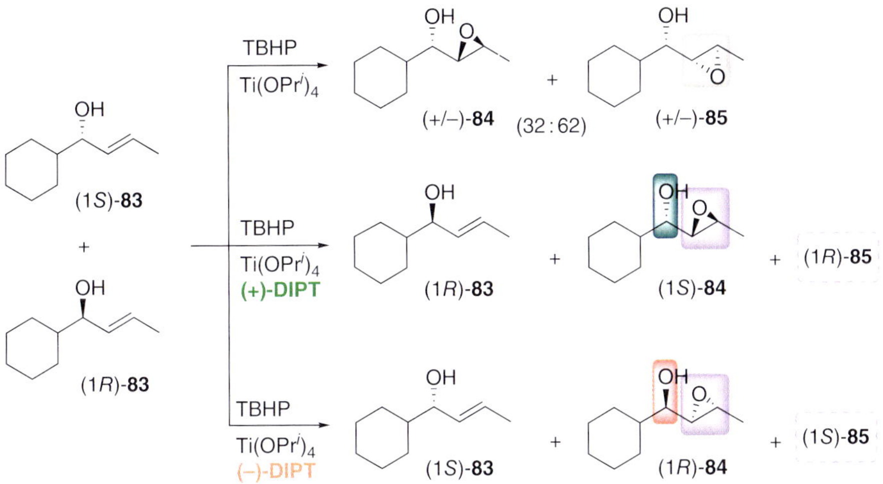

**Figure 10.14** Stereoselective epoxidation of a racemic secondary allylic alcohol catalyzed by a chiral titanium complex. The preferred enantiomers (green and red) and the new stereogenic elements formed in a diastereotope-selective mode (gray with achiral catalyst and purple with chiral catalyst) are highlighted. Upper row: Without chiral catalyst (only moderate diastereotope selectivity). **Kinetic resolution:** Middle row: In the catalytic system, the (*S*)-alcohol reacts faster and the (*R*)-alcohol remains unchanged. Bottom row: In the catalytic system, the (*R*)-alcohol reacts faster and the (*S*)-alcohol remains unchanged.

secondary allylic alcohols, the *Sharpless* epoxidation with chiral catalyst takes place almost exclusively from the "*anti*" face. In addition, the chiral catalyst will complex only a particular enantiomer (Figure 10.14).

Without chiral catalysis, the preferred direction of epoxidation is *syn*, that is, epoxide **85** will be obtained in 62% yield.[3] In kinetic resolutions performed with a chiral catalyst, 0.60 equivalents of the oxidant is given to the reaction mixture[4] and the reaction was usually stopped at 50% conversion. Enantiomeric excess of recovered (1*S*)-**83** and (1*R*)-**83** allylic alcohols (yield 45%),[5] respectively, were 100% within analytical error. The epoxides **84**, also isolated in 45% yield, were contaminated with the diastereomeric epoxide **85** (about 3%). This contamination can be traced back to the fact that the preferences of substrate and catalyst on diastereotope selectivity are opposite: while substrate selectivity favors the formation of "*syn*" epoxide **85**, the chiral catalyst that of the "*anti*" epoxide **84** (Figure 10.14).

3) Without chiral catalysis, racemic starting materials give rise to racemic products. Only one enantiomeric product is shown in the figure.
4) Kinetic resolution would theoretically require 0.5 equivalents of the reagent. An excess of 0.1 equivalent was applied; otherwise, the reaction time would have been exceedingly long.
5) The maximum yield of a kinetic resolution is theoretically 50%.

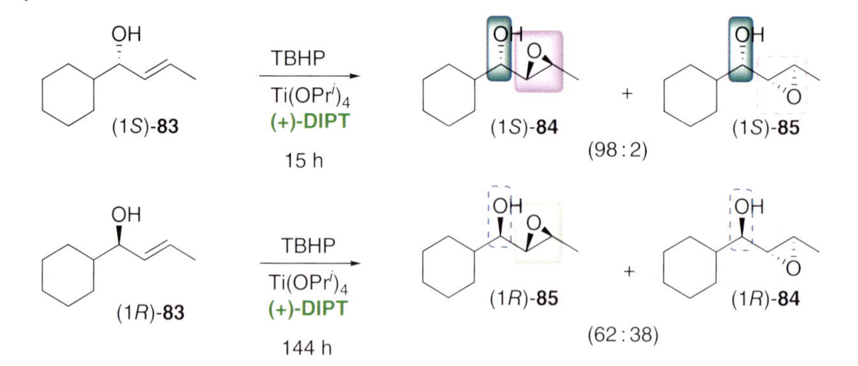


**Figure 10.15** Stereoselective epoxidation of chiral secondary allylic alcohols (**83**) catalyzed by chiral titanium complexes. Newly formed stereogenic elements formed in a diastereotope-selective manner (purple for the fast reaction and gray for the slow reaction) and the conserved ones (green for the preferred enantiomer and dotted green for the less-reactive enantiomer) are highlighted. Upper row: Fast reaction (f), bottom row: slow reaction (s). Ratio of reaction rates: $k_f/k_s = 104$.

The ratio of reaction rates underlying kinetic resolution was studied in experiments when enantiopure allylic alcohols, that is, (1*S*)-**83** and (1*R*)-**83**, were oxidized with the chiral catalyst to full conversion. When the enantiomer (1*S*)-**83** conforming to the preferred selectivity of a catalyst based on (*R*,*R*)-DIPT was reacted, the reaction was complete within 15 h and gave an epoxide of 96% *de*. When the slower reacting isomer (1*R*)-**83** was reacted with the same catalyst, the reaction was only complete in 6 days and with a diastereotope selectivity similar to that of the reaction with achiral catalyst[6] (Figure 10.15).

The Sharpless method fails with the epoxidation of olefins that lack functional groups capable of complexation with the catalyst. In this case, the *Jacobsen's* catalyst (**86**) is useful, which mimics both in its structure and mode of action the metal porphyrin prosthetic group of the enzyme cytochrome P-450 (Figure 10.16).

Oxidized forms of the catalysts – which are generated *in situ* in the reaction mixture from the commercially available catalyst of $C_2$ symmetry and an oxidizing agent (e.g., PhIO) – are performing the oxidation of the olefin. The larger the difference in the bulkiness of the substituents situated at the two terminals of the double bond, the more stereoselective is the reaction due to formation of diastereomeric transition states. Understandably, the best results were achieved with the *cis* isomers of olefins. Selectivity is even higher if the double bond is part of a ring of limited mobility (Figure 10.17).

6) Experimental data allowed the conclusion that the chiral catalyst formed a complex only with the (1*S*)-**83** isomer and thereby enhanced the rate of transformation to the preferred product [(1*S*)-**84**] by a factor of about 100. On the other hand, with (1*R*)-**83** – isomer complexing only weekly or not at all with the catalyst, reaction was slow and resulted in a product distribution corresponding to the product distribution of the reaction with achiral catalyst.

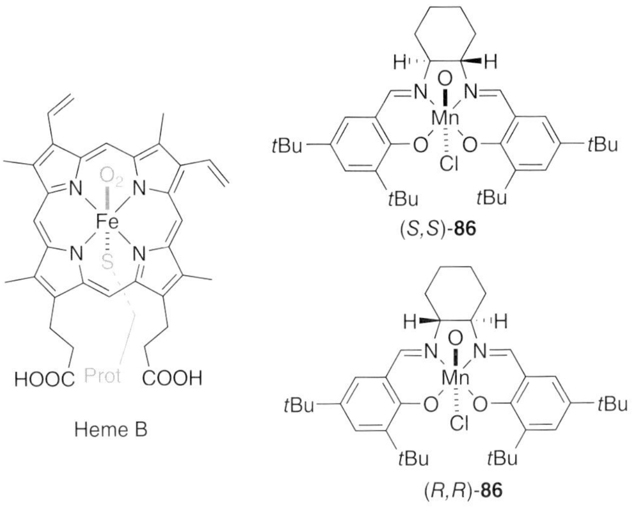

**Figure 10.16** Structures of the Heme B prosthetic group of a cytochrome P-450 enzyme and of two enantiomeric salen–manganese complex catalysts. (In the figure, the oxidized forms of (*S*,*S*)- and (*R*,*R*)-[*N*,*N*′-bis(3,5-di-*terc*-butylsalicylidene)-1,2-cyclohexandiamino]manganese(III)-chloride catalyst are shown.)

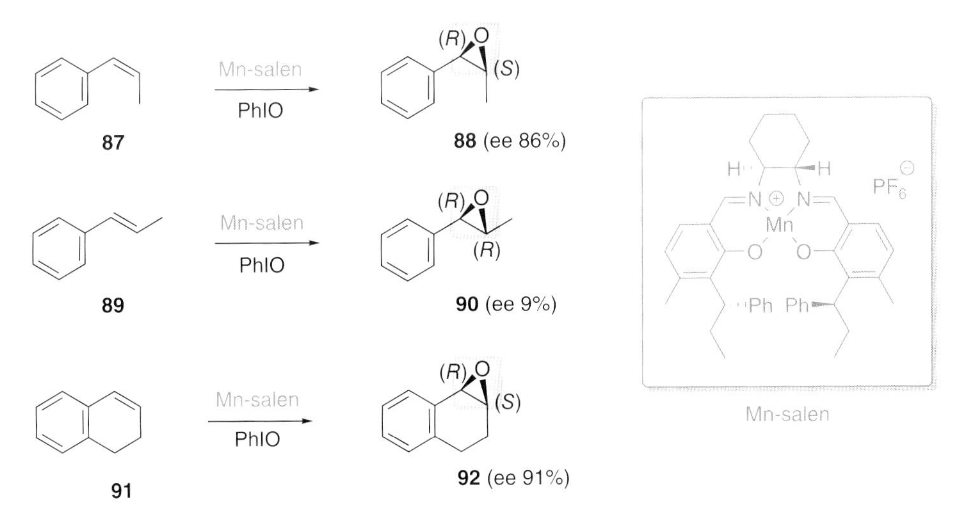


**Figure 10.17** Stereoselective oxidation of olefins (**87**, **89**, and **91**) catalyzed by a chiral manganese complex. Stereogenic elements generated in an enantioselective manner are highlighted.

In the course of ring opening by nucleophiles of epoxides obtained by oxidation of olefins, additional complex problems of selectivity (regio- and diastereoselectivity) may arise. Using the appropriate catalyst, such reactions can yet be conducted in a regio- and diastereoselective manner. Selective ring opening of epoxides of epoxy alcohols prepared by the *Sharpless* method can be catalyzed by titanium tetraisopropoxide.[7] Under the effect of the catalyst, the nucleophile attacks in a regio- and diastereoselective way with inversion at the bridgehead atom more remote from the hydroxyl group (Figure 10.18).

Enantiotope-selective ring opening (**desymmetrization**) of achiral *meso*-epoxides can be accomplished with chromium or cobalt containing *Jacobsen's* catalysts (**96**, **97**, and **98** in Figure 10.19) (analogous to **86** in Figure 10.16).

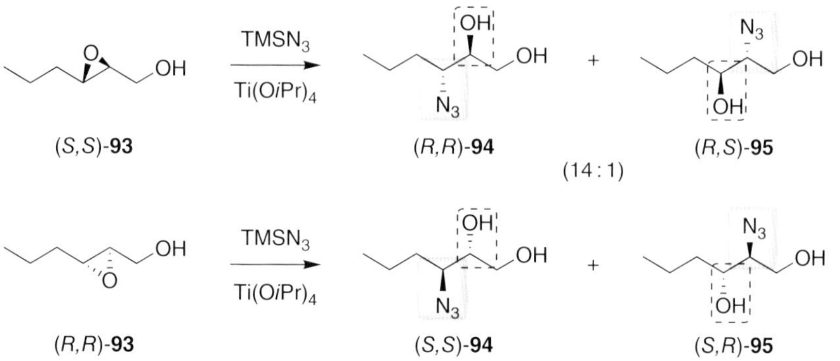

**Figure 10.18** Regio- and stereoselective reaction of epoxy alcohols (**93**) with azide ions as nucleophile with the aid of a titanium catalyst. Inverted (gray) and conserved (dotted box) stereogenic elements are highlighted.

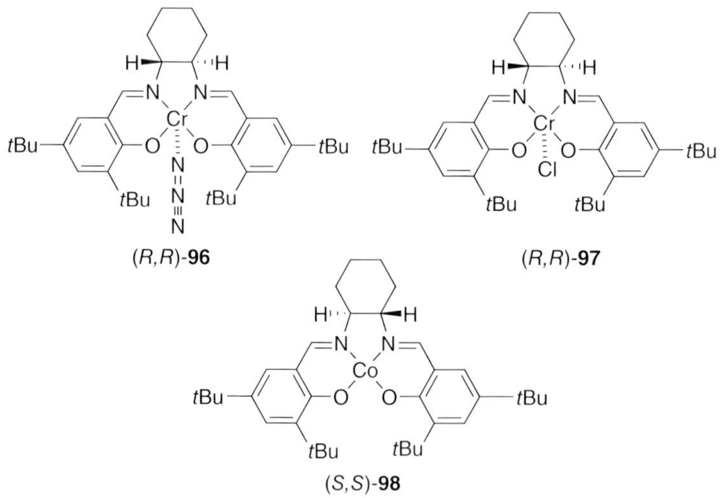

**Figure 10.19** Structures of chiral chromium and cobalt–salen complex catalysts. (Since the epoxy alcohols obtained by *Sharpless* epoxidation are chiral molecules, to achieve selectivity, there is no need for a chiral catalyst.)

7) Since the epoxy alcohols obtained by *Sharpless* epoxidation are chiral molecules, to achieve selectivity there is no need for a chiral catalyst.

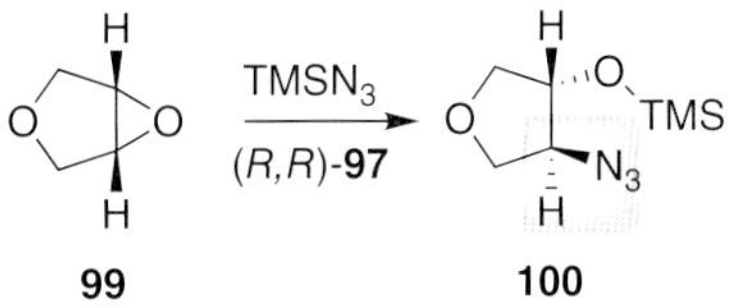

Figure 10.20 Enantiotope-selective ring opening of a *meso*-epoxide catalyzed by a salen–chromium complex. The inverted stereogenic element is highlighted.

For example, ring opening of 3,6-dioxabicyclo[3.1]hexane (**99**) with trimethylsilyl azide catalyzed by the complex (*R,R*)-**97** furnished the azide **100** of ee 97% (Figure 10.20).

Ring opening with a benzoate anion was most efficiently catalyzed by the complex (*S,S*)-**98:** the phenanthrene *meso*-epoxide (**101**) was converted to the monoacylated diol (**102**) of ee 94%. The analogous *meso*-cyclohexene epoxide (**103**), gave, however, a product (**104**) but in ee 76%, which could be upgraded to ee 98% by triple recrystallization[8] (Figure 10.21).

Salen complexes (**96**) were also useful in the **kinetic resolution** of racemic terminal epoxides. Ring opening with trimethylsilyl azide of the racemic epoxide (**105**) took place regioselectively at the terminal carbon atom of the enantiomer preferred by the chiral catalyst. Both, the azide (**106**) derived from the preferred enantiomer and the unreacted epoxide (**105**), were recovered from the reaction mixture in high enantiomeric excess (Figure 10.22).

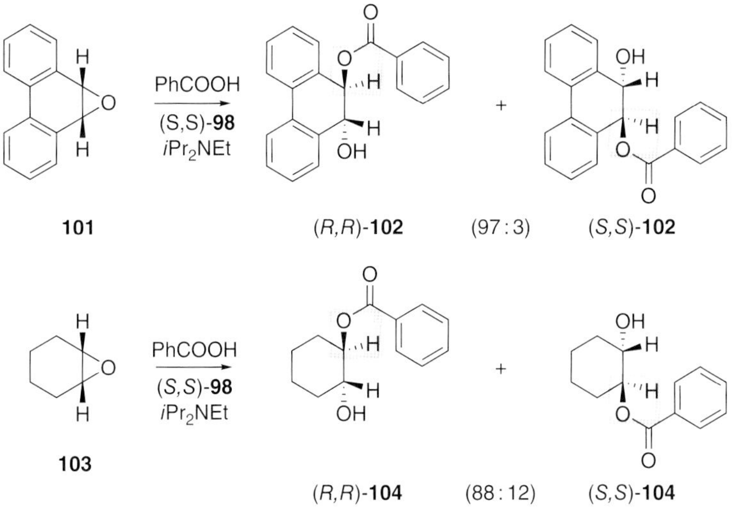

Figure 10.21 Stereoselective ring opening of a symmetrical epoxide catalyzed by a chiral salen–cobalt complex. The inverted stereogenic element is highlighted.

8) Recrystallizing three times involved a total loss of 25%, because besides a 12% of (*S,S*)-**104** in the raw product, about the same amount of (*R,R*)-**104** also remained in solution, since in an achiral solvent, solubility of both enantiomers was the same.

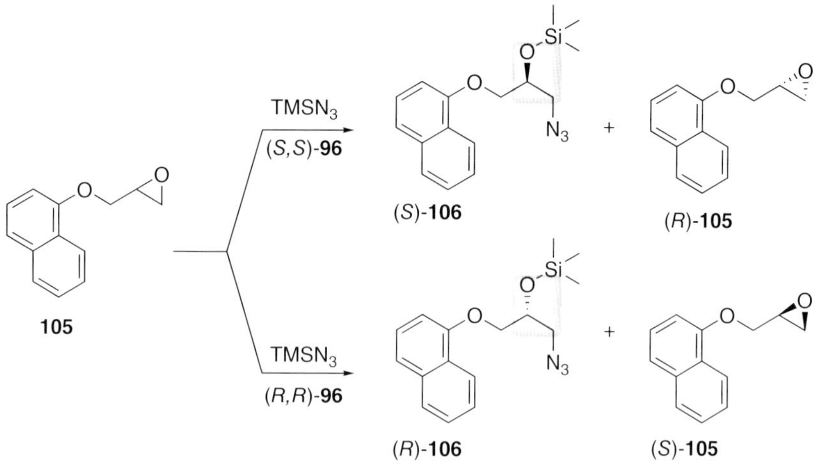

**Figure 10.22** Enantiomer and regioselective reaction of a racemic terminal epoxide (**105**) catalyzed by the two enantiomers of a chiral salen–chromium complex. The reacting stereogenic element is highlighted. **Kinetic resolution**: Top row: the catalytic system transforms the (*S*)-epoxide, while the (*R*)-epoxide remains unchanged. Bottom row: The catalytic system transforms the (*R*)-epoxide, while the (*S*)-epoxide remains unchanged.

### 10.1.3 Stereoselective Catalytic Diol Formation

Besides enantioselective epoxidation, the other often applied chiral catalytic oxidation process is the oxidative transformation of olefins to diols (**stereoselective dihydroxylation**). The method is an enantioselective extension of the *cis* specific oxidation of olefins with $OsO_4$ known for long. Among multiple oxidants and chiral ligands, most widely the system introduced by *Sharpless*, that is, potassium [hexacyanoferrate(III)] as oxidant and a ligand of $C_2$ symmetry based on quinine alkaloids [$(DHQD)_2PHAL$ or $(DHQ)_2PHAL$], is used (Figure 10.23).

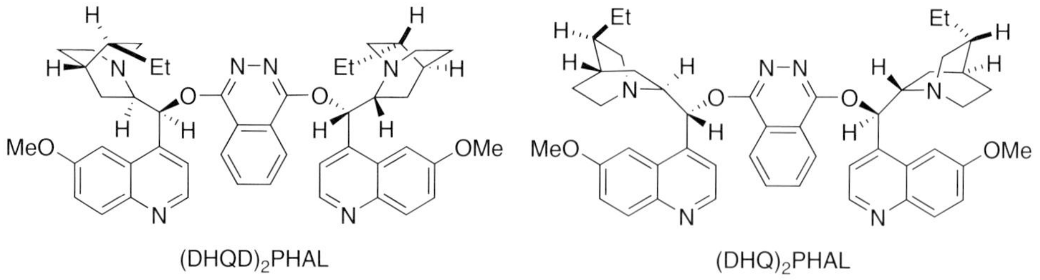

**Figure 10.23** Chiral $(DHQD)_2PHAL$ and $(DHQ)_2PHAL$ ligands (Ligands shown in Figure 10.23 are commercially available under the name AD-mix-$\beta$ and AD-mix-$\alpha$ as a mixture of the reagents {$K_2CO_3$, $K_3[Fe(CN)_6]$} and the precatalyst {$K_2OsO_4*2(H_2O)$}.) for stereoselective oxidations.

The oxidized catalyst complex performing the oxidation is formed *in situ*. Potassium osmate is oxidized by the oxidant to $OsO_4$ (step 1) followed by binding to the chiral ligand [$OsO_4$*L] (step 2). This complex binds then enantioface selectively the olefin (A), while it oxidizes it to a diolate ($AO_2$) (step 3). This complex binding the product as diolate [$AO_2OsO_2$*L] is then reoxidized by the oxidant (step 4). The resulting complex gives rise by reacting with the water content of the medium to the product diol $A(OH)_2$ and regenerates the [$OsO_4$*L] complex restarting the catalytic cycle (step 5).

1) $K_2OsO_4 + 2K_3[Fe(CN)_6] = OsO_4 + 2K_4[Fe(CN)_6]$
2) $OsO_4 + L = OsO_4{}^*L$
3) $[OsO4^*L] + A = [AO_2OsO_2{}^*L]$
4) $[AO_2OsO_2{}^*L]+2K_3[Fe(CN)_6]+K_2CO_3=[AO_2OsO_3{}^*L]+ 2K_4[Fe(CN)_6]+CO_2$
5) $[AO_2OsO_3{}^*L] + H_2O = [OsO_4{}^*L] + A(OH)_2$

The commercially available catalyst mixtures AD-mix-$\beta$ and AD-mix-$\alpha$ introduced by *Sharpless* have the advantage that they operate with opposite stereoselectivity. While using an AD-mix-$\beta$ for the oxidation of the olefin **107** gives rise to an (*S*)-glycerol derivative (**108**) in 91% enantiomeric excess, the (*R*)-enantiomer can be prepared from the same olefin using AD-mix-$\alpha$ (Figure 10.24).

Further catalysts were developed for special types of compounds. As an example, the ligand (*S,S*)-**111**, a bispyrrolidine of $C_2$ symmetry, can be quoted, which is the most efficient ligand for the *cis*-hydroxylation of *trans*-olefins. Using this system, (*E*)-stilbene (**109**) could be oxidized in 96% yield to enantiopure (*S,S*)-1,2-diphenylmethane-1,2-diol (**110**). Complete selectivity is due to the fact that the pyrrolidine rings of the ligand and the two bulky alkyl groups are situated on opposite sides of the complex, while the phenyl rings of the substrate take up an opposite orientation (Figure 10.25).

### 10.1.4
### Formation of a C–C Bond by Chiral Catalysis

The *Heck* reaction is the palladium-catalyzed reaction of aryl and vinyl halides or triflates with activated olefins. In the *Heck* reaction, in a redox reaction

OH O OH AD-mix-α O AD-mix-β O OH OH

(*R*)-**108** **107** (*S*)-**108**

**Figure 10.24** Enantiotope-selective oxidation of a terminal olefin (**107**) with a chiral osmium–tetroxide complex catalyst [$(DHQD)_2PHAL$]. The new stereogenic element formed in stereoselectivity is highlighted.

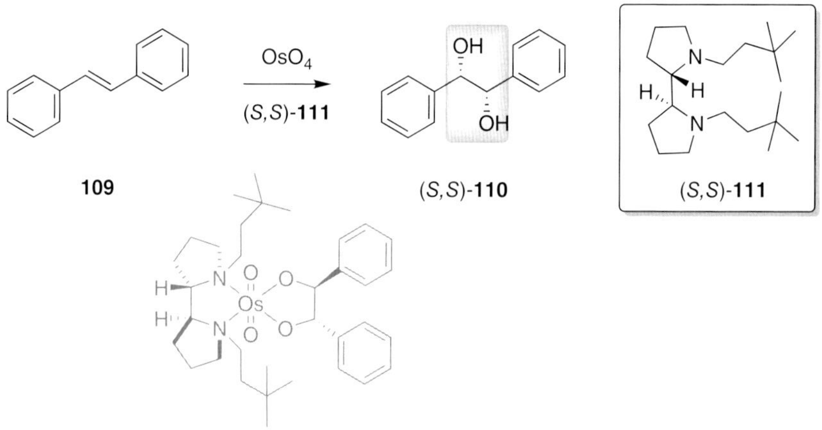

Figure 10.25 Enantiotope-selective oxidation of a *trans*-olefin catalyzed by the osmium-tetroxide complex of (*S,S*)-**111**. New stereogenic elements formed in a stereoselective manner are highlighted. In the bottom row, the complex from the chiral catalyst and the olefin is shown.

(Umpolung) between the palladium catalyst ($PdL_2$) and a halide or triflate (ArX), a complex [$ArPdXL_2$] is formed containing the nucleophilic aryl or vinyl group. This complex is added onto an olefin, and finally the palladium hydride complex [$HPdXL_2$] is eliminated. The base serves to regenerate the palladium complex (Figure 10.26).

As chiral ligands, oxazoline derivatives containing a 2-(diphenylphosphino) phenyl group (e.g., **115**) are used (Figure 10.27). The catalyst is generated *in situ*

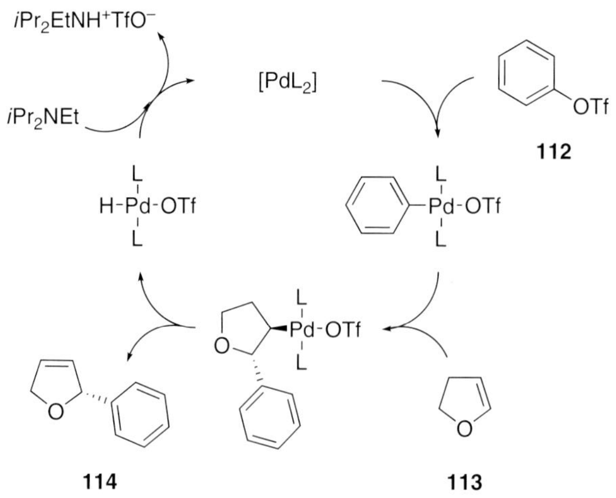

Figure 10.26 Catalytic cycle of the *Heck* reaction (L is a chiral ligand).

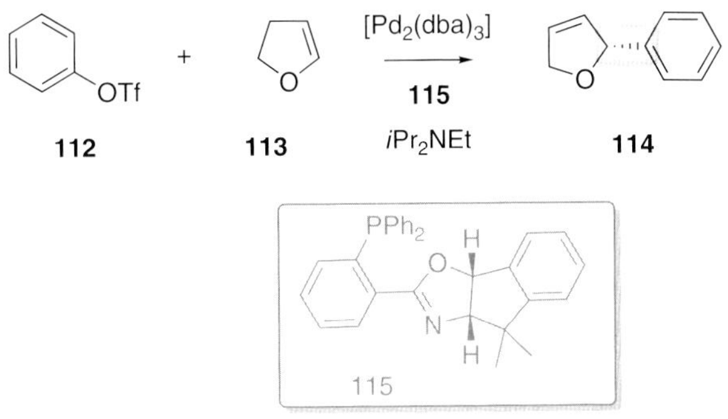

Figure 10.27 Enantiotope-selective *Heck* reaction of phenyl triflate (**112**) and 2,3-dihydrofuran (**113**) catalyzed by the chiral palladium complex of the oxazole **111**. The new stereogenic element formed in a stereoselective way is highlighted.

from commercially available $Pd_2(dba)_3$ complex[9] and the ligand **115**. In the enantiotope selectively catalyzed *Heck* reaction of **112** and **113**, the product (**114**) was isolated in 96% enantiomeric excess.

The *Suzuki–Miyaura* reaction is the palladium-catalyzed reaction of aryl halides or triflates with arylboronic acids (Figure 10.28).

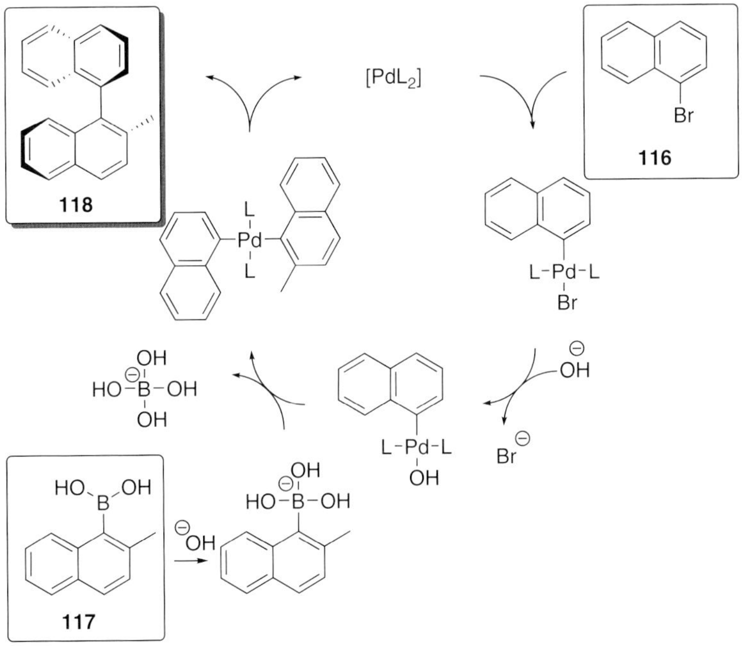

Figure 10.28 Catalytic cycle of the *Suzuki–Miyaura* reaction (L: chiral ligand).

9) dba = dibenzylidene-acetone $(PhCH{=}CH)_2C{=}O$.

In the course of the reaction, in a redox step (Umpolung) involving the palladium catalyst [$PdL_2$] and the halide (e.g., **116**) or triflate, a complex is formed [$ArPdXL_2$] in which the anionic group is exchanged for the hydroxide ion of the base applied giving [$ArPd(OH)L_2$] (Figure 10.28). This hydroxide ion is replaced in the next step by the aryl group of the *ortho*-boronate [$Ar'B(OH)_3$] (**117**) giving [$ArPdAr'L_2$]. The *ortho*-boronate anion is formed from the boronic acid [$Ar'B(OH)_2$] under the effect of the base used. From the complex containing two aryl groups, the palladium complex is recovered by concomitant formation of a biaryl (Ar–Ar′, **118**).

Analysis of the reacting centers of the components involved in this reaction reveals that both $sp^2$ carbon centers have enantiotopic faces. Thus, the reaction happens in a double-enantiotope-selective way. The reason why only two stereoisomers (i.e., two enantiomers) are forming is due to the fact that in this reaction the two prostereocenters result in not two but only one new stereogenic element (the axis of chirality, as a result of atropisomerism).

In the chirally catalyzed *Suzuki–Miyaura* reaction of **116** and **117**, the product (**118**) was obtained in 98% enantiomeric excess (Figure 10.29).

As chiral ligand, for example, a bishydrazone derivative can be used prepared from glyoxal (**119**) and (*S,S*)-*N*-amino-2,5-diphenylpyrrolidine (**120**) of $C_2$ symmetry. The palladium complex (**121**) is prepared from the ligand in a separate step and is added later to the reaction mixture of the *Suzuki–Miyaura* reaction.

### 10.1.5
### Stereoselective Catalytic *Michael* Additions

*Michael* addition is the addition of carbanions onto activated olefins, for example, α,β-unsaturated carbonyl compounds. As chiral catalysts not only chiral metal complexes, but also chiral phase transfer catalysts and organocatalysts without metals are applied. All three approaches are illustrated in the following.

Reaction of cyclohexanone (**122**) and diethyl zinc (**123**) can be catalyzed by chiral copper complexes. As chiral ligand using a phosphorus amidite derivative (e.g., **124**) of $C_2$ symmetry, the adduct (**125**) was isolated with ee 98%. The

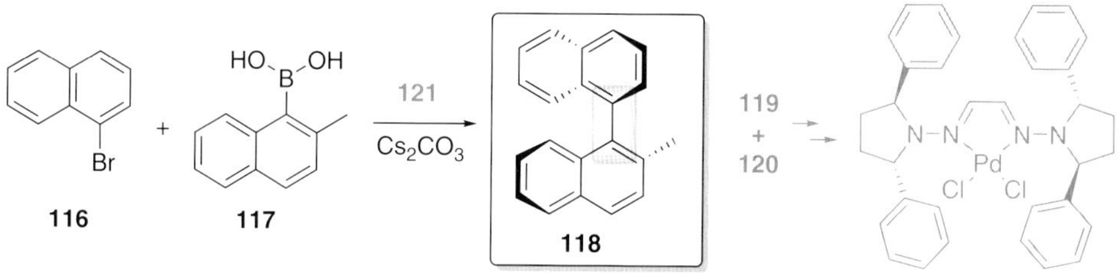

**Figure 10.29** Stereoselective *Suzuki–Miyaura* reaction of 1-bromonaphthalene (**116**) and (2-methyl-naphthalinyl)-boronic acid (**117**) catalyzed by the chiral palladium complex (**121**) prepared from (*S,S*)-*N*-amino-2,5-diphenylpyrrolidine (**120**)and glyoxal (**119**). The newly formed stereogenic element (chiral axis) is highlighted.

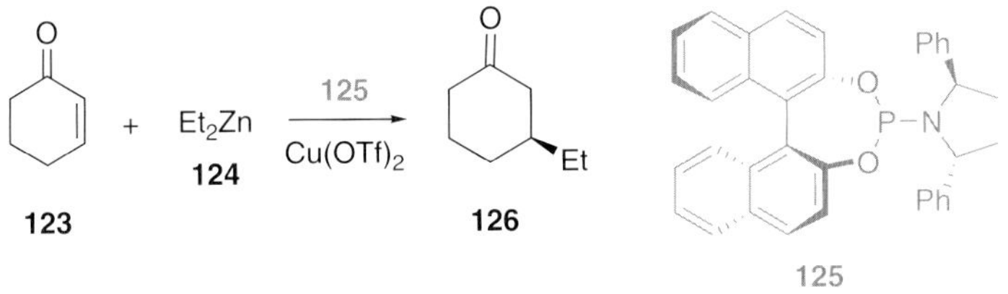

Figure 10.30 Enantiotope-selective *Michael* addition of diethyl zinc (**123**) onto cyclohex-anone (**122**) catalyzed by the chiral copper complex of the phosphorus amidite **124**.

chiral catalyst was prepared *in situ* from the chiral ligand and copper(II) triflate (Figure 10.30).

A chiral biphenyl ammonium derivative containing no metal (**126**, $R_4N^+Br^-$) has also been used as catalyst for *Michael* additions (Figure 10.31). Reaction of diethyl malonate (**128**) and 1,3-diphenylprop-2-en-1-one (**127**) catalyzed by the biphenyl **126** gave the adduct **129** in ee 90%. The operating principle of the chiral catalysts is that the cation functions as a phase transfer catalyst and transfers the basic carbonate anion into the toluene phase, followed by ion pair formation with the diethyl malonate anion deprotonated by the base. The cation shows a $C_2$ symmetry and forms therefore with the diethyl malonate anion a highly asymmetric ion pair, which attacks the two enantiotopic faces of the substrate with unequal probability.

Chiral organocatalysts can be developed on the basis of natural L-amino acids. For example, the *Michael* addition of *n*-butanal (**130**) onto β-nitrostyrene (**131**) was catalyzed by the prolinamide derivative **133** (Figure 10.32).

In the reaction in Figure 10.32, two prostereogenic elements, both in enantiotope-selective manner, are reacting and therefore two centers of chirality are generated. Thus, in principle, four stereoisomers could have been formed. Out of the four conceivable stereoisomers, the product mixture contains almost exclusively (>99%) (2*R*,3*S*)-2-ethyl-3-phenyl-4-nitrobutanal (**132**). That means

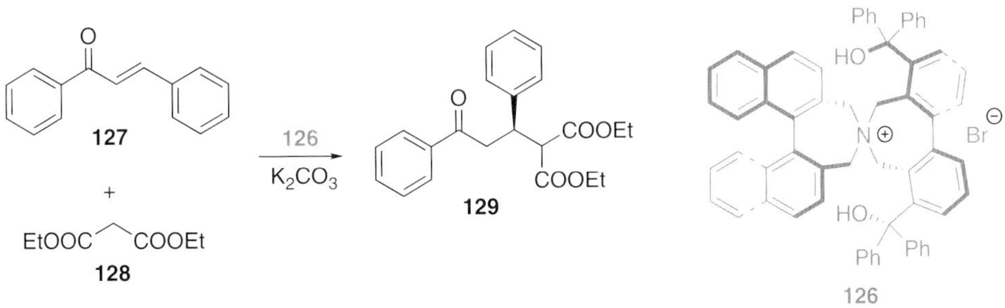

Figure 10.31 Enantiotope-selective *Michael* addition of diethyl malonate (**128**) onto 1,3-diphenylpropenone (**127**) catalyzed by the chiral ammonium salt **126**.

**Figure 10.32** Double-enantiotope selectivity in *Michael* addition of butanal (**130**) onto β-nitrostyrene (**131**) with the assistance of the chiral prolinamide-derived organocatalyst (**133**). New stereogenic elements formed in stereoselective manner are highlighted (*de* 98%, ee > 99%).

that the reaction proceeded with high enantiotope selectivity at both prostereogenic centers to result in high diastereomeric (*syn* : *anti* 99 : 1) and enantiomeric excess (ee >99% for the *syn* product).

10.1.6
### Catalytic Stereoselective Nucleophilic Addition Onto a Carbonyl Group

Enantioselective addition of metal–organic reagents onto a carbonyl group can also be catalyzed by chiral metal complexes and organocatalysts. A special case of organocatalysis is **autocatalysis**, when the product molecule exerts a catalytic effect on its own formation. Each of the three modes will now be illustrated by examples.

Reaction of benzaldehyde (**134**) and diethyl zinc (**123**) catalyzed by the titanium complex (**137**) containing as a chiral ligand TADDOL (**136**) introduced by *Seebach* gave (*S*)-1-phenylpropanol (**135**) in 99% yield and ee 98%. $Ti(O^iPr)_4$ functioning as the *Lewis* acid was applied in equivalent quantity, while the chiral titanium complex **137**, prepared separately from titanium tetraisopropoxide and the TADDOL ligand, was applied in a catalytic amount (Figure 10.33).

The explanation of enantiotope selectivity is that the chiral catalyst (**137**) forms a complex with the aldehyde and in the reactive complex owing to the axial disposition of the phenyl groups of the ligand, benzaldehyde is forced into a position that nucleophilic attack is only possible from the *Si* face. Due to the $C_2$ symmetry of TADDOL complex, addition of the aldehyde as the fifth ligand at any of the faces results in the same complex structure (Figure 10.34).

Using the metal-free amino alcohol-type organocatalyst **138** to promote the reaction of benzaldehyde and diethyl zinc also gave good results: (*S*)-1-phenylpropanol (**139**) was obtained in 97% yield and ee 98% (Figure 10.35).

A special case of organocatalysis is **autocatalysis**, when the product functions also as catalyst. It was observed that in the catalyst-free reaction of 2-(3,3-dimethylbut-1-yn-1-yl)pyrimidin-5-carbaldehyde (**140**) with diisopropyl zinc (**141**) when repeating the reaction, mixtures of various stochastic enantiomeric compositions were obtained (Figure 10.36). In successive experiments,

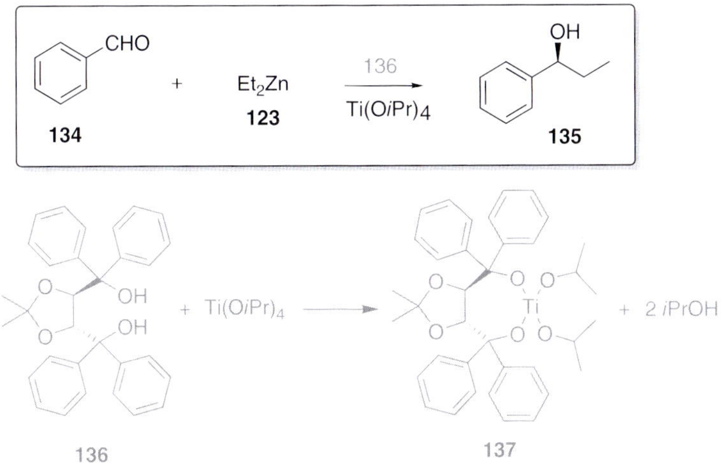

**Figure 10.33** Stereoselective reaction of benzaldehyde (**134**) and diethyl zinc (**123**) catalyzed by titanium complex (**137**) of TADDOL (**136**). In the top row, preparation of the chiral TADDOL complex (**137**) is shown.

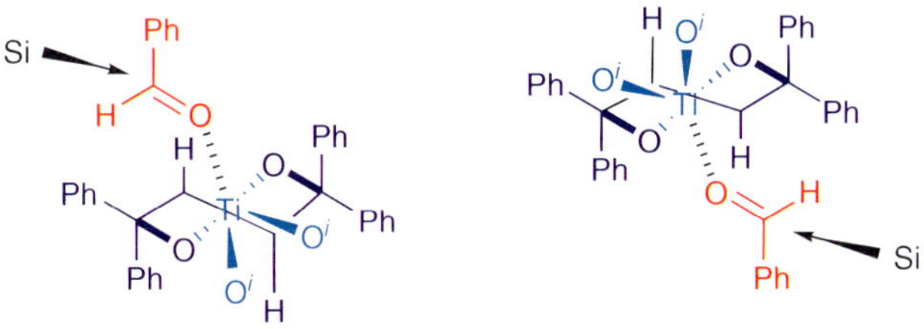

**Figure 10.34** Structure of the reactive complex composed of benzaldehyde and the TADDOL – titanium complex (red, blue, and green, respectively). Left: top view of the complex; right: bottom view of the complex. Owing to the $C_2$ symmetry of the ligand, the two complexes are identical. Addition of the diethyl zinc nucleophile takes place at the proximal *Si* face of the aldehyde. The dimethyl-1,3-dioxolane ring of TADDOL at the distal side precluding for steric reasons the attack of the nucleophile from behind was omitted for reasons of clarity ($O^i$ = isopropoxy group).

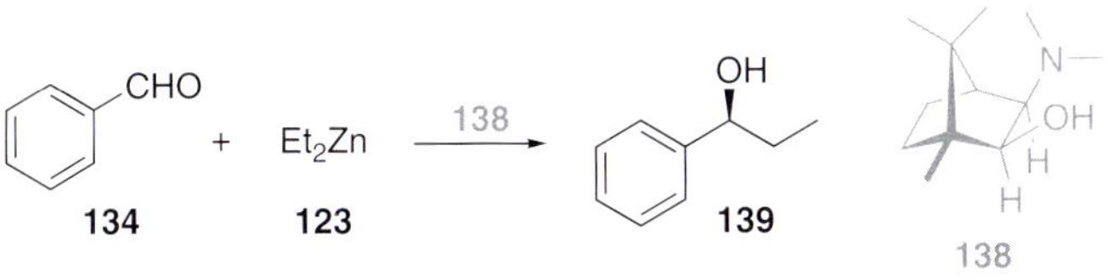

**Figure 10.35** Enantiotope-selective reaction of benzaldehyde and diethyl zinc catalyzed by (−)-*exo*-(dimethylamino)isoborneol (**138**).

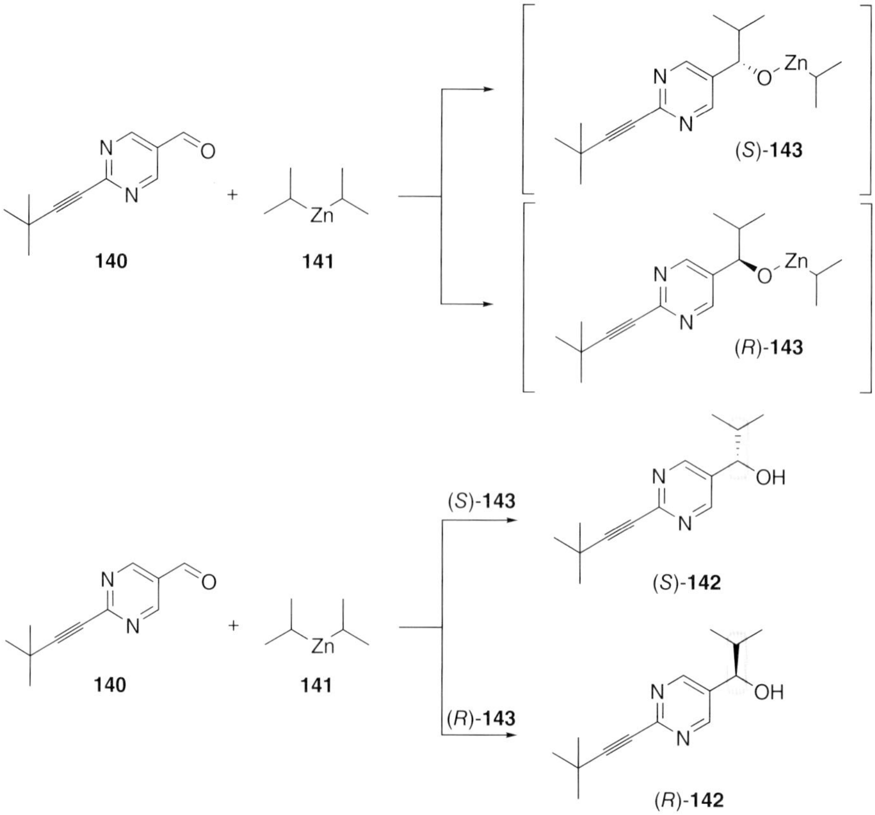

**Figure 10.36** Autocatalytic stereoselective reaction of 2-(3,3-dimethylbut-1-yn-1-yl) pyrimidine-5-carbaldehyde (**140**) and diisopropyl zinc (**141**). In the top row, formation of the zinc complex (**143**) responsible for the catalytic effect is shown.

enantiomeric composition of the product, that is, 1-[2-(3,3-dimethylbut-1-inyl)pyrimidin-5-yl]-3-methylpropan-1-ol (**142**), varied in a wide range ($R:S = 4.5:95.5-93:7\%$).

Since both reaction partners are symmetric, without autocatalysis, racemic mixtures should have been produced. In fact, the results of experiments varied in the range of ee = 33–91%, but racemic mixtures were never formed. In a longer series of 37 individual experiments, enantiomers were formed with equal probability: in 19 instances (*S*)-**142** and in 18 experiments (*R*)-**142** were the major components (Figure 10.37). Summation of the results of the complete series gave nearly racemic composition.[10] Surprisingly, the presence of one enantiomer of the product accelerated the formation of the same enantiomer at the expense of

10) On increasing the number of experiments, the aggregated results approximated the racemic composition.

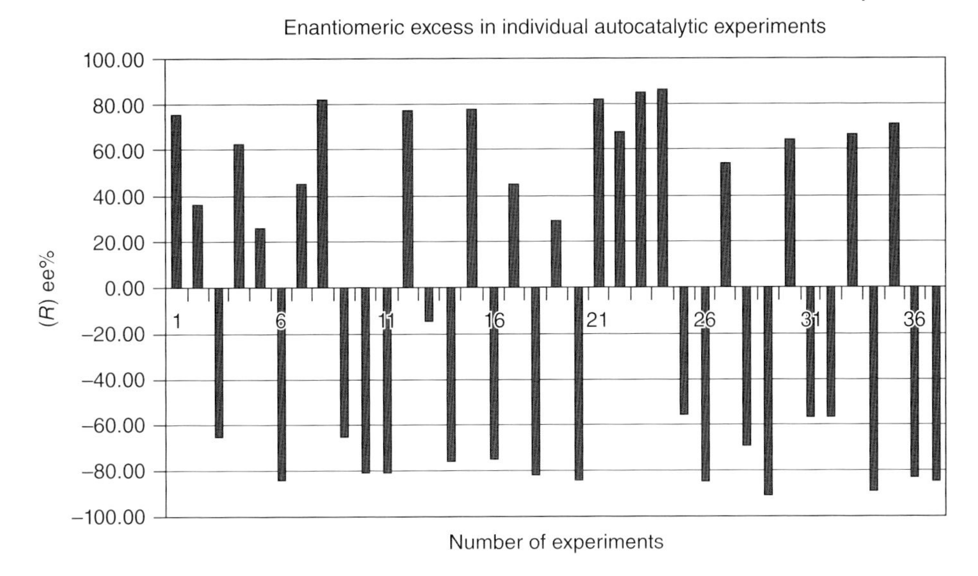


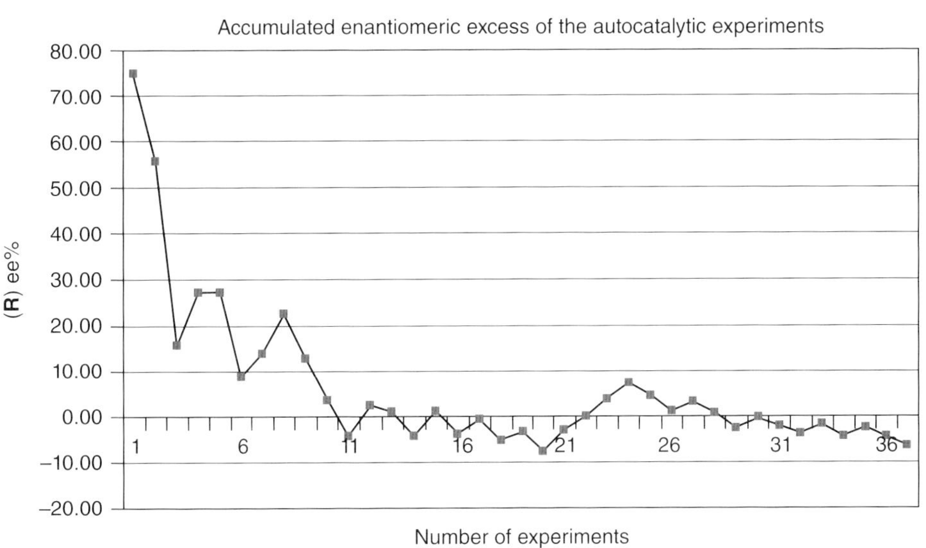


**Figure 10.37** Autocatalytic enantiotope-selective reaction of 2-(3,3-dimethylbut-1-yn-1-yl) pyrimidine-5-carbaldehyde (**140**) and diisopropyl zinc (**141**). Left: Enantiomeric excess of the individual experiments; right: accumulated optical yields of experiments.

its antipode.[11] The catalytic effect is ascribed to the zinc compound of the product alcohol (Figure 10.36).

11) If prior to the reaction, a catalytic amount of the product alcohol **142** of $5 \times 10^{-5}$ ee was added to the aldehyde **140**, a product of 99.5% ee was obtained having a configuration of that of the catalytic sample.

Mathematical modeling has demonstrated that reaction between symmetrical molecules producing mirror-image pairs of the individual enantiomers are formed stochastically. It follows that in the initial phase of such reactions,[12] the number of enantiomers may show a significant difference. For example, after transformation of 100 000 molecules most probably such mixtures are formed in which there is an excess of 212 molecules of one enantiomer over the other (the ratio is 50 106 to 49 894, i.e., 0.21%). In due course, with the number of experiments rising, the difference in the number of enantiomeric molecules approximates zero.[13] Since the zinc complex **143** is exerting its autocatalytic effect, even at ee = 0.00005%, it can be accepted that in the case of a stochastic behavior of the system in the initial phase of the reaction, even this infinitesimal enantiomeric excess is sufficient to initiate autocatalysis. The discovery of such an autocatalysis may be important for the understanding of the development of the asymmetry in living organisms.

### 10.1.7 Double-Enantiotope-Selective Catalytic Aldol Reactions

In the course of chirally catalyzed aldol reactions elaborated by *Mukaiyama*, a separately prepared silyl enolether (e.g., **144**) is reacted in the presence of a chiral catalyst with an aldehyde (e.g., benzaldehyde). The chiral catalyst is prepared *in situ* from three components: tin(II) triflate, as a *Lewis* acid, a chiral diamine (e.g., **145**), and tributyltin fluoride. In the reactive complex (**147**), the aldehyde component is bound to the central metal atom of the *Lewis* acid containing the chiral diamine ligand, while one of the oxygens of the triflate group of the *Lewis* acid is attached to the silicon atom of the silyl enolether. The fluorine atom of tributyltin fluoride attacks the silicon atom of silyl enolether, increasing thereby the nucleophilicity of the enol. The π-electron cloud of the enol attacks the carbonyl carbon atom of the aldehyde closing a cycle producing the reactive complex (Figure 10.38).

In the course of the catalyzed aldol reaction, it is not only the silyl enolether that decomposes under the attack of fluoride but also the two tin compounds are falling apart. Thus, both the tin reagent and preferably also the chiral diamine have to be put into the reaction mixture in equivalent amounts. The catalyzed aldol reaction is completely *syn* diastereoselective and the single product (**146**) is obtained in 98% enantiomeric excess.

For the stereoselective catalysis of aldol reaction, multiple metal complexes and organocatalysts based on amino acids have been elaborated. Among them, now processes catalyzed with chiral ionic liquids, as chiral media are discussed.

Ionic liquids are stable ionic compounds composed of stable aromatic heterocyclic cations and stable anions, for example, 1-ethyl-3-methylimidazolium hexafluorophosphate is liquid at room temperature. The advantage of a two-phase

12) This extremely short period of about 1 fs ($10^{-15}$ s) cannot be measured by present-day techniques.
13) In reality, the number of the molecules of the two enantiomers is never exactly the same, for example, after the transformation of 1 mol of material ($6 \times 10^{23}$ molecules), a difference of a couple of hundreds is causing an immeasurable small deviation (~ee $10^{-19} - 10^{-20}$%) from the racemic composition.

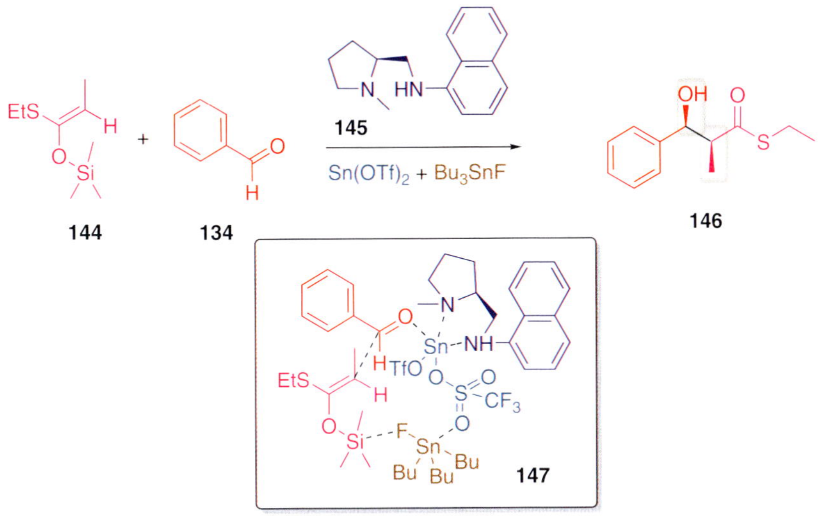

Figure 10.38 Double-enantiotope-selective aldol reaction. New stereogenic elements formed in a stereoselective way are highlighted. In the offset box, the reactive complex (**147**) formed from *S*-ethylpropanethioate silyl enolether (**144**, pink), benzaldehyde (**134**, red), *N*-[(*N*-methylpyrrolidin-2-yl)methyl]naphthalene-1-amine (**145**, blue), tin (II)-triflate (green), and tributyltin fluoride (brown).

system composed of an ionic liquid and an apolar organic solvent is that the separation of the catalytically active ionic liquid and the apolar organic solvent is simple and the ionic liquid can be recycled. If the organocatalyst exerting stereoselective catalysis is bound covalently to the molecules of the ionic liquid, recycling of the catalyst is also easy.

Using a chiral ionic liquid (**148**) as organocatalyst, reaction of hydroxyacetone (**149**) with various aromatic aldehydes, such as 2-chlorobenzaldehyde (**150**), was carried out (Figure 10.39). Regarding the two groups of hydroxyacetone, regioselectivity was 100%, *syn*/*anti* (**151**/**152**) diastereoselectivity was 94%, and *syn* product formed in ee 97%.

### 10.1.8
### Enantiotope-Selective Catalytic *Simmons–Smith* Reaction (Cyclopropanation)

The *Simmons–Smith* reaction is the reaction of olefins with diiodomethane under the action of metallic zinc giving cyclopropane derivatives. The reaction is formally cycloaddition of a carbene generated *in situ* from diiodomethane and the olefin. In fact, no carbene is formed, the active reagent is a zinc-organic species arising *in situ* from diiodomethane [reaction (a)]. Such reactive zinc-organic compounds have been prepared by reacting diiodomethane with diethyl zinc as

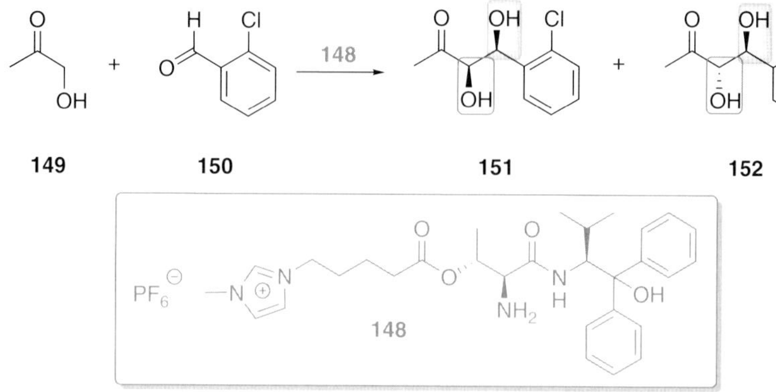

Figure 10.39 Regio-, diastereo-, and enantiotope-selective aldol reaction. New stereogenic elements formed in a double-enantiotopic-selective way are highlighted.

well [reaction (b)]:

$$I{-}CH_2{-}I + Zn \rightarrow I{-}CH_2{-}Zn{-}I \quad (a)$$

$$I{-}CH_2{-}I + Et{-}Zn{-}Et \rightarrow I{-}CH_2{-}Zn{-}Et + Et{-}I \quad (b)$$

As a chiral ligand, the bis-sulfonamide **154** of $C_2$ symmetry can be used, which by reacting with diethyl zinc produces the catalytic complex (**156**). The catalytic system can be utilized for cyclopropanation of allyl alcohol derivatives [e.g., (*E*)-3-phenyl-prop-2-enol (**153**)]. For the generation of the reactive complex, it is necessary that a zinc compound be formed from the olefin component and diethyl zinc (**123**) as well (Figure 10.40).

In the course of the procedure, first the mixture of the allylic component (**153**) and the chiral ligand (**154**) is reacted with diethyl zinc (**123**),[14] and to the solution of the complex is given the solution of the reactive zinc compound prepared separately from diiodomethane (**155**) and diethyl zinc (**123**).[15] It follows from the mechanism of the reaction proceeding with 99% yield that it is completely *syn*-diastereoselective, and owing to enantiotope-selective catalysis, the predominant enantiomer (**157**) was isolated in ee 89%.

### 10.1.9 Stereoselective Catalytic Diels–Alder Reaction

Stereoselective *Diels–Alder* reactions employ a wide range of catalysts. In the following, the efficiency of some catalytic systems of $C_2$ symmetry is compared using the *Diels–Alder* reaction of *N*-(*E*)-crotonoyl-oxazolidin-2-one (**159**) and cyclopentadiene (**160**) as model system (Figure 10.41).

When the *Diels–Alder* reaction of *N*-(*E*)-crotonoyl-oxazolidin-2-one (**159**) and cyclopentadiene (**160**) was catalyzed with the titanium TADDOL complex (**137**)

14) The solution of the complex marked blue and pink in Figure 10.40.
15) Solution of ethyl-iodomethyl zinc marked green in Figure 10.40.

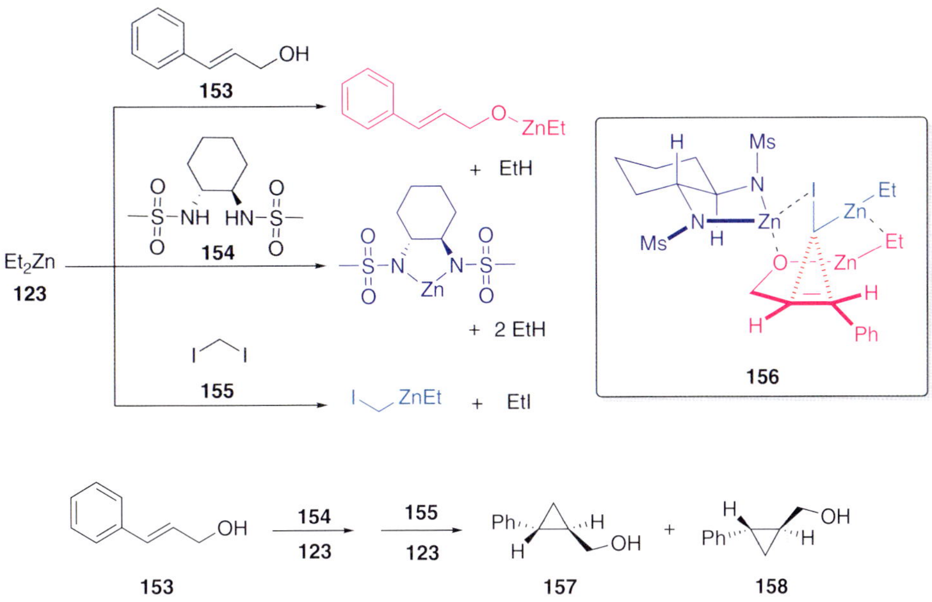

**Figure 10.40** Diastereospecific and enantiotope-selective *Simmons–Smith* reaction. Newly formed stereogenic elements are highlighted. Top left: The *in situ* processes necessary for the generation of the zinc-organic compounds. Top offset: The structure of the reactive complex. The new bonds forming the cyclopropane ring are shown as heavy dashed lines.

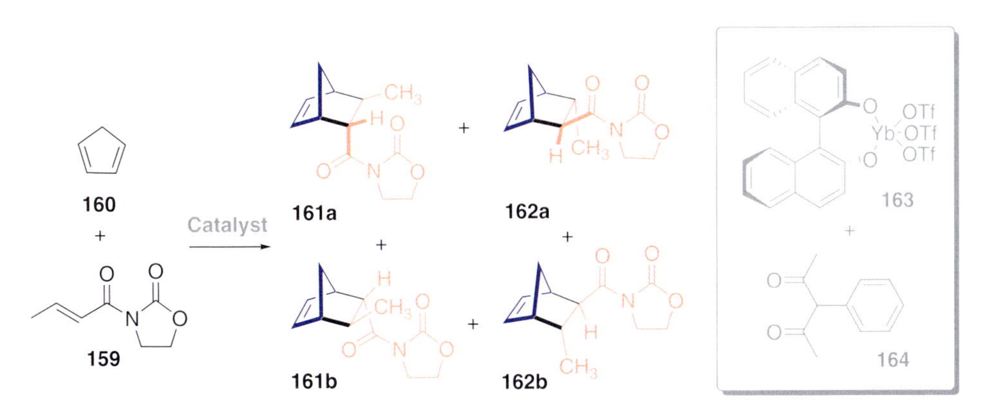


**Figure 10.41** Possible products of the *Diels–Alder* reaction of *N*-(*E*)-crotonoyl-oxazolidin-2-one (**159**) and cyclopentadiene (**160**): *endo*-adducts (**161a** and **161b**) and *exo*-adducts (**162a** and **162b**).

effective in other stereoselective reactions (see Figure 10.33), *endo*-selectivity was found to be *de* 82% associated with ee 91% and 90% yield for the main product (**161a**).

The above reaction could also be catalyzed by a complex of 3-phenylpentan-2,4-dione (**164**), and a lanthanide salt prepared from the chiral (*R*)-1,1′-binaphthyl-2,2′-diol and ytterbium-triflate (**163**). The main adduct (**161b**) was obtained in 83% yield with 86% *endo*-selectivity and ee 81%. It is interesting that the enantioselectivity of this lanthanide catalyst strongly depends on the structure of the achiral ligand. With another achiral ligand, the main product was the enantiomer **161a**.

In another variation of the *Diels–Alder* reaction of *N*-(*E*)-crotonoyl-oxazolidin-2-one (**159**) and cyclopentadiene (**160**), the catalyst was the aluminum compound (**166**) prepared from the chiral bis-sulfonamide **154** and trimethylaluminum (**165**) (Figure 10.42). As the main product, the *endo*-adduct **161a** was isolated in 88% yield, 92% *endo*-selectivity, and ee 94%.

Finally, the efficiency of two similar catalytic systems, also of $C_2$ symmetry, will be compared based on the results of *Diels–Alder* reaction of *N*-acryloyl-oxazolidin-2-on (**167**) and cyclopentadiene (**160**) (Figure 10.43).

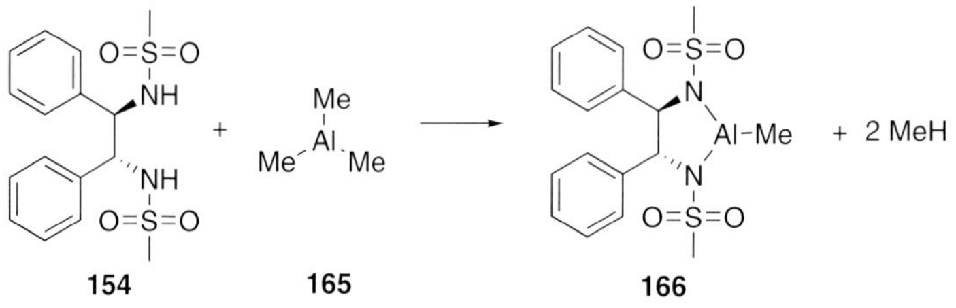

**Figure 10.42** Preparation of the chiral aluminum catalyst (**166**) used for the *Diels–Alder* reaction of *N*-(*E*)-crotonoyl-oxazolidin-2-one (**159**) and cyclopentadiene (**160**).

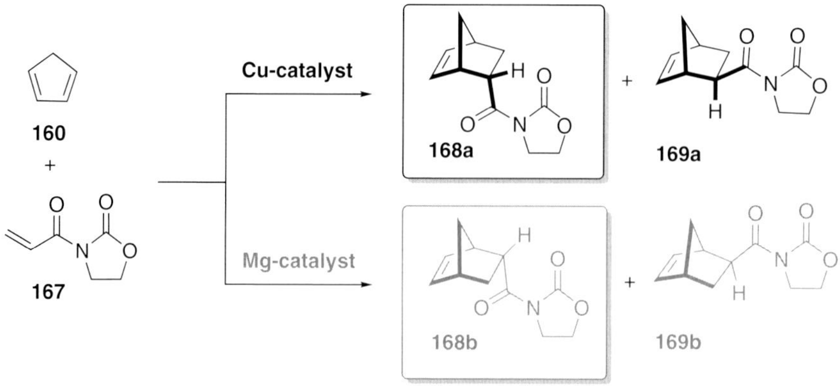

**Figure 10.43** Possible products of the *Diels–Alder* reaction of *N*-acryloyl-oxazolidin-2-one (**167**) and cyclopentadiene (**160**).

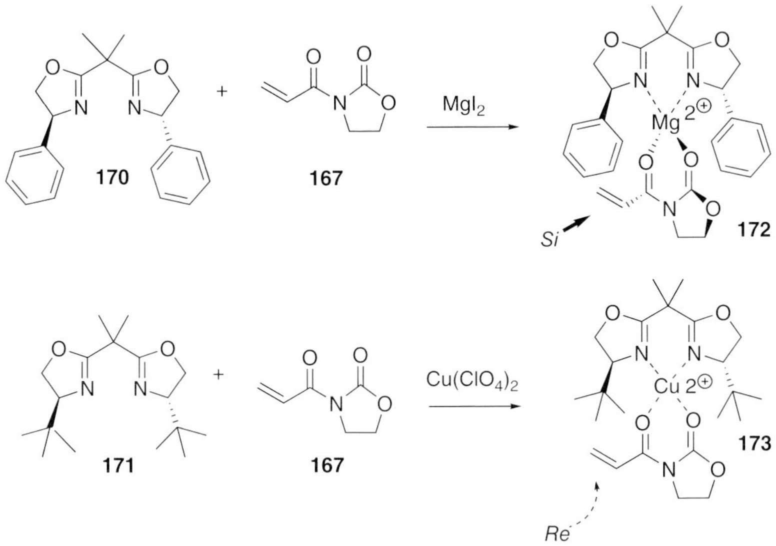

Figure 10.44 Chiral ligands (170 and 171) used for the catalysis of the *Diels–Alder* reaction of *N*-acriloyl-oxazolidin-2-one (167) and cyclopentadiene (160). Structure of the reactive complex.

The structure of the two chiral ligands is very similar and differs only inasmuch that the phenyl group in **170** is replaced in **171** by a *tert*-butyl group (Figure 10.44).

The essential difference is not this, but the bonding system of the central metal atom. When the above reaction was catalyzed by the magnesium complex of ligand **170** (i.e., **172**), the main product (**168b**) was obtained in 82% yield, 94% *endo*-selectivity, and ee 91% for the *endo*-product. On the other hand, when the reaction was catalyzed by the copper complex of ligand **171** (i.e., **173**), selectivity reversed and the main product became **168a** with 98% *endo*-selectivity and ee > 98% for the *endo*-adduct.

Magnesium(II) forms a tetrahedral complex in which the oxazolidine ring of the substrate (**167**) is oriented perpendicular to the plane defined by the nitrogen atoms of the heterocyclic ligand. The oxazolidine ring is in the proximity of the phenyl group oriented in the figure toward the viewer, while the olefinic bond is near to the phenyl group pointing to the far side. The latter phenyl group is screening the *Re* face of the olefinic bond, and thus the cyclopentadiene molecule can approach the olefinic bond preferably from the *Si* face.

Copper(II) is forming, however, a square planar complex, the heterocyclic ligand and also the substrate lie in the plane defined by the nitrogen atoms of the heterocyclic ligand and the magnesium ion. In this way, the olefinic bond of the substrate gets close to the *tert*-butyl group pointing toward the viewer and the latter blocks the *Si* face of the double bond. In this way, cyclopentadiene (**160**) can approach the double bond preferably from the *Re* face. Owing to the $C_2$

symmetry of the ligand, it can be seen that an association of the substrate (in a sense opposite to the situation depicted in Figure 10.44) leads to the same result.

## 10.2
## Biocatalysis

Enzymes are protein-based catalysts of biochemical processes proceeding in living cells. The skeleton of enzymes is similar to other proteins: a polypeptide chain built up from 20 different α-amino acids by polycondensation.

Since out of the 20 natural α-amino acids except the simplest one, that is, glycine are all chiral, the polypeptides built up from them are typically asymmetric molecules. The primary structure of polypeptides (i.e., the sequence of amino acids), the secondary structure [i.e., the local conformation of the peptide chain: aperiodic or periodic (α-helix, β-sheet)], tertiary structure (the 3D structure of the peptide chain), or quaternary structure [agglomeration of several structural units (peptide chains and prosthetic groups)] are not discussed in detail in the present work.

The part responsible for the catalytic activity of enzymes is the *active center*. Characteristic features of the catalytic activity of enzymes are as follows:

- By binding the substrate specifically (foremost by secondary chemical bonds), an enzyme–substrate complex ordered in space is formed, being the basis of substrate specificity of enzymes.
- Formation of a reactive enzyme–substrate complex provides kinetic advantage of the reaction, the free enthalpy of activation is lowered.
- The chemical reaction takes place at the part of the bound substrate that intrudes into the active center of the enzyme. Here, it is surrounded by other parts of the enzyme and is isolated partially or totally from the ambient medium.
- In the active center the substrate, eventually a coenzyme, and the reactive amino acid side chains participating in the reaction are ordered in space whereby enzymatic reactions are characterized by high regio- and stereospecificity or selectivity.
- In the active center, the amino acid side chains and eventually the coenzyme may trigger redox reactions, nucleophilic substitutions, and or acid-/base-catalyzed processes, partly or totally independent of the medium surrounding the enzyme. Occasionally, the substrate or part of it may establish a transient covalent bond with a suitable reactive side chain of the enzyme.

The internationally accepted classification of enzymes is based on the chemical reaction catalyzed and not on the structure of the enzyme. The system created by the "*Enzyme Commission*" uses a four-tier classification: EC X.X.X.X, in which X is a whole number. The first number indicates to which of the six main enzyme classes of activity the enzyme belongs. The other numbers indicate subdivisions within the classes. The six main classes of enzymes are as follows:

1) *Oxidoreductases*: catalyze redox reaction
2) *Transferases*: catalyze the exchange of structural units between two substrate molecules
3) *Hydrolases*: catalyze hydrolytic processes
4) *Lyases*: catalyze addition–elimination processes
5) *Isomerases*: catalyze isomerizations
6) *Ligases* (synthetases): catalyze synthetic reactions utilizing bioenergy (e.g., ATP).

Enzymes can be classified whether they require for their functioning the cooperation of a coenzyme (cofactor, prosthetic group of catalytic effect) or not.

(i) Enzymes that operate without a coenzyme (cofactor). The reaction is catalyzed by reactive amino acid side chains of the enzyme (being regenerated after completion of the catalytic cycle).
(ii) Some enzymes contain tightly bound cofactors. Such cofactors change their chemical features in the course of the reaction. For example, the central metal ion of oxidoreductases may change its oxidation state, which is, however, restored autocatalytically at the end of the reaction cycle.
(iii) A third group of enzymes works with a transiently bound external coenzyme. These coenzymes, for example, $NAD^+$, NADH, FAD, and $FADH_2$ change their structure during the reaction (functioning strictly speaking as reagents) and have to be reverted to their original form generally in a separate process catalyzed by another enzyme. Application of such enzymes in extracellular procedures is usually impractical either because regeneration of the coenzyme has to be carried out in a separate enzyme-catalyzed process, or the expensive coenzyme cannot be recycled at all.

The advantages of biocatalysis are as follows:

- Reactions run under mild conditions: room temperature, neutral or near-neutral pH, and generally in aqueous media.
- High selectivity (substrate-, regio-, and stereospecific operations) since enzymes are all chiral
- A wide range of chemically versatile catalysts is available.
- High efficiency, acceleration of rate by a factor of $10^{12}$ may be realized.

Disadvantages of biocatalysis:

- Enzymes are sensitive toward the ambiance (temperature, pH, solvent, etc.).
- Enzymes are sensitive toward contaminations inhibiting enzyme activity.
- Enzymes require dilute solutions, volumetric yields are poor.
- Enzymes are usually more expensive than simple chemical catalysts.

Details of the mechanism of action of enzymes and the kinetics of enzymic processes are outside the scope of this book.

Preparative biocatalytic processes can be carried out using the following methods:

- Fermentation with living cells
- Partially purified enzymes isolated from digested cells
- Synthetic enzymes prepared by solid-phase peptide synthesis
- Semisynthetic enzymes synthesized by genetically modified microorganisms
- Enzymes modified by gene technology (produced by a microorganism into which a suitably modified nucleic acid was introduced).

Isolated enzymes can be introduced into a chemical reactor

- in a free (dissolved) form;
- bound to the surface of a carrier by adsorption or chemical bonds;
- aggregated with a suitable material or entrapped into a polymeric matrix, respectively; and
- enclosed into a microcapsule, microemulsion.

The last three modes facilitate the recovery and recycling of the enzyme from the reaction mixture. Note, however, that activity of a nondissolved enzyme may be altered, depending on the mode of immobilization. Diffusion of the substrate may also play an important role in the kinetics of the reaction.

Use of immobilized enzymes permits the realization of continuously operated biocatalytic processes in tube reactors filled with the enzyme preparation.

### 10.2.1 Substrate Selectivity (Enantiomer Selectivity) – Kinetic Resolution

Substrate selectivity of enzymes is manifested not only between compounds of different constitution but also between stereoisomers. This capacity can be exploited for kinetic resolutions.

One of the first enzymatic kinetic resolutions reported was the hydrolysis of *N*-acylated racemic amino acids, for example, **174** using L-*aminoacylase*. The product of hydrolysis, that is, the L-amino acid (**175**) could be easily separated by simple extraction from the unchanged *N*-acetyl-D-amino acid (**176**). Enantiomeric excess of the products at 50% conversion was 100% (Figure 10.45).

**Figure 10.45** Kinetic resolution of an *N*-acetylamino acid (**174**) with an L-*aminoacylase*.

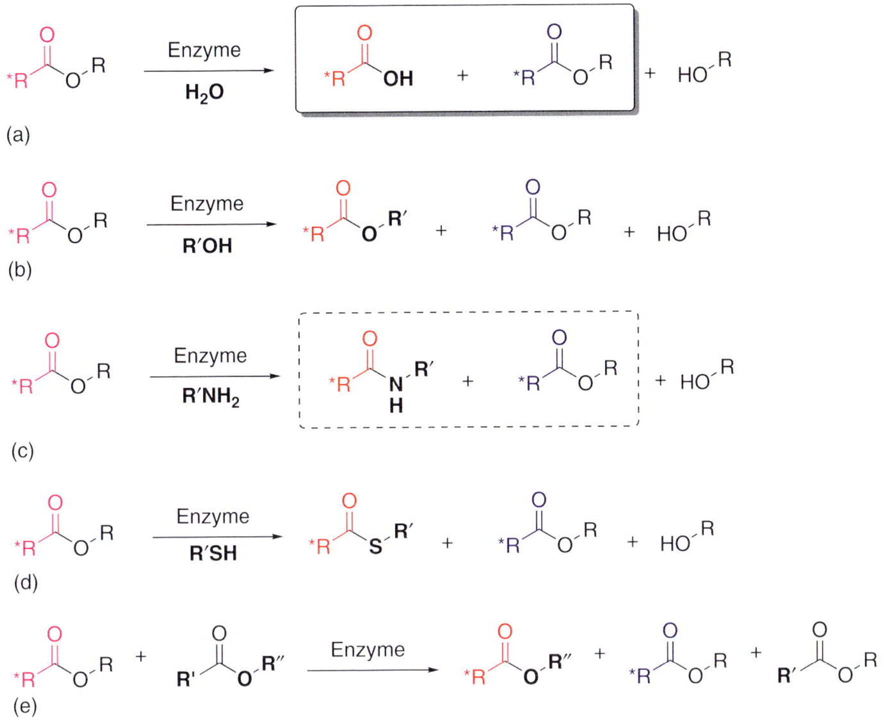

Figure 10.46 Main types of enzyme-catalyzed kinetic resolution of esters.

Enzyme-catalyzed kinetic resolutions are most often involved in reactions with various acid derivatives (Figure 10.46):

a) Ester hydrolysis
b) Ester alcoholysis (transesterifications)
c) Ester aminolysis (amide formation)
d) Ester thiolysis (thioester formation)
e) Interesterification.

Due to easiness of separation of the forming products, methods (a) and (c) are favored. In the case of methods (b), (d), and (e), the products are difficult to separate, and these processes usually led to an equilibrium.

Another version of transformations is that instead of the acyl group, it is the alcohol component that contains the stereogenic element. In such case, the following reactions depicted in Figure 10.47 can be applied:

a) Ester hydrolysis
b) Ester alcoholysis (the chiral part is in the ester)
c) Ester alcoholysis (the chiral part is in the alcohol)
d) Interesterification.

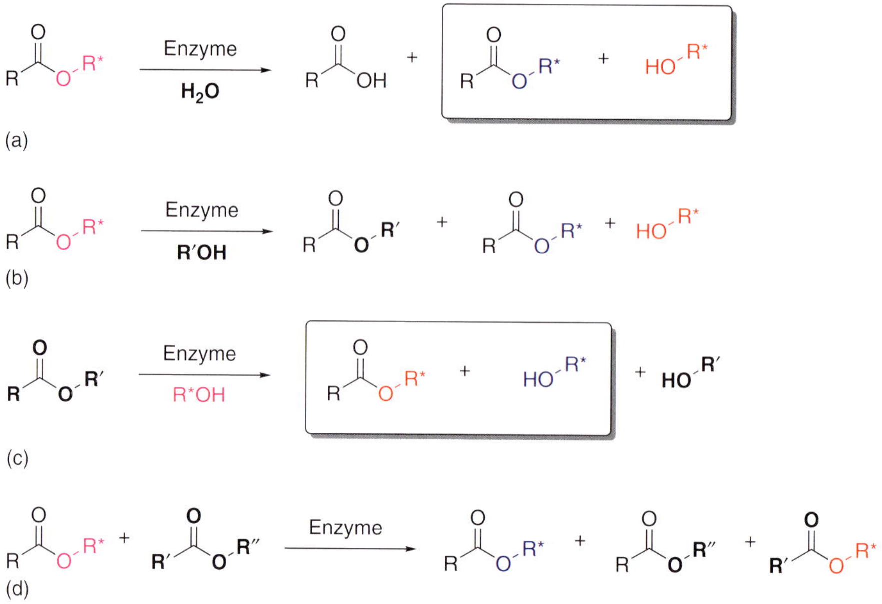

Figure 10.47 Main types of the enzyme-catalyzed kinetic resolution of alcohols.

For biotransformations involving chiral alcohols, methods (a) and (c) are favored because of the easy separability of alcohols and esters. In the case of methods (b) and (d), the product separation is more difficult, and these processes usually led to an equilibrium.

Among the widely used procedures as an example, the kinetic resolution of 4-phenylbut-3-yn-2-ol (**177**) catalyzed by *Candida antarctica* lipase B (*Ca*LB) is presented (Figure 10.48). Note that if an enzyme is used to transform nonnatural substrate, its stereoselectivity depends on the structure of the substrate.16)

In the above reaction, (*S*)-4-phenylbut-3-yn-2-ol [(*S*)-**177**] was isolated in 42% yield and ee > 99%, while (*R*)-4-phenylbut-3-yn-2-yl-acetate (**179**) was isolated in 54% yield and with ee 71% (Figure 10.48).

16) Complete stereoselectivity (stereospecificity) of enzymes is characteristic only to *enzymes involved in primary metabolism* with their natural substrates. These enzymes are usually more specialized and have relatively narrow substrate tolerance. Thus, they form with their natural substrates an enzyme–substrate complex of definite structure with a very high complex formation constant. Such a complex cannot come into existence with the enantiomer of the natural substrate. For nonnatural substrates, it is not sure that they would fit exactly into the binding site of the enzyme, and therefore its complex formation constant will be smaller. Thus, it is possible that even the enantiomer of the nonnatural substrate is capable to bind to the active site of the enzyme albeit with a variable binding constant. In the case of *enzymes involved in secondary metabolism or defense*, the substrate tolerance is wider since it is more economic if they can act on multiple kinds of substrates. It is understandable that a wider substrate tolerance sometimes comes with lower degree of stereoselectivity.

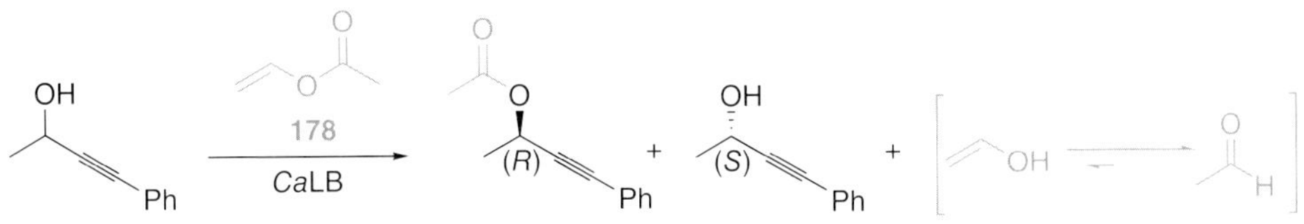

Figure 10.48 Kinetic resolution of 4-phenylbut-3-yn-2-ol (**177**) by esterification with vinyl acetate (**178**) catalyzed by *Candida antarctica* lipase B (CaLB).

Various other biological systems catalyzing chemical reactions can be utilized for kinetic resolution. The **catalytic antibody** *Ab38C2* catalyzes in an enantiotope-selective manner aldol reactions between aldehydes (e.g., **180**) and acetone (**181**) producing the β-hydroxyketone (*S*)-**182**. The reaction is reversible and when the racemic hydroxyketone (±)-**182** is added as substrate to the antibody, in conformity with its selectivity it catalyzes the retro-aldol reaction of (*S*)-**182**, while the *R* enantiomer (*R*)-**182** remains unchanged. Thus, by applying a single catalytic antibody, (*S*)-**182** can be produced by the aldol reaction and its enantiomer (*R*)-**182** by the retro-aldol reaction, both in high enantiomeric excess (Figure 10.49).

### 10.2.2 Substrate Selectivity (Enantiomer Selectivity) – Dynamic Kinetic Resolution

Transformation of the total amount of a racemic substrate to a given single enantiomer can be achieved when the unreacted enantiomer undergoes racemization in some chemical process. In this way, the reacting enantiomer is continuously

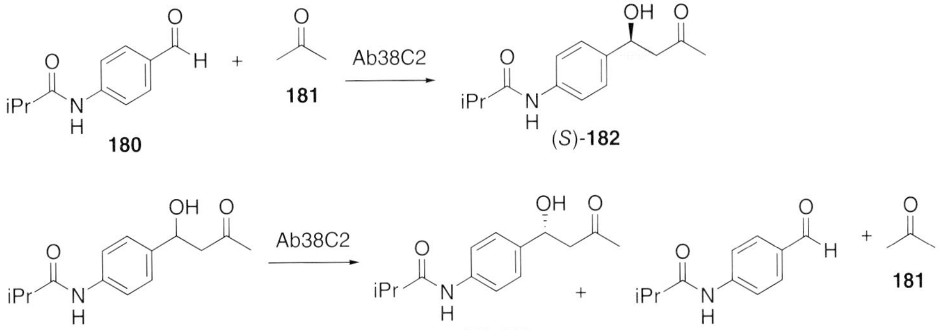

Figure 10.49 Stereospecific aldol reaction of 4-(isobutiroylamino)benzaldehyde (**180**) with acetone (**181**) and the retro-aldol reaction of racemic **182** catalyzed by the catalytic antibody Ab38C2.

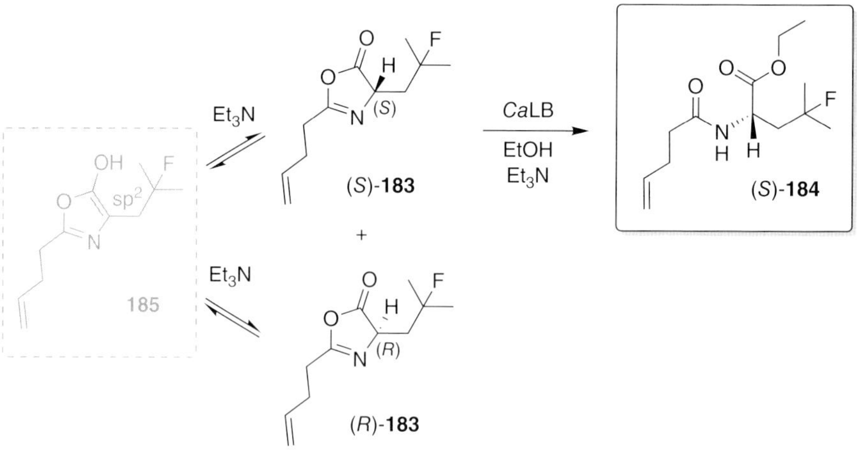

Figure 10.50 Kinetic resolution of 2-(but-3-enoyl)-4-(2-fluoro-2-methylpropyl)-4,5-dihydro-1,3-oxazol-5-one (**183**) catalyzed by lipase B from *Candida antarctica* (CaLB).

replenished until the total amount of the starting material is transformed. This process is termed **dynamic kinetic resolution** (cf. Section 8.6.4). In the example shown in Figure 10.50, equilibrium is established between two enantiomers of the racemic starting material (**183**) in the alkaline medium via the common enol form (**185**) in which the stereocenter is destroyed due to the $sp^2$ hybrid state of the $C_5$ carbon. The substrate of the esterification reaction catalyzed by lipase B from *Candida antarcticaCa*LB is the lactone (*S*)-**183**. Conducting the reaction to 90% conversion, the single product, the (*S*)-ethyl ester [(*S*)-(**184**)], was obtained in ee >96% (Figure 10.50).

Enzyme-catalyzed dynamic kinetic resolution has also an industrial application for the manufacturing of enantiopure D- and L-amino acids at 1000 tons/year scale.

In a first version of the procedure for both phenylalanine enantiomers, the racemic phenylalanine (Phe, **189**) is converted to a hydantoin (**186**) that racemizes at pH > 8 via a common enolate (**187**) (Figure 10.51).

In the course of dynamic kinetic resolution to the solution of the hydantoin derivative (**186**) racemizing in the alkaline medium, L-*hydantoinase* or D-*hydantoinase* is added. In this way, the complete amount of the substrate is transformed to the D- or L-*N*-carbamoylated Phe [(*R*)- or (*S*)-**188**]. The carbamoyl group can then be removed by adding nitrite and hydrochloric acid.

A purely enzymatic process has also been developed when the removal of the carbamoyl group was accomplished with D-*carbamoylase* or L-*carbamoylase*, while racemization of the hydantoin was carried out with a *hydantoin racemase* enzyme. In this variant, all three necessary enzymes (*racemase, hydantoinase,* and *carbamoylase*) were added simultaneously to the solution of 5-benzylhydantoine (**186**) (Figure 10.51).

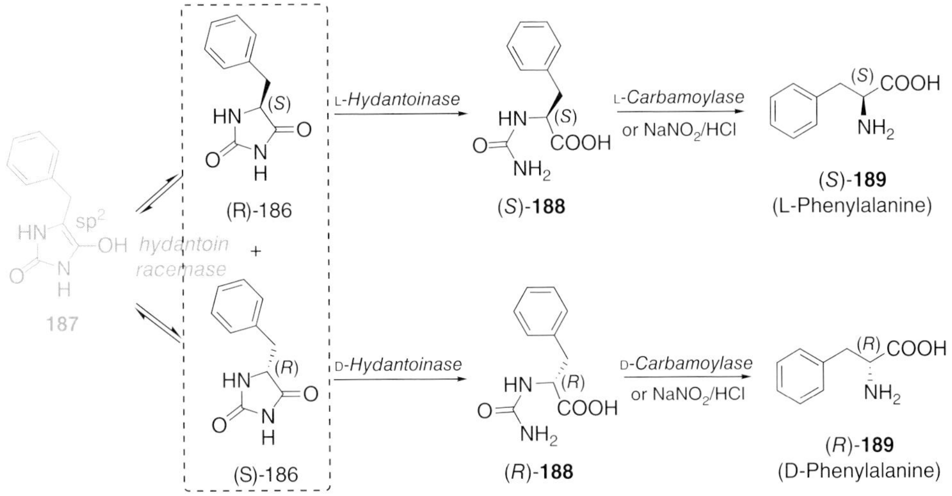

Figure 10.51 Dynamic kinetic resolution of 5-benzylhydantoin (**186**) with the aid of L- or D-hydantoinase and hydantoin racemase.

10.2.3
## Enantiotope-Selective Biotransformations

Utilization of enzymes in enantiotope-selective synthesis is enabled by their ability to differentiate between enantiotope faces and groups in achiral molecules. In the next part, typical examples of enantiotope selectivity of certain types of enzymes are presented.

### 10.2.3.1 Oxidoreductases

A broad group of oxidoreductases uses as hydride acceptor and hydride donor, respectively, the coenzymes $NAD(P)^+$ and NAD(P)H.[17] In the case of biotransformation with whole cells, there is no need to deal with the regeneration of the spent coenzyme. When using isolated enzymes, regeneration of the spent coenzyme is an important task (Figure 10.52).

In the course of the enzymatic reaction with the majority of alcohol dehydrogenases, the nicotinamide moiety of the coenzyme is approaching the substrate molecule in an orientation in which its amide group is situated in front of the less-bulky substituent "S" of the substrate. As a consequence, the carbonyl group of the substrate is attacked by the hydride ion from the top (from the *Re* face, assuming that according to the CIP rules the large substituent "L" is of higher rank than "S").[18] Most oxidoreductases follow this rule named as *Prelog's* rule, but there are known "anti-*Prelog* enzymes" as well.

17) $NAD(P)^+$: nicotinamide adenine dinucleotide ($NAD^+$) and its phosphate ($NADP^+$) are the oxidized forms of cofactors essential for many oxidoreductases. Their reduced forms are abbreviated as NADH and NADPH.

18) There is no correlation between the bulk of a group and its CIP rank.

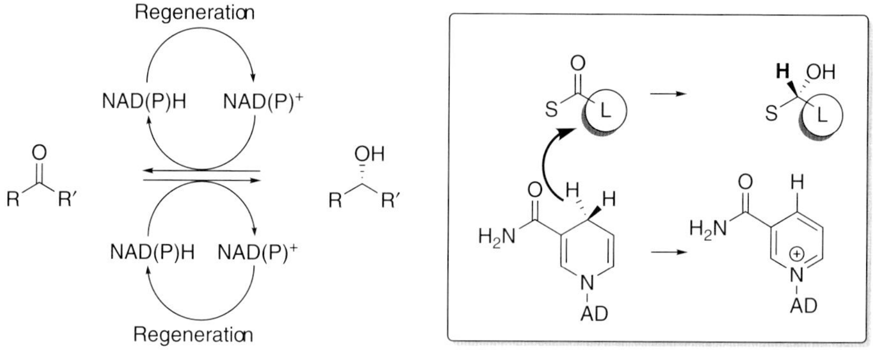

Figure 10.52 Schematic representation of the operation of oxidoreductases. Left: The reductive and oxidative reactions with the regeneration process of the coenzyme. Right: Stereochemical representation of hydride ion transfer for the majority of alcohol dehydrogenases: *Prelog's* rule.

Oxidoreductases are able to reduce not only oxo groups but also carbon–carbon double bonds[19] conjugated with a carbonyl group as well as carbon–nitrogen double bonds. As an example, the preparation of the nonnatural amino acid L-*tert*-leucine (**191**) widely used in the pharmaceutical industry is quoted. The achiral substrate of the enzymatic reaction is trimethylpyruvic acid (**190**) (Figure 10.53).

Trimethylpyruvic acid (**190**) is reduced in the presence of ammonium formate as ammonia source with *leucine dehydrogenase* (LDH). For the reduction, the enzyme is using NADH, and therefore it is necessary to reconvert $NAD^+$ formed in the process.[20] This is done with the aid of *formate dehydrogenase* (FDH) that

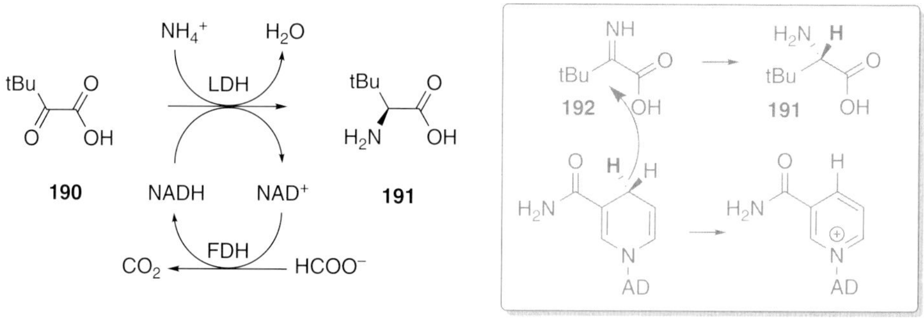

Figure 10.53 Reductive amination of trimethylpyruvic acid (**190**) with leucine-dehydrogenase (LDH) combined with the regeneration reaction with a formate dehydrogenase (FDH). Right: Stereochemical representation of hydride transfer.

19) Enzymes participating in fatty acid biosynthesis reduce the β-oxo group with the participation of NAD(P)H coenzyme, and subsequently the conjugated double bond formed after water elimination.

20) The high price of NADH precludes its stoichiometric application.

uses as hydride ion source the formate anion. Regarding the overall process, trimethylpyruvic acid (**190**) and the imine (**192**) generated from ammonium formate are reduced by the formate anion. The process is thus essentially an enzyme-catalyzed transfer hydrogenation (Figure 10.53).

An interesting example of enantiotope-selective oxidation is performed using the *dioxygenase* enzyme of the microorganism *Pseudomonas putida*. This organism is capable to utilize even benzene as nutrient source. The first step of the process is an enantiotope-selective and *cis*-specific dihydroxylation catalyzed by the *dioxygenase* enzyme. With substituted benzenes, the reaction is both regio- and stereoselective. In the course of the reaction, not only the benzene derivative (e.g., bromobenzene), but one molecule of NADH is also oxidized. Therefore, there is no need for the regeneration of $NAD^+$ in a separate process, since fermentation is operating with whole cells and the organism is using $NAD^+$ for the oxidation of ethanol and acetic acid used as energy sources. The product, (1*S*,2*S*)-3-bromocyclohexa-3,5-diene-1,2-diol (**193**), is used by the chemical industry (Figure 10.54).

An enzymatic variant of the *Baeyer–Villiger* oxidation is also known. By expressing a gene of *cyclohexanone monooxygenase* (CHMO) from the microorganism *Acinetobacter* in baker's yeast, a recombinant strain including CHMO enzyme was created. Oxidation of 4-isopropylcyclohexanone (**194**) with this recombinant whole-cell CHMO biocatalyst gave with 60% yield (*S*)-5-isopropyloxepan-2-one (**195**) of ee >98% (Figure 10.55) . In case of 2-substituted cyclohexanones, oxidation by this biocatalyst cleaves the bond between the substituted carbon atom and the carbonyl group of the (*R*)-substrate. When the substituent is bulkier than the methyl group, the enzyme strongly prefers the (*R*)-enantiomer and the reaction proceeds in a regio- and enantiomer-selective manner. Thus, oxidation of 2-isopropyl-cyclohexanone (**196**) gave with 41% yield (*R*)-7-isopropyl-oxepan-2-one (**197**) of ee 96%, while the unreacted (*S*)-2-isopropylcyclohexanone (*S*)-**196** was recovered in 46% yield and ee 96% (Figure 10.55).

#### 10.2.3.2 **Lyases**

One representative of lyases is the enzyme *oxynitrilase*, also called *hydroxynitrile lyase* (HNL), which transforms aldehydes to cyanohydrins. This enzyme is also

Br
$O_2$
Br
OH
OH
**193**
NADH
$NAD^+$
$CO_2$
EtOH/AcOH

**Figure 10.54** Regio-, diastereo-, and enantiotope-selective oxidation of bromobenzene using the microorganism *Pseudomonas putida*.

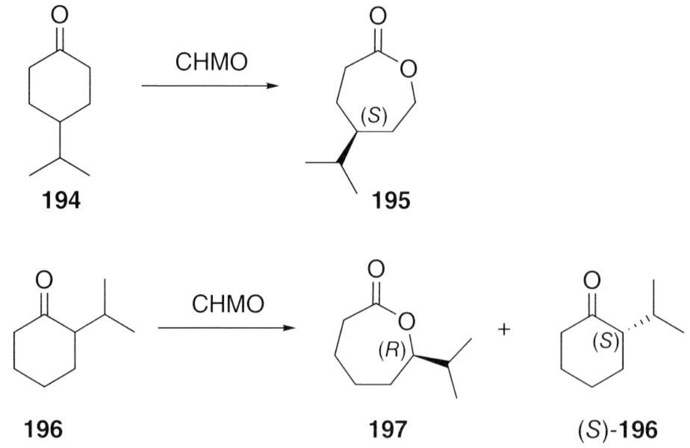

Figure 10.55 Enantiotope-selective oxidation of 4-isopropylcyclohexanone (**194**) with recombinant *Saccharomyces cerevisiae* containing the CHMO gene of *Acinetobacter*. Note that the similar oxidation reaction of 2-isopropylcyclohexanone (**196**) is regio- and enantiomer selective.

used in the industry to the stereoselective manufacturing of mandelic acid (**198**). Both *Re*- and *Si*-selective enzymes can be isolated from various organisms, for example, from plants or microbes (Figure 10.56).

An *R*-selective enzyme can be isolated from bitter almonds (*Prunus amygdalus*). This permits to obtain the cyanohydrin (*R*)-**197** in 95% yield and ee 99%. On the other hand, an *S*-selective enzyme can be isolated from the South American cauotchouc tree (*Hevea brasiliensis*). with which the (*S*)-cyanohydrin (*S*)-**197** can be gained in 94% yield and ee 99%. Performing the reaction, care must be taken to suppress the noncatalyzed reaction between benzaldehyde and hydrogen cyanide, which impairs stereoselectivity.

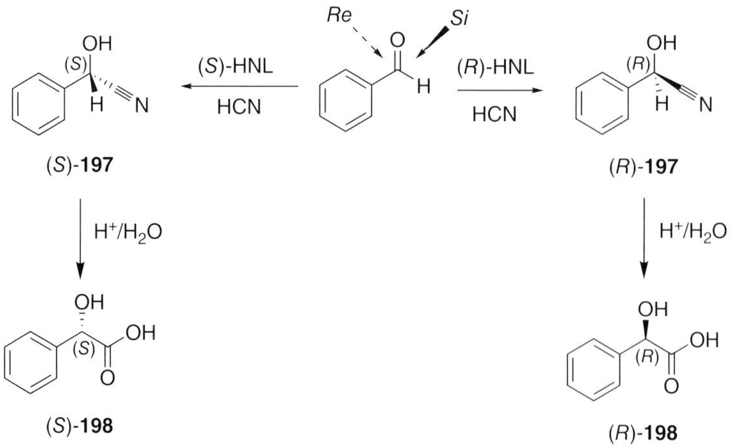

Figure 10.56 Enantiotope-selective preparation of (*R*)- and (*S*)-mandelic acid (**198**) with the aid of hydroxynitrile lyase (HNL) enzymes.

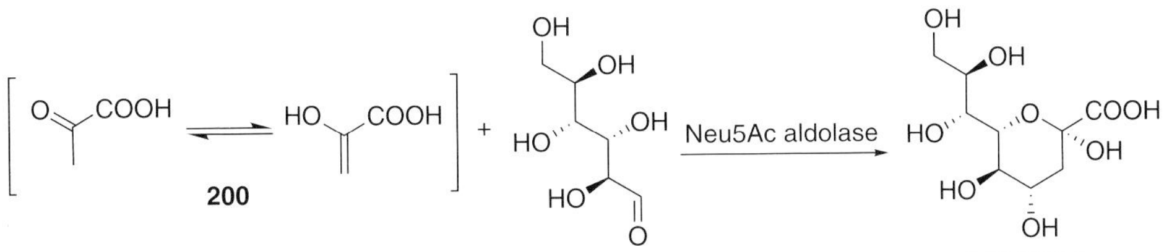

Figure 10.57 Enantiotope and diastereotope-selective preparation of KDN (**201**) catalyzed by the aldolase enzyme Neu5Ac aldolase.

Finally, we present an enantiotope and diastereotope-selective process carried out in a continuous reactor catalyzed by an aldolase enzyme that can catalyze aldol reactions between aldoses and ketoses. By membrane-bound *Neu5Ac aldolase*, enzyme 2-oxo-3-deoxy-D-*glycero*-D-*galacto*-nonopyranulosonic acid (**201**, KDN) was prepared by reacting D-mannose (**199**) and pyruvic acid (**200**). In the present example, the role of ketose is played by pyruvic acid. In the course of the catalyzed reaction, the enol form of pyruvic acid (with enantiotopic faces) attacks the carbonyl group of the aldose (with diastereotopic faces), thus in this reaction, enantiotope and diastereotope selectivity are manifested by *Neu5Ac aldolase* at once (Figure 10.57).

# Index



*Stereochemistry and Stereoselective Synthesis: An Introduction*, First Edition.
László Poppe and Mihály Nógrádi.

Companion Website: www.wiley.com/go/poppe/stereochemistry

## p



## q



## r



## s

***t***